COMMUNICATION AND EMPIRE

AMERICAN ENCOUNTERS / GLOBAL INTERACTIONS

A series edited by Gilbert M. Joseph and Emily S. Rosenberg

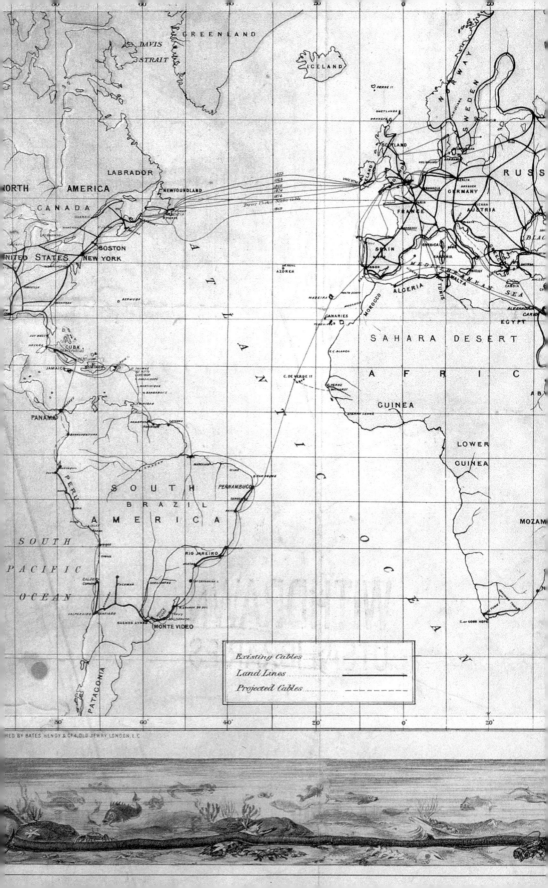

GREENLAND

ICELAND

DAVIS STRAIT

NORWAY

SWEDEN

RUSS

LABRADOR

NORTH AMERICA

CANADA

Newfoundland

GERMANY

AUSTRIA

FRANCE

SPAIN

BOSTON

UNITED STATES

NEW YORK

Direct United States Cable.

ALGERIA   TUNIS   MALTA

SEA

AZORES

MADEIRA

CANARIES

MOROCCO

SAHARA DESERT

AFRIC

EGYPT

CUBA

HAVANA

JAMAICA

GUINEA

PANAMA

LOWER

GUINEA

MOZAM

PERU

SOUTH

BRAZIL

AMERICA

PERNAMBUCO

RIO JANEIRO

SOUTH

PACIFIC

OCEAN

VALPARAISO

SANTIAGO

BUENOS AYRES

MONTE VIDEO

C. of GOOD HOPE

PATAGONIA

Existing Cables _____

Land Lines _____

Projected Cables _ _ _ _ _ _

BY BATES, HENDY & Cº, OLD JEWRY LONDON, E.C.

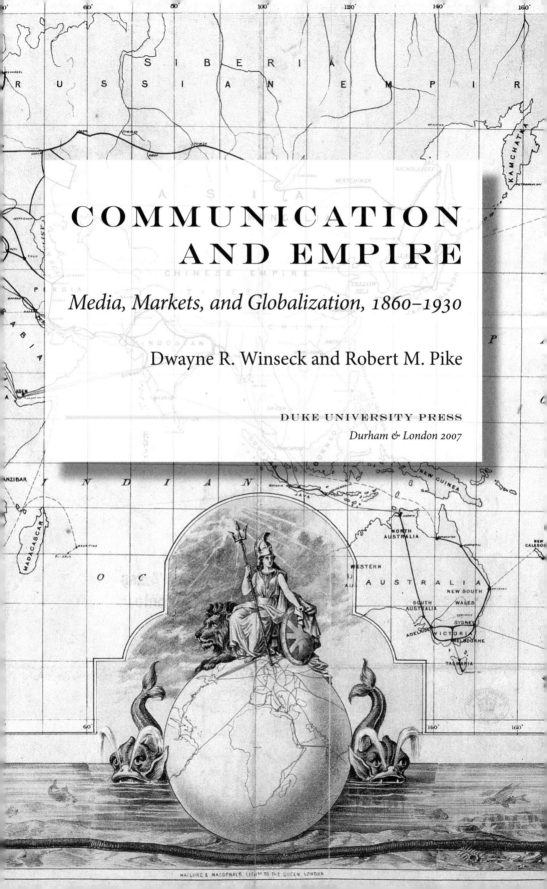

COMMUNICATION
AND EMPIRE

Media, Markets, and Globalization, 1860–1930

Dwayne R. Winseck and Robert M. Pike

DUKE UNIVERSITY PRESS

Durham & London 2007

© 2007 Duke University Press   All rights reserved
Printed in the United States of America on acid-free paper ∞

Designed by Jennifer Hill.
Typeset in Minion Pro by Keystone Typesetting, Inc.

Library of Congress Cataloging-in-Publication Data
appear on the last printed page of this book.

To Forest, the most beautiful and wonderful daughter a father could wish for. Thank you for all your patience and inspiration while the following pages were brought to fruition.   Dwayne

To my wife Faye and children Eleanor and Christopher. Thank you all for your love and your patience with my mental absences.   Bob

Contents

About the Series

This series aims to stimulate critical perspectives and fresh interpretive frameworks for scholarship on the history of the imposing global presence of the United States. Its primary concerns include the deployment and contestation of power, the construction and deconstruction of cultural and political borders, the fluid meanings of intercultural encounters, and the complex interplay between the global and the local. American Encounters seeks to strengthen dialogue and collaboration between historians of U.S. international relations and area studies specialists.

The series encourages scholarship based on multiarchival historical research. At the same time, it supports a recognition of the representational character of all stories about the past and promotes critical inquiry into issues of subjectivity and narrative. In the process, American Encounters strives to understand the context in which meanings related to nations, cultures, and political economy are continually produced, challenged, and reshaped.

Illustrations

Tables

Preface and Acknowledgments

There is an important new field emerging under the rubric of global media studies. This can be seen in the formation and widespread appeal of new journals that both herald and help to propel its arrival, such as *Global Media and Communication* (est. 2005) and the *Global Media Journal* (est. 2002). Such trends are also visible in the annual conferences of the Global Media Studies Group held since 2001, where there is concern with vital substantive topics as well as a drive to stimulate research into the global media. And at other major conferences and in the literature more generally, such research looms large.

Our work falls within this framework of studies, and we hope that its extensive use of archival sources and contextualization within current interdisciplinary scholarship on the problems of empire, international relations, history, and global political economy will help to establish a more robust historical foundation upon which these trends can be further developed. It is reasonably safe to say that much of the current research on global or, as some would say, international communication has relied upon a caricatured version of the past upon which to establish an understanding of contemporary developments in world communication. That caricature, we argue, is one in which the closest prelude to contemporary trends lies in the period we cover, from roughly 1860 to 1930. Thus, as found in much current literature, the key and distinguishing feature between the past and the present is that the advent of the global media system—consisting of intercontinental cable, and, later, wireless telegraph, communications networks, and global news agencies such as Reuters, Agence France-Presse, Associated Press, and Wolff—was driven mainly by the logic of imperialism and rivalry among the imperial superpowers (Britain, France, Germany, and later Japan, the United States, and Italy). In this view, the conquest of people and territory provided the dominant logic of international relations while the means of communication were, above all else, adjuncts of great power strategy. The leitmotiv of this era, then, is what Jill Hills refers to as "the struggle for control of global communication." And in this grand narrative, the mantel of hegemony over global communication passed from Britain to the United States after World War II, albeit with some shifts in that direction already having occurred much earlier, in the second and third decades of the twentieth century.

The analysis that we develop departs from this conventional knowledge in several ways. First, we suggest that the notion of imperial rivalry has been overplayed at the expense of deeper patterns of interdependence characteristic of the times. This was especially pronounced with respect to the global media, where large multinational communications companies were less closely allied with national capital and governments than commonly assumed. This could be seen in terms of the ownership and investment patterns behind the global cable and (though less so) wireless companies after World War I, as well as in the global news agencies. This could also be seen in the collaborative agreements these companies used to share global markets among themselves or, in more straightforward terms, the cartels they devised to divide and share global markets. The "news ring" agreements among the "big four" global news agencies, first created in 1869 and revised and renewed periodically thereafter until the mid-1930s, are relatively well known, while similar arrangements among the leading cable and, later, wireless communications companies though less well known are no less important.

Second, the web of interconnections and interdependence constituting the global media system of the late nineteenth century and early twentieth could also be seen in governments' willingness to rely on foreign firms to meet their foreign communication and military security needs. That is, during much of this period foreign governments readily relied upon British communication companies, most notably those falling under the colossal Eastern Telegraph Company umbrella, to serve their needs in the Far East, Africa, and parts of Latin America, while the British government was equally agreeable to relying on American companies such as Western Union and the Commercial Cable Company, one of the forerunners to the International Telegraph and Telephone Company, for network surveillance purposes and military security needs during World War I as well as for some of its communication activities in the Caribbean as early as the late 1880s. These patterns, we argue, were the norm rather than the exception, although we will also show that such cooperation came increasingly under strain in the run-up to World War I and thereafter.

Our third departure from conventional knowledge stems from the view that the globalization of capitalism was actually a stronger influence on the organization and control of global communication than was imperialism. Indeed, while the white settler colonies of the British Empire, as cases in point, were generally well served by communications, most zones of empire were some of the least connected, worst served places on the planet. The cases of Africa and the Caribbean, we show, stood as ample demonstrations of this neglect. Beyond these areas, however, this era consisted of unprecedented and still, in some cases, unsurpassed interconnections between foreign and domestic markets and global flows of capital, trade, people, and information. Communication networks and information flows, simply put, were densest in areas where world markets were most developed. And in this, the global media system was crucial in two ways: first, the media firms operating in world markets were some of the largest multinational companies of their times, and second, these firms provided the networks and supplied the informational and news resources upon which capitalism depended and thrived.

This does not mean that the importance of imperialism should be underrated. However, as a fourth point, we stress that there is a need to revise conventional conceptions of imperialism in two crucial ways. First, rather than focusing exclusively on interimperial rivalry, we need to focus on the extent of interimperial collaboration among the "great powers." That, as we have already indicated, could be seen at the level of ownership and control. It could also be seen in the manner in which the so-called global superpowers cooperated to jointly manage the international and imperial system among themselves. While there were dominant imperial powers, most notably Britain, the global system of the late nineteenth century and early twentieth depended on a kind of shared hegemony rather than on a single hegemonic power. This was especially visible in global communication, where cooperative agreements, elaborate multilateral organizations such as the International Telegraph Union, and a mounting body of international law found particularly fertile ground. The conception of empire that we work with in the pages ahead also required a further theoretical revision. In this sense, territorially bound concepts of imperialism that see the world as having been carved up into mutually exclusive zones of control needed to be revised to accommodate a more diffuse systems view of empire where collaboration, international law, and reasonably coherent and ambitious projects of modernization pursued in core and peripheral countries alike functioned to create and extend global markets as well as to stabilize the world system as a whole. This is an understanding recently advanced by David Harvey, among others, and we show how it applies quite well to the study of global media history.

A related aim of our study is to interject into contemporary debates swirling about the idea of empire, in particular to challenge those that argue (for example, Naill Ferguson, Michael Ignatieff, Robert Kaplan) that empire has been and could still be a good thing. Along these lines we document how communication and globalization during the period covered mutated so easily into problems of empire and in ways that offer insights into trends in our own times. Much along the same lines that contemporary analysts of empire such as Ferguson, Ignatieff, and Kaplan suggest apply today, this could be seen, for example, in widespread perceptions in the late nineteenth century and early twentieth that "weak states" and the failure to modernize the "backward regions" of Latin America, Africa, Asia, and the Middle East generated political and economic instabilities which threatened the viability of the global system altogether. Seen from this vantage point, imperialism and empire were cast as antidotes to the failures of modernization and the resultant economic instability in the global system that ensued, as evidenced in the economic crises of 1873 and 1890 as well as in the end-of-the-era mega-crisis of 1929. And of course, the intensification of imperial rivalry was real enough amid the mounting nationalism and militarism which helped to unravel the sinews of economic, technological, political, and cultural integration in the long slide toward the abyss of the Great War. In our view, the root causes of imperialism in this period and the collapse of that system in the conflagration that was World War I ought to stand as a stark reminder to those who would in our times take a rose-colored view of imperialism.

Finally, our study may claim to be unique insofar as we give prominence to the role of modernizers and reformers in the history of the global media. In doing so, we highlight the influence of two main groups of individuals on the politics of the global media and the larger processes of modernization. The first was an instrumental group of global media reformers who tenaciously fought to eliminate the cartels and to challenge the governments that supported them. Some of them, such as the Canadian engineer and visionary Sir Sandford Fleming, did so largely within the confines of an imperial sensibility, while others, notably British Member of Parliament John Henniker Heaton, came under the sway of the rising tide of liberal internationalism that characterized the times. Somewhat later, and just after the war, there were others, in particular a small coterie of individuals such as Walter Rogers, Ernest Power, and Breckinridge Long, who were closely associated with the U.S. State Department and worked tirelessly to reshape the global media system and to harmonize the regulation and legal framework for that system under the auspices of the ill-fated League of Nations. And such reformers were not alone or limited just to the centers of imperial power. Indeed, they were joined by a second group consisting of others—for example, the Chinese intellectual

Kang Youwei and the Persian diplomat Mirza Ja'afar Khan—whose efforts to create constitutional government and more liberal societies within the faltering anciens régimes of China, Persia, and Turkey as well as in the postimperial nations of South America also had an impact on the politics of the global media. The role of these reformers and their impact throughout the era are given ample play in the pages ahead.

Of course, a work of this kind could not have been accomplished without the generous support and assistance of a vast number of colleagues, friends, and research organizations. And while we can hardly do justice to them all, we would especially like to acknowledge, with thanks, the major financial support given for our project by the Social Sciences and Humanities Research Council of Canada. Financial assistance in both the early and finishing stages of our research was also given by Carleton University in Ottawa and, at the earlier stages, by the Advisory Research Committee of the School of Graduate Studies and Research at Queen's University. Particular thanks go to Dean Katherine Graham of the Faculty of Public Affairs and Management, and Dean Feridun Hamdulahpur, Vice President, Office of Research and International at Carleton University.

Our appreciation goes also to Lynn O'Malley, administrative assistant in the Department of Sociology at Queen's for managing the books, to Joan Westenhaefer, senior secretary in the department, for always being there when needed, and our gratitude goes to Michelle Ellis, secretary in the department, for her patient handling of time-consuming printing jobs. Merci beaucoup, Michelle! Dwayne would also like to acknowledge and thank two special friends, Doug and Patti Smith, who provided a sympathetic ear and moral support throughout. He also expresses his appreciation to Jocelyn Christie, who provided exceptionally loving company, warmth, and generosity in the final stages of this project.

We are especially grateful to Valerie Millholland of Duke University Press, who saw the merit in our book early on and encouraged its publication.

The following British institutions and individual staff also gave us valuable assistance: the staff of the British Telecommunications Archives; the staff of the Special Collections of University College [University of London] Library for help with documents on the Western Telegraph Company; the staff of the Guildhall Library, City of London, for material on the Globe Telegraph Company; the staff of the Cable and Wireless Porthcurno Trust and Museum of Submarine Telegraphy, and especially the director, Mary Godwin, for her patient assistance with archival sources and appropriate images; the staff of the India Office Collection at the British Library in London, and notably

Andrew Cook, for documents on early cable communications to India. In the United States, we thank the staff at the Library of Congress and Alexander Magoun of the David Sarnoff Library at Princeton for help with American source material. In Canada, we thank the staff of the Jordan Library Collections and the Archives at Queen's University in Kingston and the staff of the National Archives of Canada in Ottawa.

The following scholars provided helpful resources: Michael Krysko, who provided us with a copy of his valuable doctoral thesis from the State University of New York at Stony Brook on the topic of American radio in East Asia, 1919–41; Daqing Yang of the Department of History at George Washington University, whose then-unpublished paper on the history of early Japanese cable expansion was extremely helpful; and Marilyn Levine, the chair of the Division of Social Sciences at Lewis-Clark State College who took the time and effort to find and send us a series of images of Chinese notables. Other scholars who provided encouragement and advice were Jorma Ahvenainen, Oliver Boyd-Barrett, John Britton, Gillian Cookson, Rhoda Desbordes, Daniel Headrick, Karim Karim, Graham Murdock, Peter Putnis, Tehri Rantanen, Emily Rosenberg, Daya Thussu. Also, Amanda Laugesen, then a doctoral student at the Australian National University, provided meticulous help as a research assistant in combing the Australian source material. Several graduate students in the School of Journalism and Communication, Carleton University, also gave valuable research assistance: Ning Du, Nadine Kozak, Paul Magee, and Michael Santiani.

In the labyrinth of image collection and international copyright laws, the following deserve our gratitude: Kurt Jacobsen, director of the Centre for Business History at the Copenhagen Business School for assistance in obtaining photographs and permissions from GN Great Nordic; Marcia Thieme, coordinator of Licensing Services at the Canadian Copyright Licensing Agency for unstinting advice and assistance; Ian McDonald, senior legal officer at the Australian Copyright Council for advice in dealing with reluctant copyright holders; Andre Gailani, picture researcher with Punch Cartoon Library for his long-term friendly and personable assistance; Luci Gosling and Marcelle Adamson, researchers with the Illustrated London News Picture Library for their patience with our requests.

We have made every effort to trace copyright holders and to obtain their permission for the use of copyright material. We apologize for any errors or omissions in our list of permissions and would be grateful to be notified of any corrections which could then be inserted in future reprints or editions of this book.

COMMUNICATION AND EMPIRE

Introduction: Deep Globalization and the Global Media in the Late Nineteenth Century and Early Twentieth

There is much debate over just when globalization began. Much recent writing assumes that it was at the end of the twentieth century, whereas other scholars argue that such processes go back for millennia, if the focus is on trade, migration, and cultural contact. Still others point to the rise of capitalism in the fifteenth century or the constellation of factors—the ascendance of nation-states, industrialization, displacement of a religious cosmology by science, etc.—known as modernity in the nineteenth century as the moment in time in which the modern global order took shape. Our starting point of 1860 reflects a number of factors, including the extra-European focus of capital flows that emerged in the last few decades of the nineteenth century, the parallel rise of multinational corporate and financial institutions, the emergence of new technologies and business models that became the basis of what we call the global media system, and the advent of modernity, not just in the core of the global system, but also among certain political, intellectual, and commercial classes in the postimperial nation-states of South America and those struggling to reform the decaying edifices of anciens régimes in China, the Ottoman Empire, and Persia, among other places. And the end point of our study in 1930 reflects the backlash that mounted against globalization beginning in 1914 with the onset of World War I, but with the final nail in its coffin occurring over a decade later as strident nationalism and the failure of efforts to reconstitute a new era of internationalism based on greater technological, economic, political, and cultural interdependence led to the demise of the long global era covered in the following pages.

Globalization during the late nineteenth century and early twentieth was not just shallow and fleeting, but deep and durable. The growth of a worldwide network of fast cables and telegraph systems, in tandem with developments in railways and steamships, eroded some of the obstacles of geography and made it easier to organize transcontinental business. These networks supported huge flows of capital, technology, people, news, and ideas which, in turn, led to a high degree of convergence among markets, merchants, and bankers. This was especially so in the transatlantic economy, but the effects were also significant in the settler colonies of the British Empire, the River Plate zone of South America (Buenos Aires and Montevideo, but also Rio de Janeiro), and several Asian cities (Hong Kong, Shanghai, and Tokyo). And as governments grew more convinced of the power of communications to move markets, affect foreign policy, and shape public opinion they strove to subject them to greater control, as instruments of imperial politics and by regulation, at the national level and through international agencies, such as the International Telegraph Union (ITU, formed 1865).

The first aim of this book is to look at the rise of what we call the global media system. Of course, media historians have written extensively on long-distance communications networks and the global news agencies during this period. The existing research, however, has tended to divide the field into separate areas, with one body of writing focusing on the evolution, organization, and control of the *network industries* and a second which takes the global and national news agencies (the *content industries*) as its focus. In contrast, we look closely at the interconnections between the network and content industries by way of ownership, as customers and sources of revenue for one another and through business alliances. We also argue that the "global media system" was much more multinational and far less governed by considerations of national interests than usually assumed.

Our second aim is to contribute to an expanding body of literature which argues that globalization and the information revolution are not new but have their closest parallels in the period we cover.[1] In some ways, we are taking our cue on this score from Oliver Boyd-Barrett and Tehri Rantanen, who argue that "the links between modernity, capitalism, news, news agencies and globalization are an outstanding but neglected feature of the past 150 years."[2] We agree and relate our analysis of the global media to this larger constellation of issues.

Our third goal is to recast the relationship between communication and empire. We do so by, first, suggesting that while the new imperialism of this era (1880–1910) was crucial, it was not synonymous with either the broader processes of globalization or international communication. Our work also interjects into contemporary debates around the idea of empire, in particular to challenge those who argue that empire has been and could still be a good thing. We do so by addressing how globalization so easily mutated into problems of empire, not just as a problem for imperialist nations and their targets but as a problem of world order that ultimately contributed to the demise of globalism. The parallel between the crisis of globalization historically and the trends today, along with the efforts to legitimate imperialism in both periods, suggests that we are well advised to shine the light of historical knowledge on the issues of our own times.

EARLY GLOBALIZATION AND THE MEDIA

During the early to mid-nineteenth century, journalists and publishers were quite skeptical about the impact of the telegraph and cables on the news, but in relatively short order these new media technologies became indispensable adjuncts of the press. Contemporaries such as Julius Reuter invested in some of the most important ventures to "wire the world," founded the news agency bearing his name, and initially served as a director in the Globe Telegraph and Trust Company, an agency which strove to unify administrative control over the world's cable systems. However, Reuter soon divested his interest in these areas to focus solely on establishing news bureaus in important cities reached by the rapidly expanding cable system, as in Bombay (1870), Hong Kong (1872), Shanghai (1873), and Buenos Aires (1874), among other places. The French agency, Havas, also followed a similar practice and at times did so in alliance with Reuter. American agencies followed suit, but did not achieve a significant presence until the 1910s, although thirty years earlier James Gordon Bennett, the publisher of the largest circulating newspaper in the United States, the *New York Herald*, and Jay Gould, the owner of the *New York World*, controlled two important cable firms: the Commercial Cable Company and Western Union, respectively.

For the press, cable communications offered an opportunity to gain an edge in the increasingly competitive field of mass communications. The news wire services soon tied the urban press into a network that began, in effect, to broadcast foreign news to domestic audiences. Indeed, so great was this shift in orientation that the *Times* (London) allocated more of its front page to foreign news stories in the 1890s than it does today. The advent of wealthy papers such as *La Nacion* and *La Prensa* in Buenos Aires in the 1860s and 1870s also offered new markets for the cable companies and news agencies. Competition among the South American press also played a vital role in attempts during World War I to realign the balance of power between the American and European news agencies in the region. We draw attention to these examples to introduce one of our larger goals: namely, to show how national and global media institutions intersected with and affected one another's development.[3]

Communication markets were also influenced by complex interactions between private companies and governments. As business historian Alfred Chandler Jr. notes, the consolidation of industrial capitalism within the United States and Europe featured the rise of corporations of unprecedented scale, and the communication industries were at the forefront of these trends.[4] And nowhere was this more visible than at the global level, where the British-based Eastern Telegraph Company was the Microsoft of its age. Despite significant changes in markets, politics, and technology, the Eastern Telegraph Company continued to tower over its rivals, consistently accounting for half of the world's cable networks, even as late as 1929 when the British cable and wireless industries were merged and the company ultimately transformed into Cable and Wireless.

The leading companies had highly complex organizational structures, with a plethora of subsidiaries nestled behind their corporate umbrellas. Indeed, the communication business was so Byzantine that there is a tendency among writers to cast companies as extensions of nationally based capital and the policies of individual governments, that is, generically as British, American, German, French, Japanese, and so forth. Indeed, Jill Hills, in *The Struggle for Control of Global Communication*, claims that the nationality of these companies was essential, given their role in advancing the strategic interests of their respective governments. While there is no doubt that governments were keen to foster national companies, we believe it is a mistake to refer to firms by their national affiliation, with the implied inference that they were "tools of the state." Indeed, we argue that governments were quite ambivalent about the nationality of ownership before World War I.[5]

The disregard for the nationality of ownership could be seen, for instance, in the character of the financial syndicates that backed many of the companies. Firms such as the French Atlantic Cable Company, the Direct U.S. Cable Company, the Indo-European Telegraph Company, the Great Northern Telegraph Company, and the Western and Brazilian Telegraph Company all relied on capital from a variety of British, European, and American sources. Indeed, before the war, the industry could not be understood without grasping its ties to the largest British, European, and, although somewhat later, American financial agencies in the world.

If finance was one mechanism of structural integration in the communication industries, the formation of cartels was another. Indeed, from the outset, the global communications and media business developed as a series of cartels. Thus, the "ring circle agreement" of 1869 created a news cartel consisting of Reuters (British), Havas (French), and Wolff (German) (revised in 1893 to include the Associated Press). The cable companies quickly followed suit by creating their own cartels in the Euro-American, Euro-Asian, South American, and Indo-European markets in 1869, 1872, 1874, and 1878, respectively. The big four commercial wireless companies—Marconi, Radio Corporation of America (RCA), Compagnie Générale de Télégraphie Sans Fil (TSF), and Telefunken—did the same immediately after World War I.

These cartels were revised, expanded, and challenged periodically, but one thing is for sure: they were solidly entrenched well before the 1920s, in contrast to Jill Hills's suggestion that they only emerged in the network industries in response to overcapacity and economic depression in the 1920s and 1930s.[6] Foreign policy and business historian Michael Hogan refers to them as "private structures of cooperation" and suggests that their role went far beyond just managing competition to affect the foreign policy goals of nations. Our research strongly supports Hogan's views and actually extends them by indicating that these cartels emerged much earlier than 1918 and encompassed far more than just Anglo-American interests than he allows.[7]

The most interesting multinationals saw national identity as something that could be changed pretty much on an as needed basis. The U.S.-based Commercial Cable Company was the master at this, owning a stake in the German Atlantic Telegraph Company, claiming to be British as it sought subsidies for two "British" companies that it owned—the Halifax and Bermudas Company (1890) and the Direct West India Company (1898)—and "all-American" when standing before the U.S. Congress to promote why it should be chosen to lay a U.S.-owned cable across the Pacific (1904) and as it fronted for another firm—the U.S. and Haiti Telegraph Company (1896)— that was registered in New York but in reality owned by French interests. The company also formed alliances with European companies to offer services in the Euro-American market. Given that such relationships were the norm rather than the exception, it seems questionable to assume that the nationality of corporate actors was easy to establish or highly relevant.

COMMUNICATION AND EMPIRE:
KEY CONCEPTS AND THEORETICAL PERSPECTIVES

Global communications were overwhelmingly privately owned before the 1930s and the dominant approach assumed by governments, even if selectively, was that of free trade. Yet, that doctrine seemed rather incongruous with rampant cartelization and the companies' other favorite device for avoiding "ruinous competition": monopolistic concessions. These concessions afforded the companies incredible power to restrict access to markets, affect the development of national media systems, and, by the 1910s, to shape the advent of wireless. They also spawned intense debates within Argentina, Brazil, Chile, Uruguay, China, and, while no longer a developing country, also Japan, over the impact that foreign multinationals were having on national telegraph systems and their links to the outside world. Concessions were also a lightning rod for a tenacious group of reformers who sought to eliminate such privileges and cartels in favor of a more open, public service view of global communications.

While private ownership defined the early period of globalization, the cable companies and news agencies were drawn closer to the state from roughly 1890 until just after World War I. The military historian George Wilson portrayed this shift in a 1901 lecture to the U.S. Naval College:

> Originally planned with a view to financial returns to the constructing companies, recent lay outs have been with a view to military and strategic usefulness. . . . This policy of construction on imperialist instead of commercial grounds has arisen mainly within the last ten years. . . . This has aroused the liveliest debate in France. . . . Germany feels the necessity of meeting Great Britain by lines of cable defense. . . . Propositions for new Pacific Ocean cables . . . have recently been made in Great Britain and in the U.S. Both Governments seem inclined to support these cables with ample subsidies, if not to undertake them directly.[8]

During the era described by Wilson, several new imperial powers arose alongside the existing British and French empires, not the least of which were the United States, Germany, and Japan, while others such as Spain, Portugal, the Ottoman Empire, Persia, and China continued to decline. And as we show, it was at this time that France, Germany, and Japan rationalized their national cable industries and spearheaded a phase of significant growth, ostensibly to break Britain's stranglehold over global communications. At the same time, the British military pushed for strategic cables for defense purposes, while other British and imperial advocates such as Sandford Fleming, Edward Sassoon, Charles Bright, and John Henniker Heaton advocated an extensive system of state-owned cables to be run, essentially, as a public service network for the British Empire. This latter discourse was exemplified by Sandford Fleming, the Scottish-Canadian engineer and visionary of international standard time and global communication, in a series of essays, *To the Citizens of the Empire* (1907). In these essays, Fleming blasted the companies for their high rates and called on the governments of the British Empire to establish

> an unbroken chain of state-owned cables connecting the self-governing British communities in both hemispheres. It is believed most thoroughly that the proposal will eventually be consummated, and that by bringing the several governmental units, now separated by great oceans, into one friendly neighborhood, electrically and telegraphically, results will follow of the most satisfactory character,—commerce will be quickened, the ties of sympathy will be made more effective, the bonds of sentiment will become more enduring, and by this means, unity, strength and permanence will be assured to the family of nations constituting the Empire.[9]

Many media historians point to these trends and conclude that imperialism was one of the most, if not the most, significant factors driving the development of international communications. On the surface, there is some evidence to support such views. For one, during the "new imperialism" era (1880–1910), the volume of government subsidies given to private companies mushroomed, and the scale of state-owned cables almost doubled, from under 10 percent to nearly 20 percent. Interimperial rivalry as well as militaristic and nationalistic sentiment escalated, as evidenced by the near clash of Britain and France during the Fashoda crisis of 1898, festering animosity between Britain and Germany during the Boer War in South Africa (1899–1902), and France and Germany's near confrontation during the Moroccan crisis (1908). And while the superpowers squabbled among themselves, a huge expansion of imperialism proceeded, as over 90 percent of Africa and large swathes of Asia and Central America fell or remained under their control. Yet, in contrast to assumptions in much of the literature, most of these territories, except the settler colonies of the British Empire, continued to be among the least connected, worst served places on the planet. Instead, as we show in detail, investment in cable networks was concentrated in lucrative markets.

David Harvey describes these processes as being a part of what he calls *territorial imperialism*—a "distinctly political project . . . based in the command of territory and . . . the . . . strategies . . . used by a state . . . to assert its interests and achieve its goals in the world at large."[10] When most studies refer to the relationship between communication and empire, it is this view that they seem to have in mind. The stress tends to be on nation-states, national interests, territorial acquisition, interimperial rivalry, and a view of communication as a "weapon of politics." Yet, as John Gallagher and Ronald

Robinson put it in "The Imperialism of Free Trade," to assume that this view represents the sum total of imperialism "is rather like judging the size and character of icebergs solely from the parts above the water-line."[11] Their point being that formal definitions of imperialism based on territory fail to illuminate areas such as South America, the Ottoman Empire, and China which, while not part of any formal empire, were subject to an incredible degree of external influence and were, thus, part of "informal empires."

A more accurate measure of empire is obtained once territorial imperialism is considered together with what Harvey calls *capitalist imperialism*. According to Harvey, imperialism of this kind focuses on the flow of "economic power . . . across and through continuous space . . . through the daily practices of . . . trade, commerce, capital flows, . . . labour migration, technology transfer . . . flows of information, cultural impulses and the like."[12] Capitalist imperialism strives to make the flow of capital, commodities, and information across the different spaces of the global system—local, national, and global—as seamless as possible. These processes, in turn, require the modernization of nation-states *and* the rationalization of the political and legal underpinnings of the world system. And unlike the emphasis on the coercive power of nation-states in territorial imperialism, the harmonization of "capitalist space" relies on the "soft power" of consent and the emulation of "models of development." A further difference is that under territorial imperialism, power is exercised by single states, whereas "the collective accumulation of power is the . . . basis for hegemony within the global system" more generally.[13]

Harvey's distinction between territorial and capitalist imperialism also implies separate theoretical starting points, which we can identify as a *realist* model of power and international relations in the case of the former and a *structuralist* stance in the case of the latter. Territorial imperialism is steeped in a realist view of international relations that emphasizes national interests, balance of power politics, militarism, and the subordination of markets to the state. In our field, the result is a portrait of an unending struggle for control of communications among the superpowers—Britain, France, Germany, the United States, and Japan—with corporations and communication technology serving mainly as tools of the state. To be sure, there are numerous instances where this was the case as, for example, with the hyper-nationalistic All-America Cable Company, which liked to cast events as a "great patriotic struggle" for the control of world communication. However, that was that company's view of things (for which it had good reasons) but it gained remarkably little support, even in the United States, and matters were more complicated than an analysis on the basis of national identity, as we have already indicated, would suggest.

Whereas realists see power as anchored in states, material resources, and national interests, the structuralist view sees it as working through a nexus of institutions, as embedded in the "rules of the game," and girded by knowledge and ideologies.[14] Steve Lukes highlights the dimensions of this view by distinguishing between what he calls the "three faces of power."[15] The first "face," according to Lukes, is similar to the realist view of power. It is instrumentalist and behavioralist, meaning that actors each have their own distinct interests and power is the ability of one actor to realize its interests in the face of opposition from others. The second dimension is what he calls agenda-setting, where power is a function of actors' ability to shape the political and discursive agenda around which decisions, political change, and conflict occurs. Third, there is a systems view, in which power is the ability to shape relationships among actors by setting the "rules of the game," controlling resources (material and symbolic) and the basic goals that are open to all. To this we can also add, and this is one of Harvey's key points, the systems view does not assume that corporate interests are co-equal with or subordinate to the interests of the state or that the state is simply the handmaiden of capitalism.[16] Indeed, we show, first, that the multinational complexion of the communication industry would have made the realist assumptions about the national basis of power difficult to realize in practice; second, that governments were rather indifferent to the nationality of ownership prior to World War I; and, last, that state and corporate actors often pursued their interests independently of one another—although this was not, of course, always the case.

While a systems view of society appears to be of recent vintage, it was well developed, and in fact was a constitutive element, in early globalization processes. That this was so could be seen in the views of Charles Conant, onetime secretary of the U.S. Treasury and a sympathetic theorist of American imperialism. On the eve of the ascendancy of the American empire in 1898, Conant advanced such a view in his theory of "modern capitalist investment imperialism," which he described as follows:

> Investment imperialism required extending capitalist (or modern) property relations, and the appropriate sociopolitical environment for capitalist property relations, into noncapitalist societies. It required "modernizing" the host government's fiscal, budgetary, and taxation systems, the host societies' laws; the host societies' class structure in the direction of the commoditization of land and the creation of

a wage-earning working class. It required the introduction and spread of secular and instrumental modes of consciousness at the expense of religions and tradi-
tionalist modes, through the institutions of education, media of communication,
and otherwise.[17]

Conant's views reflected transformations in the global economy that were
taking place at the time as well as deeper intellectual currents that had begun
to migrate from European legal circles to the nascent American progressivist
movement. In particular, Conant stressed how the velocity, volume, and reach
of transportation and communication networks were leading to a high degree
of convergence among markets. Tighter integration, in turn, and the influx of
capital, much of it through foreign loans and government bonds, generated
ambitious attempts to create modern postimperial nation-states in South
America or to reform the ossifying ancient regimes of China, Persia, and
Turkey. These global forces, as we show, intersected with endogenous processes of political and cultural change within these nations to shape the development of national telegraph systems, the rise of a modern press, and
attempts to create modern constitutional nation-states.[18] Yet, while the speedy
flows of capital and information helped to synchronize markets and propelled
huge development projects, some also believed that they might be magnifying
the volatility of markets and diffusing instability throughout the world economy. At least that was one explanation of the first modern global financial
crises, namely, those which occurred in 1873 and 1890. By looking at how
crises begot efforts to further the "rational" management of nation-states
within the global system or, should that fail, to justify direct control (imperialism) over areas of chronic insecurity, our aim is to recast conventional
analyses of global media history and to shed light on the nature of similar
trends in our own time.[19]

Economic historians have highlighted the scale and impact of these crises
but say little about the role of the communications media in them.[20] Histories
of international communication, in contrast, have said almost nothing about
the relationship of the swings between financial booms and crises on one
hand and communications on the other. Nonetheless, the links were made by
Conant, in the Annual Reports of the Council of Foreign Bondholders and in
debates between that agency and Reuters conducted in the pages of the *Times*.
These views were also held in certain U.S. circles and came to underpin
Woodrow Wilson's version of liberal internationalism during his two terms in
office (1912–1920).[21] Others such as Paul Reinsch, University of Wisconsin
professor and future U.S. Minister to China under Wilson, as well as Frank

Giddings, author of *Democracy and Empire* (1900), also adopted a view of imperialism as a rational solution to crises in the world system.[22] In many ways, these were the progressive architects of empire at the beginning of the twentieth century, and their ideas probably read much like, say, Michael Ignatieff's *Empire Lite* or Robert Cooper's "The New Liberal Imperialism" do to us today.[23]

Such views also drew on a well of European experience with International Financial Commissions that were used during the late nineteenth century to deal with bankrupt countries and chronic instability in the Ottoman Empire, Egypt, Portugal, and the Balkan states, among other places. They were also part of a circuit of ideas that began to flow between European legal scholars at the Institute of International Law, on one hand, and American political and intellectual circles, on the other.[24] Of particular relevance to our study is the role that members of the institute, most notably French scholar Louis Renault and Dutch scholar Tobias Asser, played during the formative years of the ITU, and penning some of the first conventions, treaties, and regulations that came to govern global communication during this time. Others of their kind, such as Ernst Von Stephan of Germany would also use their position as agents responsible for setting national communications policies to articulate visions of very contemporary relevance, in this case a vision of a trans-European communicative space. And while the European powers dominated the ITU, the participation of Argentina, Brazil, Chile, China, Japan, Persia, and Turkey hinted at the more cosmopolitan views of communication, law, and world order, even if never fully achieved. This mixture of markets, communication, law, and a transnational circuit of ideas, nonetheless, helped to deepen and extend the sinews of globalism during this time.

Our desire to illuminate the multinational character of the global media system, the role of international law, the transnational flow of political and cultural ideas, and the systems view of globalization should not be interpreted as a Pollyannaish view of this period. Indeed, first and foremost, the shift from a realist to a systems theoretic view only reconceptualizes the nature of power rather than diminishes its role. And in that, we are keen to point out that the global media cartels, for instance, were first and foremost collective instruments of economic power as well as tools for managing the sometimes conflictual foreign policy objectives of nation-states. They were also, as we will see most notably in terms of the Asian and African cable cartels, tools used by the imperial powers to meet their collective commercial, diplomatic, and security needs.

Moreover, our attempt to recast the events in light of a systems view does not mean that the realist view was impotent. Indeed, just as today there are unilateralist hawks who deride multilateralism and international law as constraints on national prerogatives versus those who believe that such mechanisms are indispensable in an interdependent world, so too in the past were there schisms between "conservative" and "liberal" internationalists. And both camps had their imperialists, with realists prioritizing strategic, national, and racial factors to justify imperialism versus their liberal opponents who cast imperialism in a modernizing framework and a necessary, even if transitory, method for ameliorating disorder in "backward" countries. As we will show, nowhere was this more evident than in China, where the superpowers endlessly oscillated between the "single-state" model of territorial imperialism versus the collective accumulation of power and capitalist imperialism. Equally vital is the recognition that both models intersected in contrasting ways with endogenous forces within China, with the former aligning itself with commercial elites in the more dynamic city-states of Shanghai, Tianjin, Fuzhou, and so forth and the latter colluding with those striving to create a modern nation-state that was simultaneously internally cohesive and externally open to the rest of the world. Grasping this is crucial to understanding the early foreign development of cable and telegraph systems to and within China. It also helps us to comprehend proposals for the "internationalization of control" over cable communications as well as wireless communications in China in the 1910s and 1920s. It is also vital to understanding the role of the United States in China well before it had any significant material interests in the country, and the position of the United States in the world more generally, with its contradictory swings between the pursuit of "open door" policies in one region and a more straightforward "spheres of influence," territorially based model of imperialism in others.

By coming at the study of global communication with both of these models of imperialism and world order in mind—and with the realist and structuralist perspectives close in tow—we are also able to get a better grasp of at least two crucial matters: first, mounting nationalist sentiments in the run-up to World War I and, second, the postwar politics of Versailles and prospects for world order thereafter. With respect to the first point, the mounting nationalism just prior to and during the war is better understood once seen in relief against the internationalism of the preceding period and how the systems view had actually been embedded in the structure of the world economy, the global media, international law, and modernization projects. Against those standards, Germany's militarization at the beginning of the twentieth century also looks all the more remarkable. So too does Japan's aggressive territorial imperialism and its development of a vast cable system as part of that effort. And during the war, these trends could be seen as Britain, France, and Japan expropriated German cables which had previously been operated as part of the international cable cartels. Likewise, only by grasping how strong the realist view of affairs had become can we really understand how the previously weak nationalist undercurrents in the United States, which had long bemoaned the presence of British-based Marconi on American soil, became strong enough in the hands of Postmaster General Albert Sidney Burleson and key figures in the navy, such as Josephus Daniels, that Marconi's property could be, for all intents and purposes, expropriated to pave the way for the creation of RCA in 1918. During this time, communications media really did become "weapons of politics." And as weapons of politics, the belligerents turned them into instruments of propaganda, fighting for the hearts and minds of domestic populations, nonaligned countries, and ene-mies alike. This thorough transformation, we will argue, biased the orienta-tion of communication networks and tainted the operations of news agencies and journalism to such a degree that attempts at Versailles and in the early 1920s to reconstitute a modern world order along broadly liberal and demo-cratic lines were thoroughly compromised.

And this is our second point: namely, that the events after the war at Versailles and into the 1920s need to be taken far more seriously than they have been by communication scholars. In particular, we argue that proposals by the U.S. State Department and individuals such as Walter Rogers, Breckin-ridge Long, and Ernest Power represented far more than just an insular power grab by the Americans for control of world communication. Instead, we indicate that the vision they offered drew explicitly on a long line of "global media criticism" from where others before—Fleming, Heaton, and Sassoon—had left off. We also suggest that their push to have the entire system of exclusive concessions and cartels that had so defined, and clogged, the global media system reviewed and put under a more coherent international regula-tory regime had much merit. We also believe that communication scholars might see in these initiatives the origins of the "free flow of information" doctrine that has since become the bedrock principle of U.S. international communication policy, rather than, as is so often assumed, a ruse of post–World War II cold war politics. Last, we argue that once seen in the light of the systems view, it becomes easier to understand Woodrow Wilson's approach to internationalism and Versailles in contrast to those who see him as either a dewy-eyed idealist or a cunning moralist trying to cloak American ambitions in a fog of high-minded rhetoric.

Indeed, it is our view that only once we approach the issues along these lines can we grasp the fact that liberal internationalism and American imperialism were at their zenith during this time. And moreover, such a perspective also allows us to consider how questions about the political economy of the global media became central to efforts at Versailles to re-create a viable world system, in particular the idea that such a system, to be successful, not only needed to be rooted in technological, economic, and political interdependence, but personal experience or, in other words, culture. That is, rather than understanding the treatment of communication and public opinion at Versailles, in U.S. proposals, or even in the subsequent activities of the League of Nations as fuzzy idealism, the key to understanding the times is to see such themes as essential ingredients in an attempt to deepen the processes of globalization.

While we draw on previously underutilized archival material to show how these ideas were given expression by the likes of Rogers and others within the State Department, circa 1918–22, we also demonstrate that such ideas need to be treated dialectically. And by that we mean that they need to be understood on their own terms but also in relationship to the opposition they engendered among certain quarters on the world stage and within the United States. That is, the clash between the realpolitik and the "global systems" view was simultaneously a part of world politics as well as the political dynamics within the United States. Thus, while encountering opposition from certain quarters in Britain and Japan, yet with some qualified support among European powers and China, for example, there were cleavages in the United States that deeply divided the communications and media industries as well as liberal and conservative internationalists. As for the media, RCA, Western Union, the All-America Company, and the Commercial Company, they all solidly opposed the Wilsonian position at the proposed World Communication Conference, while the press and news agencies expressed far more favorable views.

In the end, though, we indicate that such matters became entangled in the broader quagmire of factors besetting the politics of Versailles and the formation of the League of Nations. As a result, matters of global communication tended to devolve back into the national arena, where they were either taken up by particular agencies in consultation with the communication industries or as part of parliamentary or congressional hearings, such as the Cable

Landing Licenses Hearings held in the United States between December 1920 and early the next year. While that event was ostensibly only about cable communication issues, it turned into one of the most expansive reviews of the global media system in the United States up to that time. And it was another moment in which cleavages in the U.S. political culture surfaced, with fiery nationalist rhetoric being spewed by a few hot-headed nationalists from the navy and the All-America Company. But it was also another moment when such views had to be weighed against those who either sought to ensure that governments and international institutions alike left the communication companies to their own devices or who sought to reform the entire global media system from top to bottom, but this time through the instrument of the U.S. government rather than the League of Nations or, as RCA derisively referred to it, a global super-regulator. As we will show, it was this climate, the swing from the postwar economic doldrums to the Roaring Twenties thereafter, and, finally, the devastation of rising authoritarianism in world politics and the Great Depression of 1929 and the 1930s that simultaneously characterized the times and, ultimately, led to the reversal of globalism altogether.

Building the Global Communication Infrastructure: Brakes and Accelerators on New Communication Technologies, 1850–70

Nearly a century and a half before the World Wide Web and a glut of fiber-optic cables were weaved around the planet, another global communication infrastructure was being put into place. This chapter highlights three key factors that drove that process: the consolidation and cartelization of telegraph industries in Britain, Europe, and North America; the cooperation between Anglo-American and European interests in establishing the first major long-distance cable systems; and the interaction between the cable promoters and speculators and the political and cultural currents within the "developing countries" of the Middle and Far East that shaped the evolution of communications systems. Overall, the chapter canvasses the visions, successes, and failures that shaped global communications between 1850 and 1870.

EXTENSIONS AND INTERCONNECTIONS: THE FIRST TELEGRAPHS

Early prototypes of the telegraph had been around since the 1810s, but it was not until the 1830s that Charles Wheatstone and William Cooke introduced the new technology in England. Samuel Morse did likewise in the United States, but not until 1843. European countries quickly followed suit, although for years strict state control in some countries prevented use of the telegraph by the general public. This was particularly notable in France, where the French Interior Minister declared in 1847 that the telegraph was "an instrument of politics, not of commerce."[1] Only after the revolution of 1848 and ensuing liberalization of the French economy was France's telegraph system thrown open for public use (1850). Likewise, the Austro-Hungarian Empire built a telegraph system in 1846, but did not permit the public to use it until 1849. Wheatstone was called on in 1845 by the Dutch government to construct a line between the stock exchanges in Brussels and Antwerp. And two years later Charles Siemens laid the basis for the communications colossus that would bear his name into the twenty-first century by constructing a network for the Prussian government. During the same period, lines were extended into Belgium, Italy, Spain, Russia, and Sweden and, over the next two decades, to Latin America, the Middle East, Africa, and Asia.

From their inception, European telegraph systems interconnected across national frontiers, and rules were put into place to govern technological standards, rates, the distribution of revenues, citizens' right to privacy, and also the prerogative of states to censor messages to ensure public morality and national security. From the outset it was clear that there were tensions around new communication technologies which would decide whether they would develop in relatively open ways or as closed systems under the strict supervision of nation-states. Beginning in the 1850s, these issues were handled by two agencies—the Austro-German Telegraph Union and the West European Telegraph Union—but in 1865 they were merged to create the International Telegraph Union (ITU). As the first multilateral organization, the ITU was on the forefront of efforts by legal scholars and diplomats to forge multilateral and cooperative solutions to world problems through the rule of international law.[2]

BRITAIN: THE FORMATION OF A GLOBAL COMMUNICATIONS POWERHOUSE

As the telegraph business took off, Charles Wheatstone turned his effort to developing this business, while his partner, William Cooke, receded into the background to focus on the science of the new technology. Wheatstone formed in 1846 what was to become one of the two dominant companies in Britain, the Electric Telegraph Company. Five years later its main rival emerged, the British and Irish Magnetic Telegraph Company. The Magnetic, as it was known, was packed with a "who's who" roster of engineers, financiers, and politicians, and over time these people became *the* global telegraph and cable barons of the age, notably its managing director, John Pender, a Manchester textile entrepreneur, and two communication technology experts, Charles T. Bright and John W. Brett. The only other firms of any significance in Britain during the 1850s and 1860s were Reuters' Telegram Company, the Universal Private Telegraph Company, and the London District Telegraph Service.

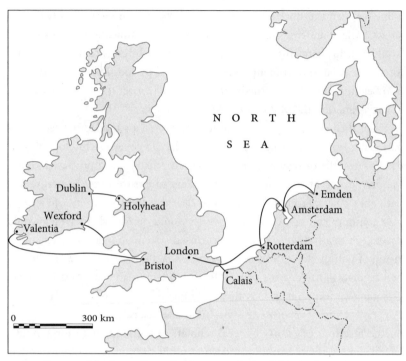

NORTH

SEA

Dublin

Holyhead

Wexford

Valentia

Emden

Amsterdam

London

Rotterdam

Bristol

Calais

0   300 km

MAP 1   European Atlantic coastline and main communication links, circa late 1850s

Pender's Magnetic Telegraph Company and Cooke's Electric Telegraph Company quickly consolidated their grip on the British telegraph industry by the mid-1850s, and as they did they diversified into the news business and expanded internationally. By 1854, the Electric Telegraph Company had its own journalists collecting and distributing news, stock market prices, and market information, and more than 120 newspapers subscribed to its services. Not to be outdone, the Magnetic began to do the same. By decade's end, however, both companies stopped competing in the news business in favor of combining with one another, as well as Reuters, in a deal that gave them sole distribution rights for Reuters's news service in British cities outside London —which remained the exclusive preserve of Reuters. For an annual fee the equivalent to US$1,000,[3] the companies provided the *Belfast Newsletter, Glasgow Herald,* and *Manchester Guardian,* among other newspapers, a daily service of approximately 6,000 words. In short, the companies leveraged their control over networks into control over content. A decade later, the cartel's grip tightened with the admittance of the United Kingdom Telegraph Company into its fold. But the cartel was a steady source of irritation, and in their campaign for state-owned telegraphs, the British provincial press quickly made major issues of the "telegraph news monopoly" and its high rates.

While the Electric Telegraph Company and the Magnetic cooperated in the British telegraph and news market, they competed with one another in European and international markets. John Brett, a director at the Magnetic, led the way by establishing the Submarine Telegraph Company in 1850. A year later, and with the backing of British and French capital, the firm laid the first cable from Dover to Calais in 1851, thereby putting the London Stock Exchange and Paris Bourse into near real-time contact with one another. The company laid five more cables to Europe over the next decade. Brett also created two new affiliates—the Mediterranean Electric Telegraph Company (1854) and the Mediterranean Extension Telegraph Company (1857)—in a bid to wire-up the Mediterranean region and to extend a system of cables from there to India—a point we return to later in the chapter. The Magnetic also expanded into Europe with two cables to Ireland in 1853 and joined an alliance with Brett's Submarine Telegraph Company in 1859. But such moves also drew into the fray the Electric Telegraph Company, which had two new cables of its own to Amsterdam and Rotterdam in 1852 and extensions to Germany, Austria-Hungary, and St. Petersburg during the next four years. In short, cooperation at home was offset by competition between the two new firms in European markets.[4]

ANGLO-AMERICAN RELATIONS AND
EARLY EFFORTS TO CREATE A TRANSATLANTIC
COMMUNICATION MARKET, 1854–58

The telegraph business in North America was more competitive from its beginnings than either the oligopolistic British telegraph industry or the state-controlled telegraph systems of Europe. By the mid-1850s, however, Western Union began to absorb its competitors at a furious pace and to establish affiliates in Canada. Close links were also forged with Cyrus Field, a New York publisher and printer who plunged into the telegraph business during the 1850s. Of particular importance to the ties between Field and Western Union was the acquisition of three small companies created by the Scottish-Canadian speculator, Frederick N. Gisborne, in the Canadian maritime provinces: the British North American Telegraph Company; the Newfoundland Electric Telegraph Company; and the New York, Newfoundland, and London Telegraph Company. Together, these firms constituted the nucleus for a planned transatlantic communications system that would reach from New York, Boston, and Halifax on one side to London on the other.

The consolidation of telegraph systems along the Atlantic seaboard was spurred on by rivalry among telegraph companies and news agencies for control over the flow of news and information between Europe and North America. American news agencies were obsessed with capturing European news as it arrived in Halifax by ship and speeding its delivery to the press and news services in New York and Boston. Using the telegraph instead of waiting for ships to make the journey from Halifax to Boston or New York shaved two days off the time for European news to be published in the North American press.[5] However, those potentials had been compromised by unscrupulous news agents who would bribe telegraph officials and even resort to cutting the lines of their rivals to gain an advantage. Recognizing the advantage of bringing telegraphs under unified control and the need to cooperate with the main American news agency, the Associated Press, Field acquired Gisborne's Newfoundland Electric Telegraph Company and the New York, Newfoundland, and London Telegraph Company in the mid-1850s.

The acquisitions, in turn, became the pillars of two new firms: the American Telegraph Company (1854), which would operate along the Atlantic seaboard of North America, and the Atlantic Telegraph Company (1856), which was intended to operate the planned transatlantic cables. With these acquisitions, Field gained monopolistic rights to operate telegraphs in the aforementioned areas for the next half century, which played a decisive role in shaping the nascent Euro-American communication market for decades. Stitching up rights along the maritime coast also gave Field the clout to deal with Western Union on a more equal basis, as was reflected in their decision to form the North American Telegraph Alliance in 1857. Although there were six members in the cartel, it was a creature of the American Telegraph Company and Western Union.[6]

As in Britain, consolidation within North America was a prelude to expansion into global markets. Between 1854 and 1856, and to howls of protest from rival telegraph companies and news agencies in the United States, Field and Western Union struck several deals that left the North American market exclusively to Western Union while reserving the yet-to-be-created Euro-American market to the Atlantic Telegraph Company. The deals also transferred the American Telegraph Company's inland system to Western Union and, in return, gave the Atlantic Telegraph Company exclusive access to West-

1   "The Atlantic Cable Projectors," Daniel Huntington's retrospective painting (1895) of a meet-
ing of the Atlantic cable projectors at the residence of Cyrus Field in 1854 to found the New
York, Newfoundland, and London Telegraph Company. Left to right: Peter Cooper, David Field,
Chandler White, Marshall Roberts, Samuel Morse, Daniel Huntington (the artist), Moses Taylor,
Cyrus Field, Wilson Hunt. Note Field's globe in the foreground and the cable fragment on the
table. Reproduced by permission of the New York State Museum, Albany.

ern Union's vast North American distribution network for the next fifty
years.[7] Rivals charged that these arrangements created an international cartel
and petitioned the U.S. Congress to intervene on behalf of maintaining competition, but to no avail.[8]

Field registered the Atlantic Telegraph Company in London in 1856, re-
flecting the fact that the American group behind the company was closely
allied with the British-based Magnetic Telegraph Company directors, notably
John Pender, John Brett, Charles T. Bright, and a few others. In fact, the vast
majority of the capital and nine of the new firm's directors were from the
Magnetic Company. The British connection was also obvious in the fact that
Field relied on two British firms—Glass Elliot and the India Rubber Gutta
Percha and Telegraph Company—to manufacture the cables to be used in his
venture. The British Submarine Telegraph Company was also hired to lay the

2   Glass Elliot's Telegraph Cable Works, 1857. Engraving from the *Illustrated London News* showing
     the Glass Elliot Cable Works in East Greenwich, London, 1857. The company name is partly
     visible on the roof of the building on the right. The mass of rolled cable for the first Atlantic cable
     is in the foreground. Reproduced by permission of the Illustrated London News Picture Library.

cables.[9] Finally, the Anglo-American connection was evident as the U.S. Congress and British Parliament passed nearly identical legislation in 1857 pledging naval surveys and annual subsidies of $70,000. In return, both governments received guarantees that their messages would be treated equally and given priority over all others. All in all, the arrangements revealed the Anglo-American foundations of the Atlantic economy, even though a cable between North America and Europe was still almost a decade away.[10]

The Atlantic Telegraph Company's first attempt to lay a transatlantic cable in 1857 failed. A second attempt a year later saw the cable work for a little over a month, but after a few press dispatches and some celebratory exchanges between U.S. President James Buchanan and Queen Victoria lauding the civilizing potential of cable communications, the line fell silent. A similar experience with the Red Sea and India Telegraph Company in 1859—as discussed below—raised grave doubts about the viability of cable communications altogether. As investors grew skittish, remaining projects were put on hold for several years.

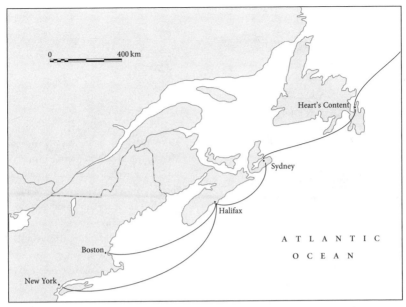

MAP 2   North American Atlantic coastline and main communication routes, circa late 1850s

In response, however, the U.S. and British governments gathered together a roster of eminent engineers and experts from the international cable communications industry in 1861 to conduct separate investigations into what had gone wrong. The results, especially of the British investigation, put things on a more solid footing and stood, as an article in the *Electrician* stated, as "the most valuable collection of facts, warnings, and evidence ever compiled concerning submarine cables."[11] The British investigation concluded with four main recommendations: improve the quality of cable technology; establish common technological standards; develop better cable laying techniques; and improve the ends of the network in terms of signaling and receiving.[12]

While that was one prong in efforts to revive the cable industry, the consolidation of two leading British cable operating and manufacturing firms a few years later (1864) was even more important. Up until this time, Britain's superiority in cable technology rested on five manufacturing companies—the India Rubber Gutta Percha and Telegraph Company, Glass Elliot, Siemens, Henley Telegraph Works Company, and R. S. Newall and Company. Britain also held a monopoly over the supply of gutta percha—the rubbery insulating material that protected cables from damage—given that it came from the Malay states within the British empire. The merging of the India Gutta Percha Company

3   Caricature of Sir John Pender, 1815–96.
Originally a Manchester textile manufacturer, his entrepreneurial skills and political
contacts, including some years as a Member
of Parliament, made him a daunting figure.
This sketch from life by James Tissot shows
the personal dynamism of Britain's cable
king at the height of his power. Reproduced
by permission of the National Portrait Gallery,
London.

and Glass Elliot created the Telegraph Construction and Maintenance Company, a company that towered over the industry well into the twentieth century. Although several smaller firms continued to operate, the scope of their
operations paled alongside the massive Telegraph Construction and Maintenance Company.[13] This new technology powerhouse was presided over by the
world's most prominent cable baron—John Pender—and woven into the fabric of business and diplomatic circles through a vast network of relationships.
The firm was also vertically integrated with the most important telegraph
company in Britain (the Magnetic) and well connected to the largest global
communication projects then being contemplated.

All of these factors gave the nascent system of global communication a
sturdy fulcrum point and this, ultimately, played a key role in fomenting
the decade-long boom in cable building efforts that took place after the
mid-1860s. Almost immediately, the transatlantic project was revived and a
new firm—the Anglo-American Telegraph Company—launched. Many of the
same interests involved in the earlier efforts were still involved, but significant
changes had also taken place. Cyrus Field and several directors of the Magnetic Telegraph Company still figured prominently, as did a bevy of British
engineers and financiers, including Thomas Brassey, Samuel Canning, and
Daniel Gooch. American capital and technology were still conspicuously ab-

sent, although exclusive rights to use the Western Union's massive North American network, the maritime concessions, and contracts with U.S. news organizations were crucial. The Telegraph Construction and Maintenance Company was at the foreground, putting up a large portion of the capital for the venture and responsible for making and laying the Anglo-American Company's new cable.

The first of the new attempts took place in 1865, but the cable broke partway through the laying. However, the following year a new cable—the fourth attempt since 1857—was successfully laid from Valentia, Ireland, to Heart's Content, Newfoundland. The 1865 cable was also found, yanked up from the ocean floor, and spliced together, so that by late 1866 two cables were in operation between North America and Britain. The cables were slow, the price was exorbitant at $10 per word, and customers were few, but after three months of service the Anglo-American Company was carrying nearly 3,000 messages and raking in $2,500 per day. The desire for more customers quickly led the company to cut rates in half, and within a short period daily revenue rose to $2,800.[14] With this success, new life was breathed into the cable industry, and a slew of new projects quickly ensued to extend a vast grid of cable and telegraph networks to the Far East

THE RED SEA CABLE AND TELEGRAPH PROJECTS: WIRING THE MIDDLE EAST EN ROUTE TO INDIA

That the world's communications infrastructure was already developing as a system could be seen in the fact that efforts to lay cables across the Atlantic coincided with similar efforts by many of the same interests to link Europe to the Middle East, India, and beyond. All told, there were four major initiatives to do so. The first two ultimately unsuccessful efforts were spearheaded by John Brett, who established two new firms for the occasion, the Mediterranean Electric Telegraph Company (1854) and the Mediterranean Extension Telegraph Company (1857), and Frederick N. Gisborne and one of Britain's smaller cable firms, Newall and Company (Red Sea and India Telegraph Company).[15] The third initiative revealed the transformation of the global cable industry that occurred in the 1860s, with several people from the Magnetic Telegraph Company—Charles T. Bright, Rowland MacDonald Stephenson, and Latimer Clark—striving, again unsuccessfully, to revive the Red Sea and India Telegraph Company in 1861. The fourth, and successful, effort was directed by John Pender and the Telegraph Construction and Maintenance

Company late in the decade. All of these efforts combined British capital and subsidies from European governments, but a decisive role was also played by the Indo-European Telegraph Department operated by the colonial government of India, the Ottoman Empire's Central Telegraph Administration, the Persian Telegraph Department, and the Indo-European Telegraph Company, the origins of which are discussed shortly.

John Brett's Mediterranean Companies and the European and Indian Junction Telegraph Company, 1854–57

The onslaught of initiatives to wire the Mediterranean region and extend communications to India was spurred on by the Crimean War (1854–56). During this time, both Brett and the Newall Company were hired by the French and British governments to quickly extend the Austro-Hungarian telegraph system toward the Black Sea and the Crimea. The efforts were significant insofar that they demonstrated a willingness for governments to jointly rely on British telegraph companies in their prosecution of the war against Russia's expansionist aims in the region. The Crimean War was also a turning point in that it was a moment in which the link between military communications, on one hand, and politicians, journalists, and the public, on the other, was tightened decisively. Along with the Mexican-American War a decade earlier, instantaneous communication introduced new actors directly into the conduct of war: politicians, the war correspondent, and public opinion. This was a substantial change in military strategy, and many tacticians lamented how the battlefield had become prone to interference by armchair warriors back at home. Military historian George Wilson expressed the concerns well in a 1901 lecture to the U.S. Naval Academy:

> The tendency of . . . the telegraph has been to centralize everything, often mixing in a sad way tactics and politics. . . . The state must declare war and . . . the central government is the best judge of the time when it should make peace, but . . . the details of the campaign should be left to those upon the field of action.[16]

While laying cables for the British and French militaries, Brett organized the Mediterranean Electric Telegraph Company in 1854. The firm had bold ambitions to create a network of cables in the Mediterranean region before adding extensions to India and had vague prospects for additional connections to the Far East. As a strong indicator of the patterns of international cooperation that were already taking shape, the new company received financial aid from the French government to lay cables to Algiers, one of its

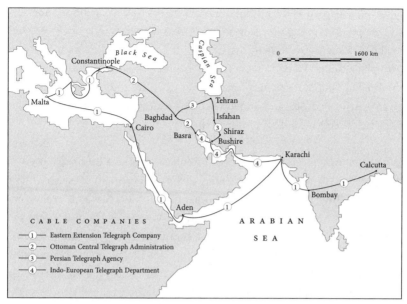

MAP 3   Proposed Indo-European communication routes, circa late 1850s

colonies on the north coast of Africa. Subventions and diplomatic support were also received from the Austro-Hungarian, British, Sardinian, and Ottoman governments. Yet, despite the breadth of support, the firm's efforts met with little success.[17] Three years later Brett formed another firm—the Mediterranean Extension Telegraph Company—to take another run at achieving his vision, but once again his efforts achieved very limited success.

Brett tried yet again in 1857 by creating the European and Indian Junction Telegraph Company, this time with even broader government support and an added sense of urgency, at least for Britain, generated by the anti-imperialist uprising in India. The Indian Mutiny shocked the security of the British Empire and galvanized the government's resolve to do whatever it took to establish communication links between London and Delhi as rapidly as possible. Subsidies were again promised by the British Treasury, the Austro-Hungarian government, and the Ottoman Empire. The Treasury also promised to lean on the East India Company for whatever commitments that private instrument of the British Empire could offer by way of subsidies, financing, or at least guaranteed use of the planned system. In addition, the Department of Indian Telegraphs (established in 1854) also endorsed the scheme as a way of linking its own network to others in Europe. The effort,

however, collapsed as the Ottoman Empire rejected Brett's petition for exclusive concessions and permission to serve areas within its territory. According
to the Ottoman government, its Central Telegraph Administration had its
own plans to build a national network "from Constantinople to the head of
the Persian Gulf."[18] Brett also learned that exclusive rights had already been
given to the Red Sea and India Telegraph Company, a firm launched by
Frederick Gisborne and his brother, Lionel.

Gisborne's Red Sea and India Company, 1857–59

After selling his Canadian companies to Cyrus Field and Western Union,
Frederick Gisborne and his brother organized the Red Sea and India Company in 1857. Much like their rival Brett, the Gisbornes planned a series of
cables from Constantinople to Cairo, telegraph lines to Suez, and a series of
cables from there through to India.[19] The fundamental difference between this
new company and earlier versions is that it had the support of the British
Foreign Office and the East India Company and, most critically, an exclusive
concession from the Ottoman Empire. The Ottoman Empire also offered an
annual subsidy of roughly $22,500 for that portion of the system that would
connect Constantinople to Cairo. The only real obstacle was the opposition of
the British Treasury to the "monopolistic" part of the company's license,
which it forcefully attempted to have rescinded, without success.[20]

But even with Brett swept aside and support from the East India Company,
the Gisbornes could not raise enough capital for the venture. This was because investors had grown skittish on account of the recent transatlantic cable
failures and Brett's parallel initiatives in the Mediterranean. The British and
Indian governments stepped into the breach by pledging annual subsidies of
$180,000 for fifty years but without making these payments contingent upon
the system actually working. While such guarantees were sure to entice investors, it was a reckless policy that the British government would come to regret.
Nonetheless, combined with the Turkish government's contribution of just
over $22,000, the Gisbornes' company now appeared poised for success. Yet,
rather than build the system, and just as Frederick had done in Canada a few
years earlier, the Gisbornes sold their concessions to the British cable manufacturing firm of Newall and Company and withdrew from the scene, at least
for the time being.[21]

In 1859 Newall and Company began its first significant international venture as it laid new cables from Constantinople to Cairo and from there to
Suez, Aden, and finally to Karachi. Finally, after nearly a half-decade of efforts,

it appeared that direct and speedy channels of communication between Europe and India were ready to begin. It was not to be. As an article in the *Times* of London recounted a few years later, "the unprotected wires rusted away and the suspended portions of the line became loaded with coral and barnacles and the whole line crumbled into hundreds of pieces by its own weight."[22] Despite legal advice to the contrary, the British government was committed to pay the subsidy for the next fifty years. This had a huge impact on British global communications policy. Burned once, the Treasury was in no mood to offer any further subsidies for the next twenty years.

The British Indian Submarine Telegraph Company, 1861–66

For three years in a row, almost every major effort to build long-distance submarine cable networks had failed, not just to India but also across the Atlantic. However, that was about to change on account of three factors already introduced earlier in this chapter: the British cable inquiry (1861), the creation of the Telegraph Construction and Maintenance Company (1864), and the success of the transatlantic cables in 1866. At an early point in these events, several individuals closely associated with Pender—MacDonald Stephenson, Charles Bright, and Latimer Clark—acquired the property and outstanding subsidies due to the original Red Sea and India Company in 1861. These assets were transferred to a new entity—the Telegraph to India Company—and plans made to start anew. Legislation enabling the new company was passed in 1862, and ships were dispatched immediately in an attempt to restore the old cable. However, the cable was irretrievable.

Eager to put India into direct contact with London, the British government decided to take matters into its own hands. Two private companies were approached to undertake the initiative, but without success, likely because of the Treasury's newfound parsimonious state of mind. However, with imperial security at stake, the government ordered its own cables in 1861 and had them laid from Malta to Constantinople, then to the north coast of Africa, and from there to Cairo. The government operated the cable for three years but then decided to lease it to the Telegraph Construction and Maintenance Company. It was this combination of leased British government cables and the old Red Sea and India Company assets that were ultimately folded into the Pender-backed British Indian Submarine Telegraph Company a few years later.[23] Table 1 summarizes the main efforts between the mid-1850s until the end of the 1860s to establish telegraph and cable connections between Europe and India.

TABLE 1   The Euro-Asian Communications Companies, 1854–68

KEY PARTICIPANTS	COMPANY NAME	DATE ESTABLISHED	SIGNIFICANT FACTORS
John W. Brett	Mediterranean Electric Telegraph Co.	1854	Partial success
	Mediterranean Extension Telegraph Co.	1857	Failure
	European and Indian Junction Telegraph Co.	1857	Failure
F. Gisborne and Newall Co.	Red Sea and India Telegraph Co.	1857	Cable laid and failed in 1859; subsidies paid by British government for fifty years
Government of India Telegraph Department (est. 1854)	Indo-European Telegraph Department	1862	Persian Gulf cables; influence on Europe to India networks
Ottoman Empire	Central Telegraph Administration	1857	Alternate land route from Europe to Persian Gulf; opened 1865
Persia	Persian Telegraph Department	1862	Two Europe to Persian Gulf lines, one via links with Ottoman lines at Baghdad, the other with Russian lines at Tehran; opened 1865
M. Stephenson, C. Bright, and L. Clark	Telegraph to India Company	1861/2	Acquire concession and subsidies from Red Sea I; attempt fails
J. Pender, C. Bright, M. Stephenson, L. Clark, et al.	British Indian Submarine Telegraph Company	1868	Use Red Sea II rights and Telegraph Construction and Maintenance Co.'s Mediterranean cables; cable to India, 1868; basis of Eastern Extension Telegraph Co. (1873)
C. Siemens, J. Reuter, German and Russian governments represented	Indo-European Telegraph Co.	1867	Unified line to Turkish and Persian lines; lines open 1870

Sources: Various, as noted in text

THE EMPIRE STRIKES BACK OR THE
DEVELOPMENT OF UNDERDEVELOPMENT?

One thing that has been intimated above but not yet directly addressed is the fact that the Ottoman Empire played a large role in each of the attempts to create a viable Indo-European communication system. As an essential intermediary power, the Ottoman Empire attracted a great deal of European finance and sought to use foreigners' plans to extend the means of communication beyond Europe to bolster its position in the world communication system and to modernize its own national economy, bureaucracy, and infrastructure, including railroads and other transportation links, ports, water systems, and, of course, telegraphs. Internally, the Ottoman Empire was also going through a significant phase of political and cultural reform, which also fed into its "modernization" initiatives. And in this regard, the agenda of reform was shaped from inside the empire by various groups with contrasting understandings of modernization and its desirability, including technocratic modernizers (the Tanzimat movement), others with a more heterogeneous conception of modernity (the Young Turks), and traditionalists who recoiled from the ineluctable forces of change.[24]

These complex cultural realities decisively shaped the Ottoman Empire's approach to new communication technologies, in particular the government's announcement in 1857 that it planned to build its own national telegraph system. Such promises notwithstanding, the project proceeded haphazardly and was finished nearly eight years later and only after the injection of fresh capital and expertise from Britain.[25] Such injections of capital and expertise came with strings attached so that in the end, what had begun as a symbol of a modernizing nation-state was largely financed and built by British and European interests.

These difficulties were further compounded in the early 1860s when decisions were made to back up the Ottoman telegraph lines between Baghdad and Basra with additional links to the Persian government's nascent telegraph system. The moves reflected British concerns regarding network security in the frontier zones of the Ottoman Empire and Persia, and separate negotiations between Britain on one side and Turkey and Persia on the other were conducted between September 1861 and December 1862 with the aim of creating a regional telegraph system linking both countries. The British proposal called for the new line to be built with British funds and operated under British supervision between the Persian Gulf cables and from there to the

Persian capital of Tehran and to Khanakin, the closest city to Baghdad near the border of the Persian and Ottoman Empires.[26]

In the initial discussions Britain offered to advance the funds and equipment to build the new Persian telegraph line. Britain also proposed that it would operate the system for twenty-five years before transferring it to the Persian government and that Persia would cover the tab for the two thousand horsemen needed to provide security for the line in the "wild zones" that were nominally part of Persia but not effectively controlled by it. However, the talks floundered and Persia halted discussions in April 1862 largely over the inability to reach agreement on who would absorb the cost of security and when the telegraph lines would be transferred to the Persian government.[27] Correspondence from Persia's Minister of Foreign Affairs, Mirza Ja'afar Khan—a modernizer keenly interested in new technology and adapting Western approaches to constitutional government—made it clear to Sir Charles Wood, Secretary of State for India, that Britain had not quite understood Persia's requirements for the new telegraph system.[28] As Khan put it, "The Persian Government is not in itself at all anxious about the construction of telegraphs in the Persian territory, because it feels that there are many other branches of public expenditure that are much more important to the country."[29]

Tearing a page from the Ottoman example, Persia restarted the negotiations by proposing to build an even more extensive telegraph network on its own from Khanakin in the north to Bushire in the south via Tehran, Ispahan, Shiraz, and other major cities, with connections to the Turkish system at Baghdad and the Indo-European Telegraph Department's system in the Persian Gulf.[30] The offer was hardly a realistic one, though, given Persia's inability to finance, build, and operate its own telegraph system without outside assistance. However, the threat of Persian independent action achieved its ends almost immediately. Worried that the Persians might follow through, British officials hastily reconvened negotiations. An agreement was signed on December 17, 1862, and ratified by the British government less than seven weeks later.

On its face, the agreement appeared equitable, with the system to be paid for out of funds loaned to Persia by the Indo-European Telegraph Department and repaid over the course of five years out of receipts. Immediately after signing the agreement, the Shah of Persia, Nasir a-Din, and Mirza Khan instructed local officials to prepare the route for the telegraph system and to construct a building at Bushire for the Indo-European Telegraph Department's offices and for connecting its cables to the domestic Persian telegraph

4   Caricature of Nasir al-Din, the Shah of Persia, as sketched for *Vanity Fair*, July 5, 1873. He began as a reform-minded leader, introducing telegraphs, postal service, secular education, Persia's first newspaper, and many other innovations designed to modernize Persia while curbing the power of the clergy. Over time, however, he grew conservative, corrupt, and resistant to popular pressures for reform. Born in 1830, he reigned from 1848 to 1896; a reign characterized in its latter years by extravagant personal expenditures and the surrender of Persian economic interests to Britain and Russia. *Vanity Fair* described the Shah's visit to the West in 1873 as being "made into one of the best advertisements ever issued in any shape to Western investors."

lines. British officials swung into action, too, ordering materials and dispatching Captain Patrick Stewart, the head of the Indo-European Telegraph Department, and two eminent engineers, Charles T. Bright and Latimer Clark, to supervise the laying of the Persian Gulf cable and its connection at Bushire and Basra. Construction of the system was to be led by British engineers but with the aid of officers appointed by the Persian government so that they could be shown "how to construct and work the line."[31] The cable was completed on April 8, 1864, and connected to the Persian and Ottoman lines in January 1865. With a Russian line to Tehran having already been finished some months earlier, the project that had begun in fits and starts eight years earlier culminated in not just one but two lines of communication between Europe and India: the main Mediterranean route via the Persian and Ottoman empires and a northern route via Russia.[32] Demand was so great that a second wire was added a year later.[33]

The accomplishment was bittersweet. The service was poor and some of the sections, as one contemporary noted, "crude, very crude."[34] Of course, the time for communication between England and India dropped considerably, but messages from Tehran to London still took 6 days, 9 hours; over the

Persian line via Moscow it was worse, at 10 days, 9 hours, and 34 minutes. Even messages over the comparatively quick and reliable Turkish route averaged 5 days, 8 hours, and 50 minutes. Communication was slow, unreliable, and expensive, at $25 for a twenty-word message. The service was, in fact, so bad that the British House of Commons organized a Select Committee a year later to look into the whole question of communication to India.[35]

STATES, MARKETS, AND THE MULTINATIONAL PROJECT: WIRING THE NEAR EAST CONTINUES

Among other things, the Select Committee recommended a direct line from London to India under the unified control of one agency. Two proposals fit the bill. The first was backed by Charles Siemens, of Germany, and Julius Reuter, who proposed an independent line from London to Tehran via Prussia and Russia. The company would be called the Indo-European Telegraph Company, as distinct from the government of India's Indo-European Telegraph Department. The new company made agreements with Russia and Prussia in 1867 that allowed it to create a system of telegraphs and cables that would run from London via Berlin, Warsaw, and Odessa, to the Black Sea, and from there on to Tiflis and then to Tehran where it would connect with the Persian lines. The company also leased lines from the state-owned Russian and Prussian telegraph administrations, and lines were constructed by Reuter between London and Berlin and by Siemens between Tiflis and Tehran. Service began in January 1870.[36]

The impact was immediate. The time for messages to go from London to Karachi dropped to 8½ hours, a third line was added to the Persian system in 1872, and an additional cable was laid in the Persian Gulf. However, rather than contributing to a drop in the price of service—which had fallen from approximately $25 for a twenty-word message in 1865 to $14 in 1868—the Indo-European Telegraph Company pressed for rate increases and the elimination of the cheaper rates offered by the Ottoman line.[37] The firm also had a big impact on international news, as Julius Reuter's rapidly growing global news agency opened its first news bureau in Bombay immediately after the new line was completed.[38]

A successive wave of agreements brokered by the British government with an ever more corrupt Shah Nasir a-Din extended the length of time before the Persian telegraph system could return to Persian control, set aside the lines built in 1866 and 1872 solely for international and European messages, and placed managerial control over both lines firmly in the hands of British and European officers. The deal also buttressed the security apparatus of the Persian state by requiring that all international messages be registered and sealed by the Persian Telegraph Office, a measure that facilitated censorship and indirectly furthered the British interest in security. The latter consideration stemmed from measures requiring the Persian security forces to protect the European staff working the lines who, by this time, were encountering much hostility from resentful Persians.[39] Consequently, the potential for the new means of communication to augment the development of Persia yielded to an apparatus of security and control. Abetted by the corrupt Shah and local elites who benefited from the status quo, the Indo-European Telegraph Company and the Indo-European Telegraph Department harnessed the Persian telegraph system to the needs of the British Empire and other Europeans. Such trends obliterated the vision of modernizers within Persia, such as Mirza Khan, a situation that remained until others like him emerged after World War I to push for the return of control over Persia's national communication system.

CABLES TO INDIA AND THE OBSTRUCTIONIST ROLE OF THE INDO-EUROPEAN TELEGRAPH DEPARTMENT

The Government of India Telegraph Department—the parent organization of the Indo-European Telegraph Department—encouraged new telegraph and cable lines so long as they extended the reach of its network and did not divert revenue from its own system. However, both agencies—the India Telegraph Department and Indo-European Telegraph Department—tried to sabotage any new venture that threatened to bypass their networks. One such example was the British Indian Submarine Telegraph Company and its backers, notably John Pender and James Anderson, who complained bitterly of how the Indo-European Telegraph Department hobbled their plans.

The British Indian Submarine Telegraph Company was created by Pender and a tight-knit group of his Manchester-based associates, many of whom we have mentioned already, such as Charles T. Bright, Charles Edwards, and MacDonald Stephenson. Other members of the new firm included Sir Richard Glass and George Elliot, both of whom had been the founding directors of the Telegraph Construction and Maintenance Company. Another notable figure who would come to loom large in these events for the rest of the nineteenth century was Sir Daniel Gooch (engineer, railway financier, and British Member of Parliament). In August 1866 the group formally registered their company and aimed to lay submarine cables from Europe to the North African coast and from there down the Red Sea to Aden, Karachi, and Bombay.

Managers and officials of the Indian telegraph departments, however, opposed the company root and branch. The Indo-European Telegraph Department's assistant director, J. Champlain, for instance, argued that service was not nearly as bad as others were making it out to be and that it would be vastly improved once the new lines of the Indo-European Telegraph Company were added to the existing system. In desperation, the department stressed the difficulty of laying cables in deep waters on account of natural disturbances to the seabed, teredo worms, fish bites, and so on.[40]

However, for each of the technical difficulties supposedly besetting the British Indian Company's plans, others pointed to myriad reasons why the existing system was wholly inadequate and beyond redemption. The reasons were simple: service was slow and unreliable, and messages were often garbled and out of sequence, if delivered at all.[41] Moreover, although the Ottoman and Persian governments were required to "appoint . . . staff possessing a knowledge of English . . . sufficient for the perfect performance of that important service," this was hardly the case.[42] Others argued that the political culture of Persia and the Ottoman Empire was inhospitable to new communication technologies. As a group of Bombay and Calcutta merchants and bankers put it, the "shortcomings of the present line are mainly due to the fact that it passes through . . . foreign territory, much of it wild and uncivilized, where European management cannot be brought to bear, and where ignorant and untrained native officers are alone obtainable."[43] The upshot was that only submarine cables bypassing these areas altogether would do. The bankers and merchants not only opposed the control of new lines by the state-owned India telegraph departments but called on Her Majesty's government to support "*any* well considered scheme for connecting India and England by direct submarine telegraph communication via the Red Sea."[44] An even more influential call for the government to subsidize the British Indian Company was made by forty of some of the most prominent figures from the realms of world finance, business, and politics, including the Rothschilds, J. P. Morgan, J. A. Gibbs, the Oriental Bank Corporation, Stern Brothers, and the mayor of London.

For its part, the British Indian Company asked for assurances that the Department of Indian Telegraphs and the Indo-European Telegraph Department would not engage in "unfair competition" or provide additional aid to the Persian and Ottoman lines. At the end of its deliberations in 1866 the Select Committee on Telegraphs to India recommended the initiative and, reaching beyond its remit, advocated that connections to Australia and China be established as quickly as possible. While the Treasury stuck determinedly to its post–Red Sea policy of no subsidies, the company still received the subsidies that had been pledged to the earlier failed Red Sea initiative. Over the course of the next half century this meant that the Treasury paid out nearly $9 million in subsidies. This was a huge benefit to the British Indian Submarine Company and left the company well placed to carve out a dominant role in the Indo-European communication system.[45]

Coming on the heels of the successful transatlantic cables, the approval of the British Indian Submarine Telegraph Company's Europe to India system signaled that the phase of trial and error was coming to an end. The momentum behind the investment boom in global communications that followed gained another boost in 1868 as the British government passed legislation to nationalize Britain's telegraph system, which it carried out in 1870. The companies' original opposition to state-ownership melted as they received extraordinarily generous payments and a pledge that they would not have to compete with the British General Post Office (GPO) in global markets. The Electric Telegraph Company and Magnetic Telegraph Company (Pender's firm), along with Julius Reuter, were the main beneficiaries, receiving some $24.5 million out of the total of $28 million initial cost of nationalization. The private international firms also benefited by being allowed to open offices in British cities. The Anglo-American Company also received a commitment from the GPO to hand over all messages bound for foreign destinations for the next fifty years, unless specifically directed by customers to do otherwise. Flush with these assurances and coffers full of cash, the British telegraph firms went global.[46]

While Britain always claimed that subsidies played an insignificant role during the formative years of global communications, the proceeds from nationalization and the holdover subsidies from the failed Red Sea venture were remarkably generous transfers of wealth from the state to private interests. Thus, it was more than coincidence that Pender and his colleagues, between the time that the Select Committee recommendations were announced and the legislation to nationalize the British telegraph system was passed, assembled four new firms—the British Indian Submarine Company (1868); the Anglo-Mediterranean Company (1870); the Falmouth, Gibraltar, and Malta Company (1870); and the Marseilles, Algiers, and Malta Company—to build a

A GREAT TELEGRAPHIC FEAT: Sending Messages Direct from London to Calcutta—6,900 Miles.

Route of the Direct Communication without Retransmission from London to Calcutta through Berlin, Warsaw, and Teheran

The portrait shows Mr. T. W. Stratford-Andrews, who has been instrumental in carrying through this telegraphic feat

MAP 4   Eastern Telegraph map of Indo-European Telegraph System, circa 1870s

series of cable lines from England to India. The cables were laid between 1868 and 1870, and in June 1870 direct service to India was opened. Two years later the four firms were merged to form the Eastern Telegraph Company—the company that would tower over the industry for the next half century. The company was also the platform for the next link in an audacious venture to push the expanding web of global communications deep into the Far East: the Eastern Extension Australasia and China Telegraph Company.

ASIAN CABLE MANIA AND GLOBAL VILLAGES

The passing of cable technology from early prototype to a mature technology, coupled with state ownership of the British telegraph system, spawned a boom in telegraph and cable investment. This was particularly true with respect to Asia. Reflecting this, Cyrus Field announced plans to lay a Pacific cable to China quickly after his transatlantic triumph.[47] C. F. Tietgen, the Danish banker, industrialist, and soon-to-be cable baron, announced another scheme to bring telegraph lines across Russia and down the Asian coast from Siberia to China and Japan. Julius Reuter, Baron Emile d'Érlanger, and a group of Manchester capitalists also competed, unsuccessfully, for Russian concessions for the same Europe to Asia route.[48] And, of course, Pender and his associates announced their own plans to extend their Europe to India system through to Southeast Asia and the Far East.

The ventures by Pender, Field, Reuter, and Tietgen, however, were neither the first nor the only such schemes. Reflecting his uncanny ability to appear wherever such activities were taking place, Frederick Gisborne had been proposing telegraph lines and submarine cables east of India and on to Australia since 1859.[49] Not surprisingly, such proposals met with a great deal of enthusiasm among the various colonial governments in the region, as evident in the Australian and Dutch governments offers to grant Gisborne annual subsidies of roughly $170,000 and $42,000, respectively, for cables connecting Australia, Java, and Singapore.[50]

Asian markets were the next big thing, and Britain and other European powers, in particular the Dutch and French, began encroaching on them steadily from the late 1850s onward, plying the region with visions of modernization, prying open markets with military force when needed or colonizing territory directly if that was deemed necessary. British trade alone in the region by the early 1860s was worth around $755 million annually and involved over 31,095 vessels. The Chinese economy accounted for the largest share of the total by far, at roughly $422 million. Not even the combined value of annual trade with Australia ($157.1 million), Singapore ($135.4 million), and the Straits Settlements ($98.2 million) and Java ($40.3 million) came close to British trade with China.[51] Altogether these images of lucrative Asian markets offered tantalizing prospects for those promoting the nineteenth-century global communication infrastructure.

As usual, Gisborne's proposals came to naught. Yet, already there were others ready to step into the breach, including the Indian Telegraph Department, which announced its own plans to extend its system throughout Southeast Asia and to China's western provinces and "open ports" immediately after Britain annexed territory in Burma and Pegu (now in Burma) in 1858 and as France did the same in Vietnam and Cambodia. As the Indian Telegraph Department put it, new lines of communication would help to further open the markets of 200 million or more Chinese to the merchants of Britain, India, and Europe and just might do so, at least so some of the naïve optimists thought, "without the employment of armies or ships of war."[52]

The visionaries pushing these schemes claimed enthusiastically that new technologies were not being foisted upon feeble regimes but were being embraced by some of the most progressive Asian monarchs of the time. In an extraordinary passage reflecting such views, Captain Richard Sprye, in a letter to the Secretary of State for India, heaped praise on the King of Ava "as . . . one of the most enlightened and public spirited Eastern monarchs of our time."[53]

Sprye was impressed by the fact that the Asian monarch had constructed a telegraph network in the capital of Ava (Mandalay) and had "expressed himself most desirous to have them connected with [the Department of Indian Telegraph lines] in Pegu, so as to place his capital in telegraphic communication with India, and the world."[54]

The king of Ava was not alone in his enthusiastic attempt to connect his little corner of the world to the expanding global communication infrastructure; so too were the chief of the Tulungagung and the king of Siam. Between 1866 and 1868, Frederick Gisborne resurrected his Asian scheme, and several other new plans were shopped around to everyone from tribal leaders to colonial administrators.[55] Two such schemes emerged in 1866, one backed by Seymour Clark and one by W. H. Read and William Patterson. All of these initiatives shared similar features. They sought to extend the Indian telegraph network beyond its terminus in Rangoon to Siam, Singapore, and areas around present-day Malaysia and Indonesia before reaching Australia and New Zealand, with branches running through Bangkok, Saigon, Cochin, and, finally, China. Each of the projects also claimed to have the support of French, British, and Dutch colonial officials. And that was another vital indicator of how far the imperialist powers remained willing to rely on a common communications infrastructure to meet their individual and collective commercial, military, diplomatic, and imperial needs. Both the Read and Patterson and Clark initiatives shared the good fortune of having struck remarkable agreements with the king of Siam and the chief of the Tulungagung in the Year of the Tiger, Siamese Civil Era of 1228, or 1866 by the Western calendar. While there were minor differences between the concessions gained by both groups, in general they had gained ambiguously worded concessions that each group claimed granted them a monopoly for the next ninety-nine years.

Despite the enthusiasm for these projects, none succeeded. Why? For one, the fact that these plans were avidly supported by the Indo-European Telegraph Department was also their Achilles heel. Indeed, while all of the initiatives planned to extend the Telegraph Department's network, the idea that all messages between Europe and Asia beyond India had to pass over the department's system collided with the Treasury Office's preference for relying on private companies in global markets. Second, the fact that all of the schemes called on the Treasury for financial support was out of step with its post–Red Sea policy of no subsidies. Thus, while Gisborne, for instance, argued that "in a matter of such national and international importance, Government aid should surely be granted, and private enterprise not be left to bear the whole

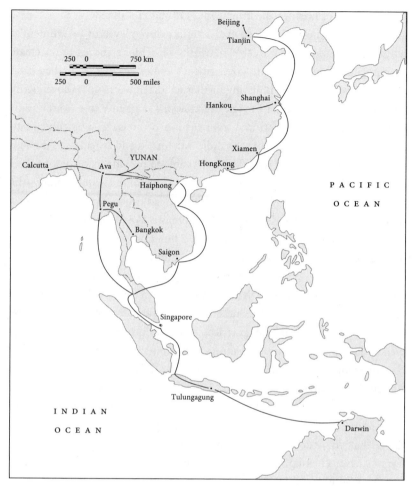

MAP 5   Southeast Asia and China, circa 1865–70

burden of the work," the plea was tersely rejected.[56] A third critical factor was the rather bizarre agreements struck between cable speculators on one hand and various Asian monarchs and the chief of Tulungagung on the other. While such things may have fueled the imagination of a world enmeshed in a web of telegraph and cable networks, they betrayed the fact that the line between reputable projects and mad schemes had blurred beyond distinction.

The last and probably most crucial factor accounting for these schemes' lack of success is the inability to gain access to China. China was, after all, the most coveted market in the region, and the government's refusal to sanction foreign telegraph and cable systems on Chinese territory effectively scuttled these projects, at least for the time being. Throughout the 1860s, the Chinese government continuously issued instructions to lower levels of government to refuse any attempt to land cable networks on China's shores or to create overland telegraph lines. The government's policy seemed to be based on three factors: its desire to limit any further encroachments on its sovereignty after nearly two decades of foreign aggression; a deference to cultural traditions insofar that the electronic transmission of messages was thought to disturb venerated processes of feng shui; and, finally, and as had been the case in Persia, concerns that it would be impossible to provide security for cables and, especially, for telegraph lines. As Prince Gong of the Zongli Yamen (the Secretary of State for Foreign Affairs), China's lead negotiator on these issues, told the British Ambassador to China, Thomas F. Wade: "should a cable . . . brought within port limits, and landed in a building on shore . . . be injured by pirates or other lawless persons being Chinese subjects, the Chinese authorities would . . . be expected to bring the offenders to justice. . . . [This would create] more mischief than can well be expressed."[57] Thus, while there had been a steady stream of requests for telegraph and cable concessions by Russian, British, Danish, and American officials since 1861, they had all been turned down.[58] The Chinese government was equally concerned that a decision in favor of one company would be seen as a precedent for all. The American position underscored such concerns, with its insistence that China's obligations under its "open door" commitments meant that rights granted to one had to be granted to all. All of this would change, and dramatically so, in the early 1870s on account of the rising influence of China's leading modernizers—such as Li Hongzhang, Ding Richang, Sheng Xuanhuai, Shen Baozhen, and Zeng Guofan—and the adoption of a more "internationalist" approach to cables and telegraphs in China among the foreign superpowers.[59] However, this set of interactions and the time frame lay outside the scope of this chapter and are the focus of chapter 4. The point for now, though, is that until "the China question" could be sorted out, the Asian cable mania of the 1860s would belong more to the realm of speculative ventures than actual accomplishments.

From the Gilded Age to the Progressive Era: The Struggle for Control in the Euro-American and South American Communication Markets, 1870–1905

> The most popular outlet now for commercial enterprise is to be found in the construction of submarine lines of telegraph.—*Times* (London), August 26, 1869

In the last quarter of the nineteenth century the market for global communications exploded. A range of transformative communication and information technologies, a boom in the world economy, and a liberal internationalist political outlook all contributed to the cause. The creation of a rudimentary framework of global communications policy and government subsidies also played their part. The combined force of these factors propelled a tenfold expansion in the global communication infrastructure from 1870 to 1900 and a doubling again in the next decade.[1]

The expansion of global communication reflected the fact that there was an integrated world economy for the first time between 1870 and 1914, in particular among the leading commercial countries.[2] Near instantaneous communication, rapid transportation, and the "dematerialization" of sources of economic value—futures trading, the greater value of symbolic goods, electronic representations of money, the news commodity, and so forth—enhanced the integration of the world economy and lent it a kind of synchronicity that helped to distinguish this phase of globalization from all those that preceded it. The global media system provided the platform upon which changes in the global political economy and cultures took shape, to say nothing of their role in the mass production and dissemination of cultural symbols that registered the ideologies and aspirations of nations and classes in this formative phase of nation-states and the search for modernity.

Britain was the cosmopolitan vortex at the center of this world political economy. As Herbert Feis put it in his seminal study, *Europe: The World's Bankers*, "London was the center of a financial empire, more international, more extensive in its variety, than even the political empire of which it was the capital."[3] Britain dominated the world economy, but the world also became more multi-polar during these years. Foreign investment from Britain nearly tripled from $6.5 billion in 1885 to $19 billion in 1914. France increased its long-term foreign investment from around $3 billion in 1880 to just under $9 billion in 1914, while German foreign investment grew from about $1.2 billion in 1883 to $5 billion in 1914. Altogether, total foreign investment by Britain, France, and Germany grew from $10.7 billion in 1885 to nearly $20 billion at the beginning of the twentieth century and to over $36 billion by 1910.[4] It was an amazing increase, and the world changed dramatically because of it. As O'Rourke and Williamson note, the volume of overseas capital investment at this time was truly staggering, and amounts during the 1980s and 1990s did not even come close to matching the rates in the last quarter of the nineteenth century.[5]

British capital turned from a European focus to an international one in the 1870s, homing in particularly on Australia, Canada, New Zealand, and South Africa within its empire and the United States, Japan, China, and the coastal regions of South America thereafter. Europe and North America benefited most from these trends, especially the United States during the railway boom and after consolidation of industrial capitalism after the Civil War. Finance and trade also augmented the power of diplomacy in a substantial way, especially in the zones of informal empire. This was particularly true in a handful of South America countries, Brazil, Uruguay, and Argentina, and specifically in the River Plate zone conjoining the latter two countries on the Atlantic coast. Britain's trade was so extensive in this region that by 1912 it nearly equaled its trade with France. While French investment maintained a European focus, it tripled from 1900 to 1914 in Asia and Latin America, although it remained marginal in both places relative to British investment. The same patterns applied to Germany.[6] In fact, the significance of the River Plate in terms of economics and as a laboratory for some of the most ambitious modernization projects of the era gave it a place in world communication all out of proportion to its compact geography. While investment poured into these regions it largely bypassed the zones of formal empire (except India and the "white dominions" of the British Empire).[7] In the main, however, cable communications and telegraph systems entered these places years after they had been introduced elsewhere, and then only after governments subsidized private companies to do so.

Just as there was a spatial dimension to the organization of the world, so too was there a temporal rhythm to the economy. Far from advancing along a single upward arc of progress, global and local economies moved in tightly synchronized cycles of economic boom and crisis, the latter occurring in 1873, 1890, and 1931. The wild swings between prosperity and crisis were registered in perceptible shifts in the movement of money and people; British investment, for instance, in Latin America during the first four years of the 1870s was $1.5 billion but plummeted to $40 million during the next five. France's foreign investment was negligible before the crash of 1873 but swung upward in the next investment frenzy during the last half of the 1880s.[8] At the height of the 1880s upswing, Argentina actually paid between one hundred thousand and a quarter of a million people per year to immigrate to the country; when the economic boom bottomed out a few years later the number plunged to less then a fifth of the peak figure and those that arrived paid their own way. Of course, suffering was borne in different degree by different classes, but these massive economic crises were a grim reaper across all levels of society.

For many, such extreme vacillations in the world economy made no sense. Greater flows of money and information should have ironed out the unevenness of the world economy and steadied its rhythms. Indeed, as Colonel Holvier, the secretary of Lloyd's of London, told a business audience, international business was being informatized: "Increased trade is not only leading to increased use of the telegraph, but the availability of the telegraph is increasing the *information component of business transactions* and providing business with knowledge about markets before they entered them."[9] But being able to communicate further and faster not only globalized the benefits of markets, it also amplified and diffused their risks. Charles Conant argued during the 1890s that improved means of communication and transportation "accelerated the pace and enlarged the size of transactions, extended and intensified competition, and facilitated the operation of industry on an increasingly larger scale. . . . As capitalism integrated national markets into a world market, the business cycle tended to become global and national cycles tended toward synchronization."[10]

To counteract these potentials, Conant called for an active approach to managing global markets and international relations. The new approach to development and foreign policy, argued Conant, should rely on economics, law, and financial experts instead of "old world" approaches steeped in power politics and imperialism. Ideally, the turn to managed markets would smooth out the business cycle, while progress in the "backward countries" would create outlets for the "superabundance" of investment capital held by the "advanced countries" and new markets for their commodities. The approach contained the elements of a new view of empire based on "shares of world economic growth" rather than spheres of influence.[11]

INTERNATIONAL LAW AND THE FOUNDATIONS OF GLOBAL COMMUNICATIONS MARKETS

Conant believed that viable global markets required new approaches to empire and the professionalization of international law. Such notions were sharpened in the field of global communication as governments turned to a more active and formal role for international law to manage the explosive growth in cable communications that followed the successful completion of the Anglo-American Company's cables between North America and Europe (1866) as well as to the Far East by the Eastern Associated Company and Great Northern Company (1870–72). The frenzy of activity that followed made it clear to governments that they had to either act fast to cobble together a legal regime for global communications or resign themselves to allowing it to develop idiosyncratically. They chose to act fast.

The fact that the United States refused to join the International Telegraph Union upon its creation in 1865 and then skipped a meeting on international communication convened by France in the same year might suggest that it was pursuing an isolationist course. Yet, despite the fact that there was no appreciable American investment or technological expertise in the early global communications and media business, the U.S. government played a key role in the late nineteenth century by proposing an expansive approach to global communication policy that set a standard against which its future policy initiatives and those of others could be measured. The new policy directions were outlined in an 1869 memo circulated by Secretary of State Hamilton Fish to twenty-three governments. The memo stated that the United States's "central position in the communication of the world entitled [it] to initiate this movement for the common benefit of the commerce and civilization of all."[12] The statement was a mixture of hubris and idealism; but it had substance. It aimed to increase the means of telegraphic communication, protect and neutralize telegraph and cable communications in time of war and peace, prohibit monopolies, and ensure the privacy of electronic communication. Louis Renault, of the Institute of International Law and doyen of international legal circles, praised the vision as "*très compréhensif.*"[13]

At the top of the list of U.S. concerns during a series of conferences during

5   Louis Renault, 1843–1918. Renault was a professor at the University of Paris and the outstanding French authority on international
law. He was a key figure in the Institute of
International Law and an expert on the legal
regulation of submarine cables. In 1907, his
work led to the award of the Nobel Peace Prize.
Reproduced by permission of the Nobel Foundation,
Stockholm.

6   Tobias Asser, 1838–1913. Asser was a Dutch
professor of international law and his country's long-time representative at the Permanent International Court at the Hague. Like
Renault, he played a major role in the Institute
of International Law and was considered by
many as a worthy successor of Hugo Grotius.
He was awarded the Nobel Peace Prize in 1911.
Reproduced by permission of the Nobel Foundation,
Stockholm.

the 1870s and 1880s was the need to adequately protect and neutralize submarine cable systems during times of peace and of war. Always at the foreground of these efforts was Cyrus Field who, with the aid of American consular officials, tirelessly advocated using international law to achieve such ends.[14]
In the views of both the U.S. government and Cyrus Field, international law
would be more reliable than the current method, namely, the inclusion of a
"communication infrastructure security" clause in the landing licenses that

the cable companies struck with a grab bag of governments to whom they offered service.[15] The proposals made little headway until the early 1880s when thirty-two countries, including the United States, finally attended an International Telegraph Union conference on the subject. By this time, the issues were propelled by a concerted push among European legal experts and their professional association, the Institute of International Law, to expand the role of international law in general and for telegraph and cable communication in particular. And in this, key figures at the Institute—especially Louis Renault (France) and Tobias Asser (Netherlands)—honed their visions by advising the International Telegraph Union and drafting the early conventions that sketched out the initial governance regime for world communication.[16]

Attempts to use international law to secure open, neutral, and protected communication networks for everyone at all times also constituted a breakthrough for an emerging "right to communicate," a broad notion then taking shape that people should be given the right to use new technologies (many governments in the early days of telegraphy tended to hoard such technologies), to use code, and to privacy. These rights, however, were nullified as governments retained the authority to intercept messages in order to safeguard state security and public morality—a thinly veiled attempt to suppress the use of new communication technologies in what were, after all, politically turbulent times. In the end, the ideal of open networks was sacrificed to national security concerns and further gutted by applying only during peacetime. In this, the United States and Britain adopted similar positions, with the U.S. secretary of state instructing the American delegate at the 1882 International Telegraph Conference as follows: "It is desirable that you and the British delegate should compare your texts so that they may be identical."[17]

FREE TRADE, CARTELS, AND THE EURO-AMERICAN COMMUNICATIONS MARKET

The rapid advent of global communications in the 1870s and 1880s compelled numerous countries to participate in the many international meetings convened during this time.[18] In addition to the above-mentioned issues, at the top of the agenda was the question of whether global communications markets should develop as monopolies or competitively. While Britain, Canada, Newfoundland, and the United States had all granted exclusive concessions to the predecessors of the Anglo-American Company, shortly after the company switched on its two transatlantic cables in 1866 things changed. In short order, the United States and Britain announced that they would no longer grant monopolistic concessions and committed themselves to a "free trade in cables" policy.

The Anglo-American Company staunchly resisted the move to free trade, arguing that monopolistic concessions were essential to gaining the confidence of financial markets for such a risky field of endeavor. So long as the world's most lucrative communication market remained a monopoly, the Anglo-American Company could also charge extraordinarily high rates. High rates also allowed the company to operate its cables at only a small fraction of their capacity while still allowing for a very handsome profit. Indeed, the company's revenues were so large that its cables were paid off in two years, far ahead of expectations. And just as obviously, such practices fomented opposition to the company and to communication monopolies, and critics charged the company with sacrificing the potential benefits of new technology and improved international communication on the altar of private gain. Even newspaper owners, such as *New York Herald* publisher James Gordon Bennett, and news agencies, such as Reuters and Associated Press, which had initially supported the company with contracts promising expenditures in the range of $3,750 to $5,000 per month, chafed under the enormous cost of cabling and soon sought to create their own cable companies.[19] Unmoved, however, the Anglo-American Company spent the next two decades defending its monopoly by lobbying the Canadian Parliament and British government, through the courts, by forming cartels, and by absorbing rivals. And in this it mostly succeeded until the Commercial Cable Company successfully established a rival company in 1884.

Competition and International Rivalry or Cartels and Multinational Collaboration? Case No. 1: The Rise and Fall of the French Atlantic Cable Company (1869–73)

The first attempt to introduce competition in the Euro-American communications market was sponsored by Julius Reuter and Baron Emile d'Érlanger. The venture was formally known as the Société du Câble transatlantique Française but popularly referred to as the French Atlantic Cable Company. The venture was well connected, given Reuter's role in the news business and the fact that d'Érlanger came from an old aristocratic family with well-placed ties to governments and finance across Europe. D'Érlanger was a significant investor in his own right with major holdings in South America and South Africa. He was also a founding director of the Council of Foreign Bondholders, a London-based agency that represented international financial institutions in their dealings with governments.[20] Reuter's status in the global news business and experience in the Anglo-European cable business and as a founder of the Indo-European Telegraph Company in 1867, as we saw in chapter 1, also boded well.

The French Atlantic Cable Company was launched on July 6, 1868, amid much fanfare and many remarks by the French Interior Minister Ernest Picard and Emperor Louis Napoleon celebrating the company as an example of France's rising commercial influence in the world. While the government of France celebrated the company's emergence and gave it a generous subsidy and exclusive rights over cable communications between France and North America for twenty years, the national identity of the firm was shrouded in mystery from the outset. Indeed, the comments of Picard and Napoleon notwithstanding, the French Atlantic Company's two principal backers—Julius Reuter and Baron Emile d'Érlanger—were more European than French. And just as important, the directors, capital, and banking arrangements (led by the Anglo-American banking firm, J. P. Morgan), and the technology and equipment used to lay the firm's cables were obtained in Britain and the United States and from those directly tied to the Anglo-American Company. Control rested with Reuter and d'Érlanger, but the Telegraph Construction and Maintenance Company made and laid the company's cable and received the next largest block of shares in it for doing so. There was considerable German capital in the company as well. The French Atlantic Company also shared two directors with the Anglo-American Company—James Anderson and Daniel Gooch—and indeed the firm's connections to Britain were so tight that it had separate French and British boards of directors.[21] Finally, the firm was registered in London, not Paris. In short, the company was more an adjunct to monopoly than a bona fide rival, a multinational in the age of capital.

The *New York Herald*, for one, pierced through this labyrinth of connections and decried the creation of a transatlantic "telegraph monopoly ring."[22] The *Herald*'s criticism also presaged the time when mounting frustration would lead its publisher, James Gordon Bennett, to enter the cable business directly so as to introduce some competition. More than just signaling Bennett's discontent, however, the paper's condemnation was emblematic of the politics of robber baron capitalism and the Gilded Age. However, it was not corruption that ensnared the company but the exclusive concessions that it had gained in France. And it was exactly on this matter that the U.S. government first announced and then honed its "free trade in cables" policy. Just before the French Atlantic Company laid its cable between Brest and

Duxbury in 1869, U.S. President Ulysses Grant informed the company that it could only enter the American market after it renounced its exclusive privileges in France:

> I refused . . . on the one hand, to yield to a foreign State the right to say that its grantees might land on our shores, while it denied a similar right to our cable to land on its shores, and, on the other hand, I was reluctant to deny to the great interests of the world and of civilization the facilities of such communication as were proposed. I therefore withheld any resistance to the landing of the cable on condition that the offensive monopoly feature of the concession be abandoned and that the right of any cable which may be established by authority of this Government to land upon French territory and to connect with French land lines and enjoy all the necessary facilities or privileges incident to the use thereof upon as favorable terms as any other company be conceded.[23]

Strictly speaking, this was not a "free trade in cables" policy but a "reciprocity policy" which leveraged access to the U.S. market to prize open foreign markets for American companies. As such, it was a powerful instrument for influencing the communications policies of other countries. Companies were also prohibited from joining cartels or combining with others to fix rates. They were also required to give priority to U.S. government messages. In principle, the policy gave the United States one of the most open communication systems in the world. Foreign companies could bring their systems to American shores and build telegraph lines to whichever cities they desired in order to directly offer their services to the public. Most other countries, except Britain and later Canada and Australia, in contrast, required foreign companies to rely solely on state-owned telegraph agencies to gather customers for them. This was an inefficient way to expand access to global communication, given that foreign firms could not directly solicit customers in, say, Paris or Berlin as they did in London, New York, or Montreal. The problem was compounded by the fact that most state-owned telegraph agencies signed exclusive deals with just one company, and in this the Anglo-American Company was the major beneficiary, which its own directors fully acknowledged, as the following quote from its managing director, James Anderson, reveals: "This Company . . . had exclusive arrangements with the English, French, and German governments, who could not deliver a message for America to any other line than this, unless the sender put upon the message that it was to go by another line; and they [the directors of Anglo-American Telegraph Company] knew by experience that not one man in a hundred [would do so]."[24]

Such measures solidified the Anglo-American Company's grip on the

Euro-American communications market. Similar arrangements prevailed on the North American side where the Anglo-American Company and Western Union had agreed to organize a cartel a decade before the first cable began operation. The impact of these arrangements on the French Atlantic Company was easily observable. Thus, even though the liberal U.S. regime allowed the company to establish its own offices in any American city it chose, it was denied access to the world's largest telegraph network and the huge volume of messages it generated by the exclusive fifty-five-year joint purse between the Anglo-American Company and Western Union. Consequently, even before the French Atlantic Company laid its cable in 1869, it joined the Anglo-American Company-led cartel. The agreement set prices and pooled the revenues of both firms, with two-thirds going to the Anglo-American Company and the rest to the newcomer.

The joint purse was subsequently modified but merely as a prelude to the firm's complete absorption by the Anglo-American Company in 1873. And when that happened, the Anglo-American Company cranked up its printing presses to create $26 million in new stock to finance the deal, with most of it going to Reuter, d'Érlanger, and the Telegraph Construction and Maintenance Company. Amid cheers and laughter at a meeting called to discuss the matter, Julius Reuter, Viscount Charles Stanley Monck (an Eastern Associated Company director), and William Montagu Hay (later Lord Tweeddale, an Anglo-American Company director) were given the task of dismantling the company. In the end, the original investors tripled their initial investment of $6 million in four years and the Anglo-American Company created enough new stock to quadruple the acquired company's capitalization.[25]

For Reuter, the French Atlantic Company was his last major cable venture. But for John Pender, James Anderson, Cyrus Field, William Hay, Daniel Gooch, George Elliot, and C. W. Siemens, among a few others, the acquisition of the French firm was only a prelude of bigger things to come. Just two months after acquiring the French Atlantic Company, these individuals became the directors of the Globe Telegraph and Trust Company. Reuter had participated in the early discussions that led to the creation of the organization but withdrew before it was incorporated on July 11, 1873. In fact, Reuter contemplated leaving the global media business altogether, as he attempted to sell his news agency to the Globe Telegraph and Trust Company in 1874. However, Pender and Anderson rejected Reuter's offer because the asking price was too high.[26]

In the years ahead, the Globe Telegraph and Trust Company suggested that there were no areas of the media business that it might not enter.[27] However, its achievements were more modest in practice. The Trust never acquired Reuters and, indeed, it never entered the news business directly. Within the cable industry, however, the aim of the Globe Telegraph and Trust Company was to "take all possible steps to amalgamate the Globe with the Eastern, Eastern Extension, Anglo-American and other Companies."[28] But even this was scaled down over time as the function became, as John Pender stated, to swap "shares in the Globe . . . for shares in submarine telegraph and associated companies," the objective being to spread the risk of cable communications over as many companies as possible.[29] Although this was a rather defensive posture, the task of protecting Trust members was executed ruthlessly against would-be rivals.

Competition versus Corporate Consolidation, Case No. 2: The Rise and Fall of the Direct U.S. Cable Company (1873–77)

The first case in point was the emergence of another rival on the lucrative Euro-American route just as the French Atlantic Cable Company was being dismantled. The new rival was the Direct U.S. Cable Company, formed in 1873, and unlike its predecessor there was no question about its independence. Its U.S. landing license, in fact, prohibited it from joining a cartel or using any other measure to restrict competition or control rates.[30] The fact that Siemens was one of the company's main backers—as a source of technology, as the second largest investor, and with directors on its board—suggested that the Direct U.S. Cable Company was part of a bid to challenge the alliance between the Anglo-American Company and the Telegraph Construction and Maintenance Company for a piece of the market in both international communication services and technology. Although the company was registered in London—hence its designation as a British company—roughly a third of its capital was held by the Siemens family and four key European banks: Société de Credit Mobilière (Paris), Banque Central Anversoise (Antwerp), Banque de Paris et des Pays-Bas (Brussels), and Deutsche Bank. The rest was held by British and European financial institutions, stock brokers, and merchants.

The firm's attempts to compete in the world's most lucrative communications market faced ruthless resistance from the Anglo-American Company and the Globe Telegraph and Trust Company from the time it began to lay its cable in 1874 until its demise two years later. As soon as ships laying the new firm's cable reached Newfoundland, they confronted the Anglo-American Company's monopoly and a court injunction prohibiting it from landing the cable. Fortunately for the Direct Company, at the same time that its operations were stalled off the coast of Newfoundland, the Canadian Parliament was tabling legislation to abolish the Anglo-American Company's exclusive concessions in the Canadian maritime provinces. Of more immediate importance was the fact that the company obtained a reprieve from the Newfoundland courts allowing it to connect its cable to a buoy in Conception Bay while finishing the rest of its network.[31]

The reprieve was crucial because two Anglo-American Company directors, Cyrus Field and William Hay, and another from the Eastern Associated Telegraph Company, Viscount Charles Stanley Monck, were busy trying to block the "anti-monopoly" legislation being considered by the Canadian Parliament.[32] Field, Hay, and Monck mixed tales of personal hardship and woe in the early days of the company with thinly veiled threats to disconnect Canada from the global communications grid altogether should the law pass. Overall, the Anglo-American Company directors hoped to either kill the bill or cause long delays that would bring about the demise of the Direct Cable Company or at least make it more amenable to their offers of amalgamation. In the end, the bill was delayed but reintroduced and passed the following year.[33] The new law gave Canada a communication policy based on free trade and the prohibition of monopolies. In practical terms, the law and the Privy Council decision gave the Direct Cable Company temporary landing rights off the coast of Newfoundland, which allowed the company to complete its system and to begin service within the year. Rates dropped and domestic telegraph companies in Canada and the United States gained a valuable new source of revenue.

However, bent on maintaining their absolute monopoly on a market that had grown to $10 million by 1873, two Globe Telegraph and Trust directors, John Pender and James Anderson, began to quietly amass a large quantity of the Direct Cable Company's shares, and by the middle of 1876 they had gained control of the company. The minutes of the Globe Telegraph and Trust reveal the outcome and its consequences:

> The Globe Company having the largest Proprietary interest in the Companies urged the adoption of such a policy as whilst protecting both Companies against unremunerative opposition would give the public the benefit of legitimate competition. . . . That the Chairman (W Lushington) and the Managing Director (W Van Chauvin) of the Direct U.S. Co met the Committee (representing AA) at this office on the 24th Oct when the relations of the . . . Companies was fully discussed. . . . During the interview . . . an intimation was given that the proprietary interests of this Company must be directly represented on the Board of the Direct U.S. Co. The Committee also intimated that before communicating with the Direct U.S. Co.

they had ascertained the sentiments of the Anglo-American Board the object of the Committee being to promote the legitimate interests of the Companies as Dividend producers to this Company.[34]

Obviously, Pender and Anderson saw nothing wrong with these activities; they were merely doing the rational thing to avoid "ruinous competition." A new slate of directors was appointed to the company and its management integrated with that of the Anglo-American Company. A new joint purse agreement adopted in 1877 further harmonized the rest of the companies' operations and dedicated both firms to obstruct any and all rivals from entering the world's most valuable communications market.[35]

The First American Global Communications Powerhouse, Case No. 3: The Rise of Western Union

The experience of the French Atlantic Cable Company and the Direct Cable Company set the mold for years as the Anglo-American Company continued to absorb newcomers into its cartel. The pattern was repeated again in 1881 with the inclusion of Compagnie Française du Télégraphe de Paris à New York (having the British nickname PQ) and, a year after that, as the Western Union–owned American Telegraph and Cable Company joined the cartel.[36] But, unlike the Direct Cable Company, both of these firms joined willingly. And once again the same terms applied: rates were set and the companies interconnected only with one another's network and not those of rivals; this state of affairs was locked in for the next forty years.[37] Paradoxically, however, as the cartel grew, the power of the Anglo-American Company began to wane.

The joint purse agreement allocated 22½ percent of the transatlantic revenues to the Western Union. In this, the cartel was no more or less beneficial to the Western Union than to the other companies involved. Yet, as Daniel Headrick puts it, the Western Union joined the cartel largely on its own terms.[38] It was able to do so because it controlled the gateway to the world's largest telegraph network. Western Union's sprawling network and over 10,000 offices located in every city of any size in the United States and east of Montreal in Canada made it indispensable to any company seeking to access customers. Reflecting the value of these connections, the agreement renewed and extended the obligation of the cartel members to exclusively use the Western Union's continental network for all their messages to North America and beyond for another thirty-eight years. Given that the Euro-American market was worth more at this time than Western Union's continental market meant that this was a benefit of considerable value.[39] In sum, Western Union

7   New York telegraph offices. The Western Union Telegraph Building on Broadway, New York City,
circa 1880. It towers above its surroundings as a symbol of the power of Western Union and of the
telegraph and cable industry. Reproduced by permission of Picture History.

was no junior partner in the cartel but a force whose interests were rapidly
assuming proportions on par with those of the Anglo-American Company.
Indeed, the Compagnie Française du Télégraphe (PQ) claimed that Western
Union, now led by the none-too-scrupulous and rapacious Jay Gould, threatened to cut the Anglo-American Company and itself off from its North American network should they refuse to these conditions.[40]

THE QUINTESSENTIAL MULTINATIONAL CORPORATION,
CASE NO. 4: THE RISE OF THE COMMERCIAL CABLE COMPANY
(1884–90)

The departure of Compagnie Française du Télégraphe de Paris à New York
from the cartel a few years later further highlighted the limits to the ability of
the Anglo-American Company and the Globe Telegraph and Trust to main-

tain, let alone expand, their hegemony over global communications, or at least its most vital market. A half decade after joining the cartel, the Compagnie Française abandoned it to join a loosely organized consortium assembled by the Commercial Cable Company. The Commercial Cable Company, a creation of *New York Herald* publisher James Gordon Bennett and mining magnet John Mackay, was launched in 1884 and was the only company in the nineteenth century to avoid the Atlantic cartel, although not entirely.

Bennett and Mackay also owned the most significant rival to the Western Union in the United States, the Postal Telegraph Company, and had strong ties to the Western Union's main competitor in Canada, the Canadian Pacific Telegraph Company (through joint purse agreements and interlocking directors). Thus, not surprisingly, one of the first things the Commercial Cable Company did upon entering the transatlantic market was tighten its control over the Postal Telegraph Company in the United States and fortify its links with its Canadian partner. The result of this tangled net of relationships was that all of the firms operating in the transatlantic communications markets were arrayed in two rival clusters. J. H. Carson, the manager of the Anglo-American Company, described the situation like this:

> All the telegraph companies are in private companies hands; the land telegraph companies have alliances with certain cable companies, and naturally through these alliances, the cable companies claim all messages that these land telegraph companies may collect. For instance, the Western Union Company for the whole of the U.S. has an agreement with the Anglo-American Telegraph Company, the Direct U.S. Cable Company, and the American Telegraph and Cable Company . . . by which all messages which come into their hands must go to those cable companies. So also in the U.S., the Postal Telegraph Company, which is the great opponent of Western Union, has an exclusive agreement with the Commercial Cable Company. The Commercial Cable Company being an American company and a part of the Postal Telegraph Company brings their messages across to England. The Canadian Pacific in Canada is allied to the Commercial Cable Company. . . . Any message those land companies collect must come to the cable companies with whom they are in alliance.[41]

Pender recognized the Commercial Cable Company as a major threat before it even laid its first cable and met with James Gordon Bennett on December 5, 1882, to discuss terms for bringing the new company into the cartel, but was rebuffed.[42] Mackay and Bennett were intent on breaking the cartel and fostering greater use of the cable by news agencies other than the

TABLE 2   Tier One Global Communications Companies, circa 1900

COMPANY	NUMBER OF CABLES	NETWORK SIZE (MILES)	ANNUAL REVENUE ($)	MAJOR ALLIANCES	MARKET VALUE (U.S.$/ MILLIONS)
Eastern Telegraph Company (ETC)	166	71,923	N/A	GNTC, AATC, TCMC, IETC, IETD	99.93
Anglo-American Telegraph Co. (AATC)	17	15,392	N/A	Western Union, DUSTC	21.82
Western Union (WU)	12	7,352 (192,075)	24,758,569	AATC, DUSTC, IOTC, CSTC, WIPTC	97.37
French Cable Co. (FCC)	27	16,053	N/A	CCC, GCC, HBTC, DWICC, USHTC	N/A
Commercial Cable Co. (CCC)	8	9,919	N/A	PTC, GCC, FCC, CPTC, HBTC, DWICC, USHTC	N/A
Total	203	121,140	N/A	N/A	

Sources: Britain, *Inter-Dept. Comm. 1st Report*, 1902, appendix J; U.S., H.R. *Pacific Cable*, 1900, 42–43

The holdings of the companies in this and the following table (table 3) are based on a tally of their affiliates and acquisitions as of 1900. Firms listed on foreign markets, but which are controlled by a parent company, are included in each of the company's main holdings. For the Commercial Company this includes the Direct West India Cable Company and the Halifax and Bermudas Company; for the Anglo-American Telegraph Company, it includes the Direct United States Cable Company. The French Cable Company was formally known as the Compagnie Française des Cables Télégraphiques after the amalgamation of the Compagnie Française du Télégraphe de Paris à New York and the Compagnie Française des Télégraphes Sous-Marins (1894). The French Cable Company also had ownership control of a U.S.-based subsidiary, the United States and Haiti Telegraph Company. The German Cable Company consisted of a commonly owned group of firms created between the mid-1890s and 1910: the Deutsch Atlantische Telegraphengesellschaft; Deutsche-See Telegraphengesellschaft; Deutsch-Sudamerikanische Telegraphengesellschaft; and Deutsch-Niederländische Telegraphengesellschaft. The firm was renamed the New German Cable Company (Neue Deutsche Kabelgesellschaft) after World War I. The Central and South American Company includes its subsidiary, the Mexican Telegraph Company. For this company, data for network size (1898), revenues (1888), and market value (early 1880s) are from All-America Cables, *A Half Century of Service to the Americas*, 22, 38, 42. Alliances include only direct ties through joint-purse agreements and ownership. The size of networks and revenues are from global markets. Western Union includes North American revenues,

Associated Press and Reuters.[43] The Commercial Cable Company's assault on the existing monopoly sparked a price war and rates declined from roughly 50¢ to 12.5¢ per word and the volume of messages rose 162 percent over the course of two years. Yet, realizing the perils of "ruinous competition" the company refused to drop prices further in the face of even deeper cuts adopted by the cartel and soon entered into an agreement with them to adopt a fixed rate of 25¢ per word, even though it never joined the cartel.[44]

The Commercial Cable Company was the most significant rival to emerge to date and soon became a magnet for foreign companies seeking shelter from the dominant cartel. The Commercial Company performed this function so well because it was the most adroit of all companies at claiming whatever national identity suited its purpose at any given moment: American at one moment, British at another, and stateless at others. It was able to do this, even though its capital was mostly American, by relying on Siemens for technology; maintaining a transnational roster of directors drawn from the United States, Canada, and Britain; and forming alliances with other firms seemingly without regard to the nationality of ownership. The first firms to associate with the Commercial Company were two French companies: Compagnie Française du Télégraphe de Paris à New York (PQ) until 1894, when all the French cable companies operating in the North Atlantic and Caribbean regions were amalgamated (as discussed below) into the Compagnie Française des Câbles Télégraphiques (Compagnie Française), which continued the association with the Commercial Cable.[45] This portrait is obviously daunting and there is no way to escape the fact that the tangled web of inter- and intra-corporate connections is so complex that it is nearly impossible to keep their identities and links to one another straight. The portrait of the first and second tier global communication companies in table 2 above and table 3 below tries to provide

---

(*Table 2 note continued*)

in parentheses, for illustrative purposes and are based on its *Annual Reports* for 1901–2 and 1910. Some of these firms are introduced in the pages that follow.

AATC=Anglo-American Telegraph Co.; CCC=Commercial Cable Co.; CPCC=Commercial Pacific Cable Co.; CSTC=Cuba Submarine Telegraph Co.; DUSTC=Direct United States Telegraph Co.; DWICC=Direct West India Cable Co.; ETC=Eastern Telegraph Co.; FCC=French Cable Co.; GCC=German Cable Co.; GNTC=Great Northern Telegraph Co.; HBTC=Halifax and Bermudas Telegraph Co.; IETC=Indo-European Telegraph Co.; IETD=Indo-European Telegraph Dept.; IOTC=International Ocean Telegraph Co.; PTC=Postal Telegraph Co.; TCMC=Telegraph Construction and Maintenance Co.; USHTC=United States and Haiti Telegraph Co.; WIPTC=West India and Panama Telegraph Co.; WU=Western Union

TABLE 3   Tier Two Global Communication Companies, circa 1900

COMPANY[a]	NO. OF CABLES	NETWORK SIZE (MILES)	ANNUAL REVENUE ($)	MAJOR ALLIANCES	MARKET VALUE (U.S.$/ MILLIONS)
Central and South American Telegraph Co. (CSATC)	18	10,757	614,489	WU, Trans Andine Telegraph Co.	5.3
German Cable Company (GCC)	N/A	N/A	N/A	CCC, GCC, HBTC, WIPTC, USHTC	N/A
Great Northern Telegraph Co. (GNTC)	24	6,982	N/A	ETC	N/A
Indo-European Telegraph Co. (IETC)	N/A	N/A	N/A	ETC, IETD	N/A
Indo-European Telegraph Department (IETD)	N/A	N/A	N/A	IETC, ETC	N/A
Total	69	32,062	N/A	N/A	N/A

Sources: Britain, Inter-Dept. Comm. 1st Report, 1902, appendix J; U.S., H.R. Pacific Cable, 1900, 42–43; All-America Cables, A Half Century of Service to the Americas, 22, 38, 42

[a]For the background to this table and definitions of firm abbreviations used in it, see the explanatory note to the previous table (table 2).

a useful reference point to sort this out and is a source to which readers can return periodically. The complexity of this global system can be overwhelming, as, indeed, to be honest, it often was for the authors.

At the beginning of the twentieth century, the total size of the global cable communications system measured 146,419 miles in length. Of this figure, the top eight firms controlled just over 90 percent. The top four tier one global communication firms alone controlled roughly 72 percent, and the biggest of them all, the Eastern Telegraph Company, just under 50 percent. The tier one players shaped the institutions of communication markets and had a global presence or a dominant position in the most lucrative regional communication markets, while the tier two firms were regionally focused and depended on tier one players for access to markets.

CARIBBEAN AND SOUTH AMERICAN
COMMUNICATIONS MARKETS

The unstable economies that became the bane of Charles Conant and others were rampant in Latin America. The drawn out implosion of the Spanish and Portuguese empires from the 1820s onward gave rise to demand for new communication technologies among officials who strove to preserve their disintegrating empires and, conversely, among those who wished to forge new nation-states and viable markets in the region. The chronic instability of the Caribbean and the modernizing thrust of the larger South American countries, notably in Argentina, Brazil, Chile, and Uruguay, begot hordes of concession hunters and speculators who acquired one concession after another for the development of telegraph and cable systems in Latin America, especially between the mid-1850s and 1870 before the major global communication firms, supported by foreign diplomacy, ironed out the arrangements among themselves. And as "normal markets" were carved out of instability one of the most discernible facts was that the development of communication markets in Latin America did not occur as a reflex of Anglo-American rivalry and a more general struggle for the control of global communication between these and other national interests, as claimed by most media historians, but rather did so cooperatively.

The International Ocean Telegraph Company was the first company to emerge from the quagmire of tangled interests in 1866 and was a paragon of nineteenth-century international collaboration. Formed by the retired American Civil War officer James Scrymser, the firm established strong ties to the leading global communication companies and relied heavily on governments, especially Britain, the United States, and Spain, for diplomatic support, subsidies, concessions, and network security in this volatile area. The company also had impeccable ties to Western Union, the Anglo-American Company, and the Eastern Associated Company. Indeed, the new company was approached before it laid its first cable in the Caribbean by Norvin Green, the president of Western Union, who sought to coordinate the firm's operations with those of Western Union in the United States and the Anglo-American Company in the Atlantic.[46]

The U.S. had not yet adopted its free trade in cables policy, and this allowed the company to obtain exclusive concessions from the Florida legislature and the U.S. Congress for fourteen and twenty years, respectively. Secretary of State William Seward also instructed the American ambassador to Brazil "to give such assistance as may be in your power . . . to obtain aid and authority from the Brazilian government to extend the lines . . . of the International Ocean Telegraph Company . . . over the West India Islands to a convenient point on the coast of Brazil."[47] American consular officials and company agents also persuaded the Spanish government to transfer several Latin American concessions held by Arturo de Marcoartu, the head of the Spanish Engineering Corps, and the famous French engineer and builder of the Suez Canal, Ferdinand de Lesseps, to Scrymser. British diplomats also appealed to Spanish officials in Cuba to grant concessions for Scrymser's new venture. While the British government refused to give monopolistic privileges in its Caribbean colonies—and, indeed, made Scrymser renounce claims to such rights around the expected location of the Panama Canal—the Treasury, Colonial Office, Foreign Office, and Board of Trade all approved the project and encouraged the colonies to contribute subsidies. By 1869, the company was on exceptionally solid ground with a host of landing rights—some exclusive, some not—and pledges from the Spanish, Danish, French, and British colonial governments for annual subsidies totaling about $85,000. It was also engaged in ongoing discussions with Columbia, Peru, Chile, Argentina, and Brazil for additional operating rights in these countries.[48]

The company laid its first cable from Key West to Havana in 1867. During the next three years additional lines were added to Jamaica, Puerto Rico, Trinidad, and Panama—although the latter cable never provided effective service. The firm's tight connections to John Pender and others at the top of the global communication system were readily apparent from the outset. At first, this was only at the shallowest of levels through the company's reliance on the Telegraph Construction and Maintenance Company for the making and laying of its cables. Banking relationships, however, were one index of deeper ties, and in this, the International Ocean Company followed in the footsteps of Western Union and the Anglo-American Company by using the firm of J. P. Morgan to handle its financial affairs. The significance of these firms' reliance on J. P. Morgan is revealed by the fact that their only rivals to date—the Direct Cable Company, the Commercial Cable Company, and the Compagnie Française du Télégraphe (PQ)—all had different banks. Morgan was a paragon of Anglo-American cooperation, with close ties to the centers of finance and political power on both sides of the Atlantic. The bank was also a renowned advocate of "managed markets" and "rational capitalism," and its position as the chief banker to the leading global communication firms meant that that logic would loom large in their approach to the business.[49]

Immediately after laying its first Caribbean cables, the International Ocean Company joined Charles T. Bright, chief engineer for the Telegraph Construction and Maintenance Company, Cyrus Field of the Anglo-American Company, and a coterie of other, mainly British, interests to create two new companies: the West India and Panama Telegraph Company (1869) and the Cuba Submarine Telegraph Company (1870). The West India and Panama Company system ran between Cuba and Puerto Rico, Panama, Jamaica, Trinidad, and a few points in Mexico and South America. It also laid a cable to British Guiana and Venezuela, which was intended as part of a much larger system that would eventually connect the east coast of South America to the United States, but this never functioned properly and over time failed completely. The company was originally conceived as an extension of the International Ocean Company, but in the end most of its capital and directors as well as its registration were located in Britain. The company's president, W. F. Smith, and another American, Moses Taylor, still played prominent roles in the company, but they were overshadowed by British banking and merchant interests, and others from interest groups in the British Empire, notably the West India Committee, were predominant.[50]

A year after the West India and Panama Company (1869) was formed it joined with the International Ocean Company to create the Cuba Submarine Telegraph Company. The new company was intended as an alternative to the existing state-owned telegraph system in Cuba, which was constantly disrupted by the Cuban independence movement's guerilla warfare against Spain. With this in mind, the company was created to link commercially important cities by way of a secure network laid underwater around the coast of the island.[51] The firm was financed by leading figures of the British aristocracy, and its directors were drawn from the International Ocean Company, the West India and Panama Company, and British diplomatic circles. And as could be expected, the three companies involved in these processes—the International Ocean Company, the West India and Panama Company, and the Cuba Submarine Telegraph Company—immediately formed a cartel. The following

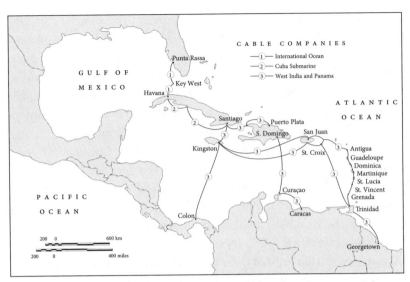

MAP 6   Caribbean showing operations of International Ocean Telegraph Company, Cuba Submarine Telegraph Company, and West India and Panama Telegraph Company, circa 1870

extract from the cartel agreement gives some indication of its impact on the organization of the Caribbean cable communication system:

> Neither company party hereto will enter into any agreement or traffic arrangement with any other company or . . . telegraphic line whatsoever which may be prejudicial to the interests of the other company hereto without the assent in writing under the seal of such company being . . . obtained. But either company shall be at liberty to send messages over their lines at the written request of the sender of such messages without solicitation or suggestion by any competing lines, provided that in such cases the other company party heretoshall share in the amount accruing in respect of such messages in proportion to the amount each compete would have received had the messages gone over the lines of both companies.[52]

Decades later, court cases declared that the agreements were still "of the most binding character."[53] The West India and Panama Company was continuously hobbled by the agreement, low demand, the poor quality of its technology, and the weak Caribbean economy. Obligations to deliver messages for the Spanish government for free that had been exchanged for concessions, and its commitments to provide a compendium of U.S., British, and French news to the Caribbean islands, also aggravated its difficulties, although the commercialization of the news compendium as a service provided to the

Caribbean press in the 1870s did establish a valuable new source of revenue. The company's weak condition led to constant complaints of poor service and threats of new rivals and, despite this, to yet even higher rates as the company sought to offset its losses.[54]

In the end, the weakness of the companies led all three—the International Ocean Company, West India and Panama Company, and the Cuba Submarine Company—to become more tightly yoked to one another between 1873 and 1875. This was illustrated as the International Ocean Company was acquired by the Western Union and a new board of directors installed. The West India and Panama Company was brought closer to the Anglo-American Company and Eastern Telegraph Company through a similar process of restructuring. In Jorma Ahvenainen's words, by the mid-1870s the links between all of these interests were so tight that they effectively comprised "a telegraph monopoly between Europe and the Caribbean: from Europe westbound traffic was handled by the Anglo-American Telegraph Company, Western Union, International Ocean and the two Caribbean companies, which were partially jointly-owned and which had mutually beneficial traffic agreements."[55]

GLOBAL COMMUNICATION AND WIRED CITIES: THE BRITISH AND EUROPEAN COMPANIES GO TO SOUTH AMERICA

As the Western Union and Eastern Associated Telegraph Company sorted things out in the Caribbean, they developed their plans for South America. Negotiations initiated by Scrymser with the Brazilian government in the late 1860s and transferred to Charles T. Bright during the creation of the West India and Panama Company culminated in the creation of several new companies between 1870 and 1874 that soon began operations in South America.[56] While there had been previous plans to link South America to Europe and the United States since the early 1860s, most notably by the South Atlantic Telegraph Company and by the International Ocean Company after 1865, none of these had been realized. However, as the outlines of the Caribbean system were put into place and global communications put on a sounder footing, definite plans, indeed a veritable scramble, to wire up the continent of South America occurred between 1870 and 1874. Together, Charles T. Bright of the Telegraph Construction and Maintenance Company and Matthew Gray, the managing director at the India Rubber, Gutta Percha and Telegraph Company, worked with well-connected South Americans—for example, Irineu Evangelista de Souza, Baron of Mauá in Brazil, Andrés and Pedro S. Lamas in Uruguay, and Carlos and Mariano F. Paz Soldan in Peru—to revive and re-register old concessions which had never been used or to acquire new ones.[57]

As Bright, Gray, and their South American colleagues set about their activities, three new and separate British and European cable companies were registered in London with the aim of laying transatlantic cables to North and South America: the New Atlantic Telegraph Company, the Direct American Telegraph Company, and the Great Northern Telegraph Company. Recognizing that there was little hope for all three to succeed, the backers of the Great Northern Company, C. F. Tietgen and Henrik Erichsen, persuaded the other groups to combine their efforts, and in April 1872 they registered a new firm in London: the Great Western Telegraph Company. The aim was to develop a transatlantic system stretching from Britain to Bermuda with two branches from there, one to New York and the other to South America by way of the Caribbean.[58]

From the outset the company was cast as the extension of Tietgen's other major cable company, the Great Northern Telegraph Company, which had just completed its telegraph and cable system from London through Europe and Russia to Korea, Japan, and China in the Far East. As such, the new venture was well resourced and poised to create a global communications powerhouse that would rival the only other major global system in existence at this time, the firms arrayed around John Pender, James Anderson, Charles T. Bright, and a few others, to whom we have already been introduced: the Anglo American Company and the Eastern Associated Telegraph Company.

The Great Western Telegraph Company was indeed taken seriously by Pender and others in the Eastern Associated Company, and they did their utmost to cast doubt on the new venture through several unflattering articles published in the *Times* in April 1872.[59] Yet, while the Great Western Company was undoubtedly a formidable rival, it lacked one indispensable thing that the Pender group of companies did have: concessions. Indeed, by this time, Bright and his associates had gained two vital sets of rights. The first were *non-monopolistic rights* to establish direct cable connections between Brazil and the United States and Europe. The second set was even more important and consisted of a "*monopoly . . . for a period of 60 years . . .* for submarine telegraph communication between Rio de Janeiro, Bahia, Recife (Pernambuco), Ceara, Maranhao, Para, Santos, Sao Paulo, Florianopolis, and Rio Grande de Sul."[60] A third set of Uruguayan rights, originally acquired by Arturo de Marcoartu and Matthew Gray of the Gutta Percha Telegraph Company nearly a decade earlier, were also thrown into the pot in 1873. Altogether, these rights cleared the way for a cable system from northern Brazil to Montevideo and then to Buenos Aires, and no one without such rights had a chance to enter South America.[61] The Pender group was clearly in an enviable position. Surprisingly, however, rather than developing the concessions, the Eastern Associated Company formed the Brazilian Submarine Telegraph Company in 1872 and simultaneously transferred the Brazilian licenses to another new company formed by Tietgen and associates: the Western and Brazilian Telegraph Company. That company, in turn, had been formed as the successor to the Great Western Telegraph Company and still involved the same actors, with the Hooper Telegraph Company holding the largest block of shares (4,801), Erichsen and Tietgen the next largest share (4,397), followed by Bischoffsheim, Goldschmidt and Company, and then John Heugh (2,035).[62]

The obvious question is why did Pender so eagerly dispose of his companies' interests to the new rival? One answer is that Pender and Tietgen had adopted a similar course of action, as we will see in chapter 4, with respect to the Asian communication markets, and the South America deal paralleled that precedent. Second, the fact that these events took place amid a volatile phase in the world economy (just prior to the crisis of 1873) may have been another reason. Third, the arrangements revealed another feature that was already becoming a mainstay of the global communication business: the cartel. Now the preference for "private structures of cooperation," to use Michael Hogan's language, was being applied to South America.[63] And in this, the details of the deal between the two companies were familiar. In return for ceding control over communications between Brazil's coastal cities, the Brazilian Submarine Telegraph Company received a third of its erstwhile rival's gross revenues and obtained commitments from the Western and Brazilian Company to use its cables exclusively for all messages between South America and Europe, and the two agreed to jointly manage the coastal network of cables and the transatlantic cable as a "single system."[64]

In some ways however, this explanation oversimplifies matters a bit too much because the Western and Brazilian Company did not have the entire coastline from the northern Brazilian city of Para all the way through to Montevideo and Buenos Aires to itself. Instead, in the most lucrative markets south of Rio de Janeiro it had to either compete or cooperate with three small firms which had been set up between 1865 and 1872: the Platino Brazilian Telegraph Company, the Montevideo and Brazilian Telegraph Company, and the River Plate Telegraph Company. The minutiae of these firms and the myriad

ATLANTIC

OCEAN

Buenaventura

from Portugal
via Cape Verde

Payta

Ceara/
Fortaleza

Recife

Callao/Lima/
Chorillos

Bahia / Salvador

Mollendo

Arica

Iquique

Rio de Janeiro

Sao Paulo

PACIFIC

OCEAN

Caldera

ATLANTIC

Coquimbo

Florianopolis

OCEAN

Valparaiso

Rosario

Santiago

Montevideo

Buenos
Aires La
Plata

CABLE COMPANIES

circa 1875

——①—— Brazilian Submarine
——②—— Western & Brazilian
——③—— Montevideo & Brazilian
——④—— River Plate
——⑤—— Transandine
——⑥—— West Coast of America
——⑦—— Tehuantepec Railway / Western Union
        (1873 Proposal)

300	0	900 km

300	0	600 miles

MAP 7   South American cables system, circa 1875 (planned and actual)

overlapping concessions they possessed are detailed by Jorma Ahvenainen and Victor Berthold.[65] Here it suffices to note that from the beginning the general thrust was for all three to work closely with the Western and Brazilian Company and the Brazilian Submarine Telegraph Company. Between 1873 and 1875, agreements were struck which allowed the three smaller companies to retain ownership of their cable and telegraph lines while permitting them to be operated as a "single system" by the much larger Western and Brazilian Company. In return, the three smaller companies received a portion of the revenues accruing to the "unified system" and shares in the Western and Brazilian Company.[66]

These complex interactions gave the Western and Brazilian Company a de facto monopoly over cable communication between all of the commercially significant cities from Argentina in the south to the top of Brazil in the north. Others could lay a cable to any one of these cities and between those not covered by the Brazilian concession, such as Para, Recife, Bahia, Rio, Sao Paulo, Florianopolis, and Rio Grande do Sul, but they could not establish a parallel national system between them. Overall, the concessions provided access to two-thirds of the Brazilian population that lived in these coastal cities. This position in Brazil, Uruguay, and Argentina was a formidable barrier to competition for the next half century and, ultimately, became a magnet for the most significant events in the politics of global communication between 1890 and 1920.

Pender's Brazilian Submarine Telegraph Company laid the first cable from Lisbon to the Cape Verde Islands off the west coast of Africa and then to Pernambuco in 1874. While the 1870 concession also permitted a direct cable connection to the United States, this was not laid until the 1920s. However, attempts to create two indirect routes to the United States were under way at the same time that the Brazilian Submarine Telegraph Company was laying its cables across the South Atlantic. The first was undertaken by Tietgen and Erichsen's Western and Brazilian Company as its cables progressed along the northern coast of Brazil to British Guiana, where it was suppose to connect with another cable laid by the West India and Panama Company. However, that company's cable, as we saw earlier, never worked and was abandoned. With it went the first of two planned routes between North and South America.

DREAMERS AND VISIONARIES: CABLE
COMMUNICATIONS, ICE SKATES, CANALS, AND
THE GLOBAL FINANCIAL CRISIS OF 1873

The second proposed cable between South America and New York was an
aspect of an audacious venture to relocate current trade routes connecting
Europe to the Far East by way of the "wild zones" of the Middle East to a new
"international zone of security" cut across the isthmus of Central America.
Although the Panama Canal was not completed until 1914, similar schemes
emerged during the 1850s, and by the early 1870s there were several plans
consisting of a mixture of British, French, and American interests that appeared to be quite viable.

One prominent version of the Panama Canal Project in the early 1870s was
spearheaded by New York capitalist and president of the Tehuantepec Railway
and Ship Canal Company, Simon Stevens. In early 1873, Stevens began discussions with William Orton, the president of Western Union, Cyrus Field,
Henry Meiggs (an American railway builder and developer in South America), and the British consul general in New York, E. M. Archimbald. Together,
the group outlined their vision of a network of cables and telegraph lines
between New York and Callao on the west coast of South America, crossing
Central America at the point where Stevens and Meiggs would link the Atlantic and Pacific by way of a canal and railways. The proposed cable system
would link up at Callao with another being organized by others closely linked
with Pender's inner circle—Charles T. Bright, Matthew Gray, and C. S. Stokes,
all of the Gutta Percha Telegraph Company—before linking into the vast
system of cables being laid by the British and European companies along the
east coast of South America.[67] Archimbald also enthusiastically endorsed the
venture in a letter calling on his counterpart in Peru to offer as much support
for the project as he could muster:

> The time has fully arrived when this great work should be carried out. The rapidly
> increasing commerce of the Pacific Ocean,—which would be increased ten-fold by
> the construction of a ship canal at the isthmus of T—, loudly calls for the con-
> struction of this important work. . . . Whoever shall complete it will . . . realize the
> grand dreams of Columbus, and eclipse the fame of Lesseps. There is, in my mind,
> no comparison between the world wide importance and value of a Canal between
> the Pacific and the Mexican gulf—and one between the Red Sea and Mediterranean.
> The former will be the *great work* of the age.[68]

Peru granted concessions for the project in 1874 and everything seemed ready to go. Then the venture disappeared. What happened? A seemingly improbable connection lies in the fact that just as the backers were ready to embark on their project others were trying to sell ice skates in Brazil and others still to peddle fine china to indigenous tribes throughout the continent. While a cable and telegraph venture backed by leading firms and the diplomatic weight of the British Empire seems unrelated to such silly schemes, the fact was that financial markets had lost the capacity to discriminate and were running up the stocks of all three—cables, skates, fine china—in a speculative bubble of global proportions and it was about to collapse. When it did late that year, along with the hucksters of fine china and skates in Brazil went Stevens's Panama Canal venture and the cable and telegraph system associated with it.[69] Whereas submarine cables had been the "most popular outlets for commercial investment" a few years earlier, they were now seen as the most "dangerously speculative"[70] and remained so for the next half decade.

Networked Nations and the Impact of Global Connections

The worldwide economic crisis did not mean, of course, that all activity ceased. Indeed, Tietgen and Huegh's Western and Brazilian Company continued to forge ahead with its vast network of cables between cities along the Brazilian coast north of Recife up to British Guiana. Concurrently, the three other smaller companies that we mentioned above—the Platino Brazil Company, the Montevideo and Brazilian Company, and the River Plate Company—were set in motion to extend the network of coastal cables in a southerly direction to the holy grail of the South American economy—the River Plate estuary that straddled the borders of Uruguay and Argentina. The first firm laid a cable from the southern city of Chuy to the border of Uruguay, at which place the Montevideo and Brazilian Company's cable and telegraph lines extended to Montevideo, while the last gap from there to Buenos Aires was finished by the River Plate Company, all as the transatlantic cable between Europe and Brazil was switched on in 1874.

Thus, despite the meltdown of global financial markets, the commercial, political, and cultural life of Buenos Aires, Montevideo, and Rio de Janeiro was revolutionized. Speedy connections to the metropolitan centers of the world—London, Paris, New York—meant that such places now loomed larger in the imaginations of local and transnational elites who looked to them for their cues on style, politics, the design of urban spaces and architecture, and their ideals of modernization and progress. News, commodity prices, and market transactions that had taken months to circulate from the metropoles in the past now took hours and sometimes just minutes. Immediately agreements were struck with two of the largest global news agencies—Reuters and Havas—and both opened joint offices in Recife, Bahia, Rio de Janeiro, Montevideo, and Buenos Aires in June 1875. Without any direct cables to the United States, the American press was left out, a point that was to grow in significance in the 1910s as the United States began to covet a much greater role in the affairs of South America in general and its leading metropolitan centers in particular.

The River Plate was emerging as a key nodal point on the world economy and as a focal point in the system of world communication. In fact, it was the most sought after hub in the transatlantic cable system south of Miami, perhaps even New York. This could be seen not only from the fact that it was the first place reached by the South Atlantic cable after Rio de Janeiro, but by the fact that Charles T. Bright, Baron Mauá, and their associates had established their first South American company there: the River Plate Telegraph Company of London in 1865. The company became another link in the coastal cable system, but before that the telegraph system that it operated in Montevideo and to the inland cities of Rosario, Canelones, San Jose, and Colonia made up Uruguay's de facto national telegraph system. The commercial importance of the River Plate zone was underscored again in 1871 as the Oriental Telegraph Company began a small telegraph system from Montevideo into the northern interior of Uruguay en route to Brazil. A year later the Anglo-Chilean firm of Clarke and Company constructed a line over the Andes Mountains from Valparaiso to Buenos Aires in anticipation of the South Atlantic cable. The Transandine Telegraph Company opened on July 28, 1872, to celebrations and speeches laced with utopian hype, as the following excerpt from the *Daily Standard* of Buenos Aires shows:

> Twenty years ago it was in its infancy, and now it forms . . . the nervous system of the universe . . . This electric fluid that comes not from below, but from above, that is so ethereal, so spiritual, and that now binds together the two great republics of South America . . . is a sure prophecy of the greatness and glory that are coming, in the near future.[71]

Nationalism, universalism, science, spirituality, and technology all intertwined in one utopian mélange. Yet, then as now, such panegyrics concealed as much about the culture and power relations of the time as they revealed. The telegraph radically changed the perception of distance and brought Chile's main cities two weeks closer to Buenos Aires and to Europe. Yet, the accelerated flows of information applied only to a thin sliver of the continent. Valparaiso, Chile, and Lima were inside the network, but only just. Plans to connect both countries to the United States, as discussed, had disappeared as quickly as they had appeared, and it was another decade before they were revived.

Geography still mattered immensely, and the companies described so far connected only the most commercially viable markets and urban centers of the River Plate and along the coast of Brazil, Argentina, Uruguay and only then proceeded to Chile and Peru, while bypassing most of the rest of the continent. Life in these cities was good by global standards, and one of their favorable qualities was the availability of rapid and reliable communication and transportation links to other cities. Thus, in Buenos Aires, an electronic message could be sent to London or New York in 90 minutes, but it was incredibly expensive.[72] Global electronic communication was a luxury, not a necessity. That it was just another circuit among the upper echelons of world commerce was clear from the fact that the lines between Rio de Janeiro, in the south, and Europe and North America at the other end principally served about ninety firms in eighteen European and North American cities during the mid-1880s: twenty-five in New York, eleven in Paris, seven apiece in Antwerp and Baltimore, six in both Hamburg and Marseilles, four in Bordeaux, two in Bremen, and one each in Montreal, Christiana, Naples, Porto, Rome, and Venice.[73] The network spreading out from Rio certainly had an impressive reach, but it was targeted at very well-defined places which, in turn, revealed just how much of an elitist phenomenon global communication really was.

The means of global communication could be as much lauded for their beneficial impact on world commerce as they could be chastised for aggravating its instabilities. The Council of Foreign Bondholders, for instance, lambasted Reuters for what it believed were inaccuracies in its reporting on South American economic conditions in 1875, and the skirmishes between the two agencies were played out in the London press, notably the *Times*.[74] But underneath these criticisms of inaccurate business news there seemed to be a sense that the sheer speed and volume of information was magnifying global insecurity so that events in South America were being connected to equally calamitous events in the markets of Europe, North America, and the Middle East, all within an incredibly compressed period of time. Within three years, economies from Argentina to the Ottoman Empire collapsed.[75] Indeed, as the global economic crisis unfolded, the Council of Foreign Bondholders sent harried messages back and forth to their offices and to British consular officials in South America badgering them to take care of business.[76]

As markets collapsed, the global communication industry was rationalized. Western Union withdrew from South America for the time being; work on national telegraph systems in Chile, Brazil, Uruguay, and Argentina was hindered or halted altogether; and those who had built up the only significant positions in these countries—Henry Meiggs (Peru), William Wheelwright (Chile), the Transandine Telegraph Company (Chile and Argentina), the River Plate Company (Uruguay)—and state-owned telegraph systems guarded them jealously. Even the biggest companies suffered badly with the Western and Brazilian Company and its network of cables along the east coast of South America characterized by regular interruptions, weak revenues, and the constant expense of repairs during their first years of operation, as shown in table 4.

The conditions were not only frustrating to the Western and Brazilian Company but to the governments whose countries it served, to clients who were forced to pay high rates for poor service, and, crucially, to the Brazilian Submarine Telegraph Company, which depended on the Western and Brazilian Company for the origination and termination of messages. Thus, the Western and Brazilian Company faced a constant barrage of criticism from governments, customers, and, crucially, John Pender, who personally began to push for changes during the 1880s. But even before that, several of Pender's allies from the Anglo-American Company, the West India and Panama Company, and the Eastern Extension Company—Henry Weaver, W. S. Andrews, C. W. Earle—were put on the firm's board of directors between 1877 and 1881.[77]

While the cable system on the Atlantic coast was in operation by 1874 this was not the case on the west coast. There Charles T. Bright, C. S. Stokes, and Matthew Gray, along with two members of an influential and elite Peruvian family, Carlos and Mariano Paz Soldan, collected concessions on behalf of the India Rubber and Gutta Percha Telegraph Company. Concessions were obtained in 1870 and the overall plan was to build a cable system from Chile to Peru and ultimately, to New York. However, when efforts began in 1873–74 they were obstructed by existing private- and state-owned telegraph agencies, which feared that a new cable would cut into their already meager revenues. In Chile legislation calling for the existing state monopoly over telegraph service to be extended to the coastal cables was introduced. The Peruvian Senate also opposed new cables on the grounds that they would ruin the state-owned system.

TABLE 4   Western and Brazilian Telegraph Company, 1876–81

YEAR	CAPITALIZATION (MILLIONS)	NET REVENUES	AMOUNT CARRIED TO RESERVE FUND	RESERVE FUND EXPENDITURES	REMARKS
1876	6.5	157,500	126,000	126,000	
1877	9	65,000	130,000	253,000	
1878	9	182,000	16,000	120,000	Relay cables
1879	9	16,300	100,000	0	Special repairs
1880	9	180,000	20,000	0	Special repairs
1881	9	230,000	0	0	Cable renewal
1882	9	266,000	0	0	Cable renewal

Source: Britain, *Inter-Dept. Comm. 1st Report*, 1902, appendix G; revenues from Ahvenainen, *European Cable Companies in South America*, 130, 179

Note: All amounts are in U.S. dollars.

These hurdles were only overcome when the Gutta Percha Telegraph Company agreed to compensate the Peruvian and Chilean state-owned systems for "lost business" caused by the new service.[78] The cables were finally laid in 1874–75, followed by another a year later, and then transferred to yet another new firm connected to John Pender's global communication empire: the West Coast of America Telegraph Company. However, the cumbersome arrangements between the West Coast of America Company on one hand and the Peruvian and Chilean state-owned telegraph companies on the other, coupled with the fact that there were no direct links to the United States, meant that service was poor and rates high. These conditions also compromised the commercial viability of the West Coast of America Company, and it was a weak vessel from the outset. And for Peruvians, although nominally "inside" the network, they still pondered whether they were linked to the global communications grid after all.[79]

Global Connections and National Aspirations

Obviously the major global players had assumed a great deal of influence over national telegraph systems, regardless of whether they were privately or state owned. Just as in the Euro-American market, their connections and revenues were crucial. Although Brazil, Uruguay, Argentina, Chile, and Peru all claimed a government monopoly over domestic telegraph services in the 1850s, they commercialized and privatized such services in the late 1860s and early 1870s. This did not mean that state-owned systems disappeared but that private companies either supplemented or competed with them. By the 1870s, regardless of whether nations had privatized their services or not, the private global communication companies were, de facto, running parallel national telegraph systems between the most commercially vibrant cities while leaving the rest of the country to be run by others as a public service. Even during the height of the economic resurgence in the late 1880s, many national telegraph systems continued to be run at a loss, and this was laid at the feet of the global communication providers.[80] The result was animosity and growing debate over private and state ownership.

In the 1880s, the Director General of the Brazilian Government Telegraph Department, Dr. G. S. Capanema, complained in a series of newspaper articles that the Western and Brazilian Company no longer supplemented the national telegraph system, but formed its nucleus.[81] Capanema's criticism was severe indeed, and soon the Brazilian government was pushing to nationalize the operations of the company altogether. And if that was not enough, as the 1880s progressed the Brazilian Submarine Telegraph Company and John Pender personally began to advocate substantial changes to the original 1872 joint purse agreements. In one such proposal, Pender even suggested that the state-owned Brazilian Telegraph Administration join the companies' joint purse, a proposal which Capanema rejected out of hand. The intervention of the Brazilian revolution and deposition of the monarchy in favor of a republic in 1889 gave the company a brief respite as the new government attempted to improve its relations with foreign companies and investors. However, it was a short-lived rapprochement and by the mid-1890s the company was again in the sights of the advocates of state ownership. Indeed, the Brazilian congress approved the takeover of the company's cable system in 1894, a move that was only aborted because of the poor condition of state finances and the broadside blow given to the economies of South America in general by the Barings Bank near collapse of 1890.[82]

While the Western and Brazilian Company faced enormous challenges in Brazil, the severest criticism of the companies was in Uruguay. There, the River Plate Company had been offering commercial telegraph service between Uruguayan cities and to Buenos Aires for two decades (1866 to 1887) despite consistent unrest and civil war. In 1887, the Uruguayan government hired the general manager of the Platino Brazilian Company, F. A. Lanza, to expand and operate the state-owned telegraph system for the next five years. Just before the contract was up for renewal in 1892, a report by the engineer L. Strauss for the Government Post Office presented a scathing indictment of Lanza and the

Platino Brazilian Company as well as "the good faith and credulity of the state."[83] The system was a disaster, claimed Strauss, because the government was paying inflated costs, there were huge deficits concealed behind an elaborate web of subsidies, and there were no direct connections to the capital. As the report concluded, all "the routes chosen were merely branches of the company's telegraph lines."[84] The defense of the South American coastal monopoly had literally been hardwired into the very architecture of Uruguay's public telegraph system. The contract, obviously, was not renewed, and over the next two years the government brought the telegraph system under post office management, extended its lines to Montevideo and other key cities, and implemented a uniform national rate system and sweeping rate reductions to which the private companies were forced to respond.

The West Coast of America Company faced similar pressures in Chile, and calls for state ownership were revived on a regular basis in the 1880s. Opposition to the company, as we will see below, gained momentum under the Balmaceda government in the late 1880s and early 1890s as it sought to offset British dominance of the national economy with investment from other countries, notably the United States.[85] In sum, after having experimented with a strong dose of privatization between the 1870s and late 1880s, many South American countries found it wanting. Consequently, they either expanded the role of government in such services or assumed a more encouraging attitude to those who sought to compete in South American communication markets. The two biggest beneficiaries of that new attitude were James Scrymser's Central and South American Telegraph Company and the Compagnie Française des Télégraphes Sous-Marins (Société Française), while existing companies were hit hardest.

THE REVIVAL OF SOUTH AMERICAN MARKETS: COOPERATION, COMPETITION, AND CONFLICT

A Vision Renewed

After Jay Gould and the Western Union took over the International Ocean Company in 1873, James Scrymser withdrew from the field. By 1880, however, things had changed. Ferdinand de Lesseps emerged with concessions for a canal across Central America and plans to take up where Simon Stevens had left off. James Scrymser also resurfaced with two new companies in hand: the Mexican Telegraph Company (1878) and the Central and South American Telegraph Company (1879). Scrymser and de Lesseps shared a vision that a combination of new cables to South America and a new Panama Canal would revolutionize the world economy, as captured in personal correspondence between the two near the end of the 1870s:

> The cutting of the isthmus of Panama . . . will give those regions an incalculable importance and will cause your system of submarine lines in the Pacific to become indispensable. . . . I am particularly struck with the advantages that will result from placing Japan, China, Australia, etc. etc. in telegraphic communication with the United States and Europe, by connecting the lines of your projected Central American cables with the telegraph system of New Zealand and Australia. . . . Henceforth to communicate from Europe eastward we will have to go West.[86]

The revival of Latin American markets and the Panama Canal project was indicated by the relative ease with which Scrymser gained the support of leading American financial institutions—J. P. Morgan, Winslow, Lanier and Company, and Drexel and Company. Scrymser's initiative also benefited from exclusive agreements with Western Union for the interchange of messages between North and South America. Three Western Union directors also sat on the new companies' boards of directors: J. Pierpont Morgan, Charles Lanier, and Percy Pyne. A joint purse agreement with the West Coast of America Company covered Peru and Chile and gave it indirect access to Argentina, Uruguay, and Brazil.

Corporate Connections: James Scrymser and U.S. Foreign Policy

From the outset, Scrymser's Central and South America Company built firm ties not just to financial institutions—as shown above and consistent with the practices of Western Union—but to people who shaped U.S. foreign policy and politics in the late nineteenth century and early twentieth. To be sure, U.S. foreign policy interests had long been represented on the board of Western Union, but the weight of such figures in James Scrymser's company was far greater and deeply affected its corporate culture.[87]

The close ties between the company and U.S. foreign policy were evident, for example, as soon as its attempt to lay cables from Peru to Chile were blocked by fighting between the Balmaceda government and agricultural landlords in the center of the country who had gained control of several Chilean cities. While the U.S. government worked behind the scenes and U.S. Navy ships floated offshore, the company's cable was brought to Santiago and placed in the service of the Balmaceda government while the insurgents' connections to the rest of the world were cut. Again during the Spanish-

American War of 1898 and the rise of the American empire, the company showed its eagerness to meet the communications needs of the U.S. military. And as we will see in later chapters, the company was always the most bellicose and nationalistic when it came to questions about the need for U.S. cables across the Pacific to the Philippines and the Far East or, later on, to South America.[88] More examples could be piled up like leaves in autumn but the point for now is that the company was far more willing then other U.S. or European companies to wear its nationalistic stance on its sleeve and eager to assist in the prosecution of whatever military and imperial ventures the United States embarked upon.

Cooperation or Rivalry: Scrymser versus Pender, 1880–1900

Nowhere were the swings between cooperation, "normal" competition, and conflict that defined the global political economy as a whole in the last quarter of the nineteenth century more apparent than in the relations between the Central and South American Company and the existing companies operating in South America, in particular the West Coast of America Company and the Western and Brazilian Company. The companies initially approached one another in a relatively cooperative fashion, and there were few obstacles in the Central and South America Company's path as it built its system. Between 1879 and 1882, the company gathered fifty-year exclusive concessions in Mexico, Nicaragua, Costa Rica, and Ecuador, while twenty-five-year concessions were obtained in Columbia and Peru. The Columbian and Peruvian concessions were interesting because the first explicitly exempted the "international zone of control" surrounding the planned site of the Panama Canal while the latter only went as far as Payta, where the Central and South American Company's network connected with the West Coast of America Company.[89]

In weak states (such as Mexico, Guatemala, Nicaragua, and Costa Rica), the company gained most-favored nation status, could open its own offices, and gave special half-rates only to presidents and ministries of foreign affairs. In contrast, stronger states (such as Ecuador, Peru, and Chile) did not grant most-favored nation status, gave monopolies of shorter duration, protected their telegraph systems from competition, and obtained half-rates for all official correspondence. The nuanced character of the Central and South American Company's privileges relative to those of the 1860s and early 1870s also suggested that Latin American nations had learned a thing or two about "communication policy" in the interim period. This was crucial because cable landing licenses were the key instruments for regulating global communication markets and, as such, reflected the micropolitics and power dynamics between the states and companies that negotiated each and every one of them, line by line.

Scrymser's two companies began laying their cables as soon as the legal and political formalities were put into place, first with a link between Galveston to Vera Cruz and an overland line to Mexico City in 1881. Later that year and into the next, the Central and South American Company laid its network of cables along the west coast of South America and by the end of 1882 its system connected New York (through Western Union lines) to every commercially significant city on the west coast as far as Callao. From there it had indirect access to Santiago, Buenos Aires, Montevideo, and Rio de Janeiro by way of its agreements with the West Coast of America Company and the Transandine Telegraph Company. The long-sought aim of direct communication from New York to the markets of South America was now an established fact.

Scrymser's arrangements with the West Coast of America Company only lasted during the formative years of the new company. The expiration and nonrenewal of the joint purse agreement between the Central and South America Company and the West Coast of America Company in 1891 signaled the change. Over the next few months, the newcomer extended its cables from Callao to Valparaiso and Santiago and, after a bidding war with the West Coast of America Company, acquired the Transandine Telegraph Company that ran from Santiago overland to Buenos Aires. The Central and South America Company now had a network running straight into the heart of the most lucrative communications market in South America: the River Plate. The early period of uneasy cooperation was over.[90] And if the most important communication markets in South America had not benefited from significantly cheaper and faster service during the 1880s, they did now. Messages between New York and Buenos Aires, for example, no longer had to be routed through Europe, and this saved time and money. Rates plummeted by at least half for Central American countries, while from New York to Peru, Chile, and Argentina they dropped from between $4.60 and $7.25 per word to $2.17 and $4.00 per word; still an extravagant cost, to be sure, but a substantial reduction nonetheless. The U.S. press also made its first tentative steps into the area, most notably by the *New York Herald*.[91]

Corporate Consolidation and the Creation of the Western Telegraph Company, 1880–99

Scrymser broadened his initiatives considerably by forging new working agreements with the only two independent telegraph companies on the east side of South America—the Oriental Telegraph Company (1871) and the River Plate Telegraph and Telephone Company (1887).[92] Scrymser also renewed requests that he had made in the mid-1880s to the Brazilian and Argentinean governments to bring his cables to Rio de Janeiro and Buenos Aires.[93] The West Coast of America Company scrambled to keep pace by bringing about its own rate cuts and soliciting closer ties with Pender. For their part, Pender and the Brazilian Submarine Telegraph Company reacted by taking several decisive steps to defend their influence over the national and global communications markets of the region. In terms of the west coast, Pender pitched in to assist the floundering West Coast of America Company by creating a new company, the Pacific and Europe Telegraph Company, in 1893 to replace the link that had been lost by Scrymser's acquisition of the Transandine Telegraph Line two years earlier. Nonetheless, the West Coast Company's revenues and capital reserves continued to fall until it was liquidated in 1896 before being resurrected a short time later as a full-fledged member of the Eastern Associated Company. While the West Coast of America Company and the new Pacific and Europe Company continued to flail about in the face of formidable competition from Scrymser, the impact on the Eastern and Associated Company was negligible since both of these subsidiaries were rather marginal vessels in the overall corporate empire.[94]

Although the west coast was largely conceded to Scrymser, neither Tietgen and Huegh's Western and Brazilian Company nor Pender's Brazilian Submarine Telegraph Company had any intentions to do the same on the east coast. There, the competitive threats became acute as the Compagnie Française des Télégraphes Sous-Marins laid a new cable from the northern Brazilian city of Salinas to Venezuela and Haiti in 1889 (which gave it an indirect link to the United States on account of the firm's connections at Haiti with the United States and Haiti Telegraph Company). But even before these moves, the Western and Brazilian Company and Brazilian Submarine Telegraph Company were already trying to make the "single system" principle more of a reality than the chimerical ideal it had been since the first joint-management agreement of 1872. These organizational changes, in turn, also reflected a recalibration of the balance of power between Pender, Anderson, and Bright on one hand and Tietgen, Erichsen, and Huegh on the other. Indeed, over the period between 1877 and 1899 the latter group was essentially forced to accept an ever more diminished role in South America.

Relations between the two groups had never been particularly cordial (and not just in South America but also in Asia, as chapter 4 shows), with Pender and other directors at the Brazilian Submarine Telegraph Company constantly worried about the poor condition of the Western and Brazilian Company's system and its unerring propensity to alienate local governments. Such concerns were magnified as the number of competitors banging at the doors of South American governments increased and as the push in some quarters for state ownership gained momentum. In the face of these threats the Western and Brazilian Company was yanked ever more tightly into the Pender corporate empire. The process began in earnest between 1877 and 1881 with the appointment of more Pender allies to its board. Pender also pushed for changes to the joint purse. The Western and Brazilian Company resisted, but steady pressure from Pender, including several court challenges, brought the company around to accepting a new agreement in 1889. The agreement earmarked 55 percent of all revenues for the Brazilian Submarine Telegraph Company, while also obligating the latter to renew its old coastal cable system.

In addition to tightening control over the Western and Brazilian Company, Pender and the Eastern Associated Company used three other measures to obstruct would-be rivals: discriminatory rates, renewal of monopolistic concessions, and new cables. Even though Scrymser's acquisition of the Transandine Company and agreements with the Oriental Telegraph Company and the River Plate Telegraph and Telephone Company gave his company direct access to some of the most lucrative markets of South America, he still depended on the Western and Brazilian Company for access to customers in Brazil. And it was this dependency that made it easy for that company to discourage its Brazilian customers from sending messages to the United States by the Central and South American Company's more direct route. Just as important, while the Brazilian and Argentinean governments were openly frustrated with the Western and Brazilian Company and generally supportive of new competitors, their options were constrained by the original licenses. Their most-favored nation clauses required that the Western and Brazilian Company be given the opportunity to complete any anticipated new schemes, so long as it did so on comparable terms. Consequently, as soon as the Western and Brazilian Company caught wind of new efforts to build competing networks, it bid to do the same thing on comparable terms. Given the ironclad nature of these concessions the governments had no choice, and just as the push for competition became most intense the concessions were renewed for another twenty years beginning in 1893.

In addition to these defensive tactics, a more proactive strategy of massive investment was pursued by the incumbent companies. Beginning in the early 1880s, the Brazilian Submarine Telegraph Company invested $2 million to duplicate the cable between Recife and Portugal. The new cable was opened for service in 1884. The company's fortunes improved, and its reserve fund soared from a modest $1.1 million just prior to the outlay on the new cables to nearly five times that amount in 1899. Several years later the Western and Brazilian Company, now much more firmly under the direction of Pender and the Brazilian Submarine Telegraph Company, embarked on a similar path. As a result, conditions in the Western and Brazilian Company reversed dramatically, and its revenues climbed steadily from a paltry $62,000 in 1888 to three times that amount four years later. Reflecting the more aggressive approach to competition and the fact that both companies now really did work as a "single system," the Western and Brazilian Company spent nearly $1.5 million between 1890 and 1892 to modernize its old cables and to duplicate the entire east coast cable system.[95]

While the threat of competition was undoubtedly driving these initiatives, they also reflected the fact that the River Plate zone was in the process of recovering from the Barings crisis, being again in the early throes of another euphoric phase of economic development, with the centerpiece a $20 million, European-financed "world class" port in Buenos Aires. Thus, despite mounting competition from the Central and South America Company and La Compagnie Française des Télégraphes Sous-Marins, revenues for the east coast and transatlantic companies were escalating. Seizing the opportunity Pender created another new subsidiary in 1891: the South American Cable Company. The new firm laid the third transatlantic cable between Europe and South America, by way of Senegal, in 1892 and immediately forged joint purse agreements with the Brazilian Submarine Telegraph Company and Western and Brazilian Company shortly thereafter.[96]

The number of messages sent over the east coast companies' networks tripled between 1884 and 1894, and revenues increased correspondingly, although rate reductions in the early 1890s tempered this somewhat.[97] The Brazilian Submarine Telegraph Company continued to fare exceptionally well, but matters were far different at the South American Cable Company and the Western and Brazilian Company. The South American Company lost money year after year and was sold to the French Post and Telegraph Authority just after the end of the nineteenth century. Likewise, the Western and Brazilian Company soon wilted in the face of competition by the Central and South American Company and Compagnie Française des Télégraphes Sous-Marins. The company's deteriorating condition only accelerated the long march toward the inevitable and in 1899 it was absorbed entirely into the

Pender family's corporate empire, as were the other east coast companies and the West Coast of America Company. Out of these changes a new continental communications colossus emerged that would tower over South America for the next four decades: the Western Telegraph Company.[98]

THE STRUGGLE FOR CONTROL OF CARIBBEAN AND SOUTH AMERICAN COMMUNICATION MARKETS: ALPHABET SOUP AND THE MULTINATIONAL CORPORATE IDENTITY CRISIS

If the web of companies and corporate affiliations was not already Byzantine enough at this point, they only became more so as time went on and as events in South America became more tightly bound to changes in the Caribbean and the North Atlantic. There is no way to avoid the fact that the economics of the global communication business and the tangled web of relationships tying it all together got thicker and messier during the 1890s. By the beginning of the twentieth century, a dozen firms in addition to the amalgamated Western Telegraph Company vied for control of Caribbean and South American communication markets, an alphabet soup of corporate interests, big and small. Four were ostensibly French, the Compagnie Française du Télégraphe de Paris à New York (PQ) (1879), Compagnie Française des Télégraphes Sous-Marins (Société Française) (1888), La Compagnie Française des Câbles Télégraphiques (Compagnie Française) (1894), and the South American Cable Company (1892). Three were members of the United States–British Caribbean cartel, the West India and Panama Company, the Cuba Submarine Company, and the International Ocean Company. Four others, the Halifax and Bermudas Company (1890), the Direct West India Company (1898), the U.S. and Haiti Company (1896), and a new German Cable Company (1894), were connected to the more loosely knit consortia being formed around the Commercial Cable Company, either by joint ownership, cross-appointed directors, or joint purse agreements.

The complicated organization of Caribbean and South American communication markets grew even more so. Société Française started operations in 1889 after having laid a new cable from Cuba to Haiti, with a few links to other Caribbean islands, and Venezuela, where it held monopolistic concessions for thirty years. From Venezuela, the company extended its cables to British and Dutch Guiana and then to the northern Brazilian city of Vizen.[99] Chafing under high rates, poor technology, frequent service interruptions, and the most punishing of all cartels, Spanish, Danish, British, and French colonial governments in the region granted Société Française concessions and subsidies totaling about $85,000 against howls of protest from the existing Caribbean cartel, especially the West India and Panama Company.

That latter company bore the brunt of competition as rates were cut, subsidies were transferred to the new company, and as whatever lingering hopes it had to obtain more subsidies vanished. The French government also committed to spend at least $35,000 a year for official communication to Société Française.[100] Yet, Société Française was only really interested in the Caribbean as a hub in a bigger system spanning from New York to the River Plate. But as it sought entry into these markets, it was not the West India and Panama Company that stood in its way, but Western Union and its allied members in the Caribbean cartel. Western Union launched a lawsuit in the American courts that aimed to block the French company's entry into the United States while the cartel reduced its rates, not so much to meet the competition, but to entice Société Française to use the cartel's networks instead of building new ones. However, cheaper rates were a poor substitute for the company's ultimate aim: offices in Manhattan.

The Transformation of French Global Communication Policy

Part of the backdrop of these tensions was the fact that France had recently undertaken significant changes in its communication policies that would not only help to batter open the electronic gates erected in the path of the French company by Western Union but also to a much enlarged role for French interests in global communication markets altogether. Some of these changes we have already seen in earlier sections of this chapter, notably the departure of the Compagnie Française du Télégraphe (PQ) from the dominant Atlantic cartel in favor of the Commercial Cable Company–led consortium in 1888. The second major change was ushered in by an 1892 report conducted by France's Ministry for Trade, Industry, Post, and Telegraphs. Among other things, the report led to the amalgamation of the privately owned French global communication companies—Compagnie Française du Télégraphe de Paris à New York (PQ) (1879) and Compagnie Française des Télégraphes Sous-Marins (Société Française) (1888)—into one new company: Compagnie Française des Câbles Télégraphiques (Compagnie Française [1894]; we will refer to it as the French Cable Company).[101] The commission also advocated allocating $21 million to restructure France's global communication industry and to extend the new company's communication network by some 20,000 miles so as to reach every commercially and strategically important center of French interest. New legislation was also advocated and subsequently introduced in 1902. In addition, the French government supported the firm with annual subsidies (revenue guarantees, to be precise) of approximately $160,000 per year for thirty years. It also merged several communications equipment suppliers into one technology manufacturing powerhouse along the lines of the Telegraph Construction and Maintenance Company.[102]

Underlying all of these changes, Ahvenainen argues, was France's desire "to overthrow the British monopoly of the cables."[103] But if that is so, the third development in this vast restructuring revealed just how far "nationalist" aims were constrained by global realities. This was apparent as soon as the new company's initiative—a cable from New York to Haiti—ran into opposition from the U.S. government and Western Union on account of its exclusive privileges in Haiti, Venezuela, and Brazil. However, in a bold move, the company laid the cable in 1896 without approval from the State Department. The ostensibly French company had circumvented the obstacles by creating a new American company: the United States and Haiti Telegraph Company. The company's stock was owned by the French Cable Company, but registered in New York and stacked with a contingency of American directors drawn from the Commercial Cable Company.[104]

By 1896, the French Cable Company's new cable between Brazil and New York began operation, its position bolstered by a secret agreement with the Venezuelan government which renewed its exclusive concession for fifty years and allowed it to lay new cables between the main cities along the Venezuelan coast. This was largely on account of the fact that Venezuela was in the throes of yet another debt crisis. The crisis fomented a political revolution that brought a new government into power in 1892 but which also left 40 percent of the national telegraph department's revenues assigned to the payment of foreign debts. With its revenues thus diverted the country turned over the development of its national telegraph system to the French Cable Company, although this would prove to create more headaches than revenue in the long term as we will see shortly.[105]

The Commercial Cable Company, 1895–1904:
The Quintessential Multinational

The rise of the French Cable Company also strengthened the consortia developing around the Commercial Cable Company. The retooled French Cable Company was now allied with that firm in the Euro-American communication market as well as between New York and Brazil. The French Cable Company also benefited from the fact that it now had access to the Postal Telegraph Company in the United States and the Canadian Pacific Telegraph Company in Canada. The simultaneous ascendance of the French Cable Company and bolstering of the Commercial Company's role as the nucleus of a nascent global consortium marked a radical change in Atlantic communication markets. Henceforth, two "super-cartels"—one allied with the Commercial Company and the other around the Western Union/Anglo-American Company— would vie for the control of communication in these areas. Coupled with Scrymser and the French Cable Company's assault on Pender's monopoly in South America, the 1890s represented a dynamic decade of fundamental changes in the global communication business. Other events at the time and just after the end of the nineteenth century only reinforced such an impression, most notably with the creation of two new "British" companies by the Commercial Company: the Halifax and Bermudas Company (1890) and the Direct West India Company (1898).

The two new companies revealed that the Commercial Cable Company was no mere tool of French imperial communication policy, but equally willing to serve the needs of any empire: French, British, German, or American. At the same time that the Commercial Company was helping the French Cable Company masquerade as an American company, it was vying with the "British" West India and Panama Company for subsidies to create a new line of cables from Halifax to Bermuda and from there to the Caribbean. The new cables were desired in order to create competition with the existing Caribbean cartel and to meet the rising British demand for strategic "all-red" lines to every point in its Empire. Proposals for the new Halifax-Bermuda-Jamaica route were entertained from both firms, that is the West India and Panama Company and the Halifax and Bermudas Company. The Colonial Office preferred the West India and Panama Company on the grounds that it was more British. But the other firm had the support of the Treasury Office and, as such, it was the Halifax and Bermudas Company that laid the first cables, with a subsidy of $42,500 per year for twenty years from the British government, in 1890.[106]

As had been the case around the initial Halifax-Bermuda leg, so likewise was the rivalry for subsidies and concessions in the Bermuda to Jamaica section being sought by the incumbent and "more British" West India and Panama Company, on one hand, and the upstart Commercial Cable Company, on the other. As had previously been the case, the Colonial Office favored the incumbent on the grounds that it was "more British" while the Treasury, unmoved by sentimental considerations, supported the latter on more narrowly drawn economic considerations. The West India and Panama Company submitted proposals to the Colonial Office that requested a subsidy of $65,000 per year for twenty-five years with a reduction in rates from London to Jamaica from $1.37 per word to $1 per word. Rejected, the firm submitted a revised proposal a year later that dropped the amount of subsidy to just over $50,000 for twenty years, but it was still higher than the deal offered by the Direct West India Company. Negotiations continued for the next two years, with the Colonial Office making the case that the promotion of trade and "imperial unity" made the West India and Panama Company the natural choice, so long as it dropped its rates to 75¢ per word. The answer came back from the company shortly: it could not be done. Consequently, the Direct West India Company was selected and registered in London, several British and Canadian directors appointed, its cable laid in 1898, and a subsidy of $40,000 for twenty years received. Henceforth, the new company was a "British company."[107]

CORPORATE INTERESTS AND THE NATIONAL INTEREST: THE CASE OF THE GERMAN CABLES AND THE REDEFINITION OF FREE TRADE

The Commercial Cable Company's loose attachment to any national identity could be seen in one last example that will be introduced before we complete this chapter. The example covered in the following pages took shape around the rise of the German Cable Company in the mid-1890s, an expansion that was facilitated as much by its tight links with the Commercial Company as it was impeded by the obstreperous tactics of the Anglo-American Company and British government.

Until the mid-1890s, most of Germany's international communication services had been handled by the Anglo-American Company. But much as the Compagnie Française du Télégraphe (PQ) had done a decade earlier, the new German Cable Company was preparing to leave the Anglo-American Company and Western Union cartel on the cusp of the twentieth century in favor of joining the Commercial Cable Company–led consortia. Such changes did not occur suddenly but culminated processes that had taken shape over the previous ten years as Germany undertook a process of "forced industrialization" at home and a stronger projection of its interests abroad. Among those involved in both activities was the firm Felten and Guilleaume.[108] Felten and Guilleaume mimicked British and French practices as it moved from being a manufacturer of communication technology and electronic equipment for domestic markets to build and operate a global communication system. Between 1890 and 1910, the company developed four regional networks: one for transatlantic communications (Deutsch-Atlantische Telegraphenges.); another for communications between cities along the European coast of the Atlantic (Deutsche-See Telegraphenges.); another affiliate for South America (Deutsch-Sudamerikanische Telegraphenges.); and finally, in collaboration with the government of the Netherlands, another network between Indonesia and China (Deutsch-Niederländische Telegraphen Ges.).[109]

The German Cable Company's efforts were strongly oriented toward the transatlantic economy. During the first years of the 1890s Germany continued to rely on the old transatlantic cartel for sending and receiving messages from North America, while in Europe the nascent German Cable Company's cables interconnected with state-owned networks in Britain and France and offered services directly to the business districts of Madrid and Barcelona. As the company expanded it increasingly collided with the dominant Euro-American cartel and the British government. This was evident from the very first concession granted to the German Cable Company by the German Post Office in 1894 for a cable between Emden, Spain, and North America. The Anglo-American Company and British government continuously opposed the German Cable Company for two principal reasons: first, that its connections with the Commercial Cable Company would divert business from the Anglo-American Company and Western Union cartel and, second, that its proposed route between Emden, Waterville, and then to the United States would bypass London. The Anglo-American Company urged the British government to refuse the new company's application for a landing license, molding its arguments into a classic statement on the perils of "ruinous competition": "The growth of the competition for this traffic is approaching the point at which it tends to become destructive of capital rather than helpful to trade, and the fact that the right to control this growth is admitted in the Government, implies that the latter has also a duty to protect the legitimate enterprise of its own subjects when that protection as in the present instance can be afforded without prejudicially affecting any public interests or advantage to trade."[110]

This was a remarkably brazen invitation for the British state to defend private monopolies. The chairman of the Anglo-American Company, F. A.

Bevan, went even further to argue that his company should have the right to review any proposed new cables to ensure that they comported with its own interests.[111] In an incredible demurral to corporate power, the British government agreed as it incorporated the essence of the company's position into its decision of the German Cable Company's landing license. Thus, on January 3, 1900, the British Prime Minister Lord Salisbury approved the German Cable Company but only if it came to terms with the "British" firms, routed its traffic through London, and avoided competing with existing companies on the basis of rates.[112] The German Cable Company and Commercial Cable Company were probably aware of the unacceptable British offer before it was formally announced. In fact, just before the British decision was issued, the two companies signed a deal that subverted almost every goal of British policy. London was bypassed in favor of the Azores as the hub of their operations and the Anglo-American Company–led cartel notified that the German company was abandoning it.[113]

CONCLUDING REFLECTIONS ON COMMUNICATION AND THE GLOBAL SYSTEM, EARLY 1900S

What the German Cable Company incident shows is not so much the extent to which imperial rivalry had swamped communication markets by the beginning of the twentieth century but the extent to which these markets, like most other capital intensive industries, continued to contain a complex admixture of collaboration, competition, and conflict, self-interest and opportunism, private enterprise and state intervention. These conditions paralleled those of international relations generally and, as stated previously, whether arranged as cartels or consortia, the organization of communication markets offered a framework in which companies continuously struggled to preserve their entrenched positions in some markets while managing their links with one another in others.

The benefits of this "system" were mainly in the preservation of "viable markets" and the avoidance of ruinous competition. Yet, at the same time that there were benefits, there were losers by the score. For one, the Western and Brazilian Company, despite having a monopoly over the Brazilian system of coastal cables handed to it on a silver plate by the Pender group, floundered continuously until it was folded into the Western Telegraph Company in 1899. And then, of course, there was the West India and Panama Company, always a weak company, which was simultaneously hamstrung by the Caribbean cartel and decimated by the rise of the French Cable Company and two firms backed by the Commercial Company: the Halifax and Bermudas Company and the Direct West India Company. But the precarious condition of Caribbean and South American communication markets did not mean that these new rivals were runaway successes, either. The latter two merely joined the West India and Panama Company as the objects of a never-ending stream of British investigations into what could be done to improve conditions in the Caribbean.

Similarly, even though the French company was solidly backed by the French state and obtained an additional infusion of $2 million in 1898, it had become more and more a tool of the state. Eventually, this "politicization" compromised its concessions in Venezuela, where accusations of meddling in the politics of that country during the revolutionary times of the 1890s led to its concessions being cancelled and its control over the "shadow national telegraph system" running between major cities along the Venezuelan coastline being expropriated in 1902. Complicating matters further, the French Post and Telegraph Authority acquired the South American Cable Company from the Eastern Associated Company in 1902 and then forged a joint purse with that firm and the German Cable Company that lasted until World War I. Although Headrick and Griset claim that the alliance between the French and German companies would have astonished French public opinion, that was unlikely to have been the case. As we have shown, multinational cooperation between corporate actors and nation-states had been strong influences on the evolution of global communications since the outset. Indeed, at least up until the beginning of the twentieth century, nowhere had communication markets developed principally as a reflex of national interests.

Indo-European Communication Markets
and the Scrambling of Africa: Communication
and Empire in the "Age of Disorder"

The Indo-European market was opened in the mid-1860s by a consortium of
state and privately owned British, European, Turkish, and Persian companies,
followed by John Pender's Eastern Associated Telegraph Company's cables to
India in 1870. Unlike elsewhere in the world, between 1870 and 1878 the
Eastern Associated Company's cables had to compete with the consortium:
the Indo-European Telegraph Department (the regional operating arm of
the India Telegraph Department), the Indo-European Telegraph Company
(owned by Siemens, Reuters, and others, with a few German and Russian
government-appointed directors), the Persian Telegraph Agency, and the Ottoman's Central Telegraph Administration. After trying in vain to obstruct
the Eastern Associated Company, these companies turned aggressively to defending their monopoly before largely giving up to join the cartel organized
by that firm in 1877–78.[1]

In the interim, however, the consortium struggled to defend its monopoly
by tightening control over the Persian telegraph system, improving the Ottoman Empire's Central Telegraph Administration, adopting cheaper rates, and
impeding the Eastern Associated Company's access to customers in major
cities. Several agreements between 1868 and 1872 transformed what had originally been a joint venture for the development of Persia's national telegraph
system into a concession of longer and longer duration, the cumulative effect
being to force the Persian government to cede control over its national telegraph system to the Indo-European Department south of Tehran and to the
Indo-European Company to the north of the Persian capital, roughly parallel-
ing the division of the country into informal spheres of British and Russian
influence, respectively. In the process, new Persian lines were created but
largely turned over to exclusive control by British and European interests and
dedicated to the transmission of international messages. By 1892, subsequent
renewals had resulted in Persia relinquishing control over its national com-

munication system until 1925 and a kind of "informal electronic empire" put in its place.[2]

A different balance of considerations led the Indo-European Department to continue aiding the Ottoman Empire's attempt to build a modern telegraph system. The Turkish lines provided a strategically important route to India and helped check the Eastern Associated Company's power to turn the Indo-European communication market into a monopoly. The assistance given to the Ottoman's Central Telegraph Administration by the Indo-European Department was one small measure in a larger foreign policy toolkit that aimed to keep Turkey within the bounds of Britain's "informal empire." Pender and other directors at the Eastern Associated Company complained bitterly of the favoritism shown toward the Ottoman's Central Telegraph Administration, but their complaints largely went unheeded, at least for the initial half decade or so.[3]

But what galled the private companies the most was that the old consortium had reduced their rates just as the new private company was laying its cables in the Mediterranean and the Red Sea en route to India. Whereas sending a message from London to Bombay had started out at approximately $25 for twenty words when the lines first opened, they were reduced to roughly $14 in 1868, and a new class of messages designed to increase the social uses of the telegraph was instituted at about $9.10 for ten-word messages. According to Pender, James Anderson, and the other global communication barons trying to compete in the Indo-European market, the rates were unremunerative and were crippling all of the companies. In a series of letters, Pender, Anderson, and William Montagu Hay, all directors of the Eastern Associated Company, called for the abolition of the special ten-word rate and requested "the aid of the Government of India in procuring the raising of the rates . . . between Great Britain and India."[4]

William Hay complained, in a letter to the secretary of state for India, that beyond "cheap rates" the Indo-European Department was blocking the new company's access to customers. The complaints included refusing to allow the company to open its own offices in Karachi, Bombay, Aden, and Calcutta; refusing to provide it with private lines to connect with its cables along the shores of India and the Persian Gulf; charging discriminatory rates; routing traffic over the Indo-European route rather than the company's cables; and, generally, "the powerful influence [that] the Indian Government seems to be exerting in favour of the Persian Gulf lines."[5] The Eastern Associated Company needed their own offices and private lines, explained Hay, to deal directly with customers, as was the practice in Britain, North America, and, later, Australia. Furthermore, access to private lines and their own offices would allow them to unify administrative control over their own system, which would enhance efficiency, reliability, the secrecy of customers' correspondence, and so on.[6]

Everything that Hay said was probably true, but equally true was the fact such appeals smacked of hypocrisy. Indeed, shortly after criticizing the Indo-European Department, Hay turned to defending the Anglo-American Company's monopoly in the far more lucrative Euro-American communications market before Canadian and British officials, as we saw in chapter 2. In the present case, however, the criticisms carried weight, and certain changes followed in 1871. The company, for instance, was permitted to offer its own services directly to the public but in offices shared with the Indian Telegraph Department, rates to use the Department's telegraphs were reduced, and, most significantly, the ten-word social rate was abolished and general rates raised to approximately $22, close to their original precompetitive levels.[7] But the Eastern Associated Company did not get private lines, their own offices, or a rate increase on all routes to Europe. Indeed, the Ottoman administration kept the old rate of $14 despite considerable pressure to do otherwise.[8]

MODERNIZATION, MODERNITY, AND CRISIS: THE NEW MEANS OF COMMUNICATION IN THE OTTOMAN EMPIRE

Beyond these considerations, however, the most significant outcome was that none of the original companies capitulated to the Eastern Associated Company's incessant efforts to dismantle the consortium in favor of a more iron-clad cartel under its stewardship.[9] By the late 1870s, however, this changed, especially as the Ottoman Empire teetered between becoming a "modern cosmopolitan liberal state," on one hand, and the brink of disaster, on the other. The precarious condition of the empire was evident as the Eastern Associated Company pressured the Central Telegraph Administration to fall in line with the new rates adopted by the other companies and by its more frequent attempts to take over the Turkish lines to the Persian Gulf altogether.[10] The cheaper rate on the Ottoman lines was leading to a considerable loss of revenue—nearly 20 percent of all revenue went to the Ottoman lines in the early 1870s—for the British and European companies, and there was much consternation as a result, although more so with the Eastern Associated Company.[11] While that company was unable to gain control of the Turkish Telegraph Agency directly, several factors led to the Turkish system's elimination as a viable competitor in the longer term.

This was most notable with the eruption of the Russo-Turkish War of 1877, but in the previous year the deep problems besetting the Ottoman Empire were signaled as the country defaulted on its foreign loans, in effect declaring itself bankrupt. This led to the administrative and fiscal machinery of the state being taken over in 1881 by the International Public Debt Administration, a body representing the interests of European financiers and, less so, governments. These arrangements were a response to the debt crisis plaguing the country and other economies as well and which hindered Turkey's ability to repay the massive foreign loans taken out in the 1860s and early 1870s to fuel its modernization efforts.[12] These modernization efforts, in turn, reflected attempts by both indigenous forces, the Tanzimat movement (technocratic modernizers) and Young Turks (liberal reformers), and European influences to revamp the Ottoman Empire and prevent its collapse altogether.

The International Public Debt Administration signaled the failure of these efforts while ostensibly aiming to deepen them by putting Europeans directly in charge of the fiscal and administrative machinery of the state. In some ways, this intertwining of global forces and local political and cultural dynamics helped revive efforts to devise a new constitutional order along liberal and cosmopolitan lines. Thus, in 1876, a new constitution was adopted and a kind of cosmopolitan Ottomanism put into place based on citizenship rights, religious pluralism, the separation of church and state, and an overall architecture for the state, economy, and even the design of urban space and public goods inspired by liberal European ideals.[13] However, while registering the ascendance of certain political and cultural trends that would shape the next half century in the modern history of the Ottoman Empire, such measures were insufficiently grounded in the local culture to provide the basis for a new political, economic, and cultural order. They collapsed in the wake of the Russo-Turkish War, the repudiation of the constitution a few years after that, and ultimately in the face of the chaotic disintegration of the empire altogether.

This aborted attempt at modernization was also apparent as the previous capacity of the Ottoman Empire's Central Telegraph Administration to keep foreign interests from directly taking over its telegraph system began to wane. At first, it was in the areas north of Constantinople and toward the Black Sea as John Pender cobbled together a collection of concessions to create the Black

Sea Telegraph Company in 1874. In and of itself, the Black Sea Company was a marginal player, but beyond revealing the enfeebled stature of the Ottoman's Telegraph Administration, it constituted another route between the Balkans and the rest of Europe and London on one side and India on the other.[14] As such, it was a potential competitor to the Indo-European Company that had emerged just a few years earlier and to European state-owned telegraph systems. During the first few years the Black Sea Company posed little threat to either the Central Telegraph Administration or the Indo-European Company, but this changed during the Russo-Turkish War. The Indo-European Company's lines were cut at the beginning of the war, and this further weakened confidence in a company that was already lacking in this regard.[15]

The situation enhanced the value of the Black Sea Company's telegraph network in the Balkans and enticed the chastened Indo-European Company to join a cartel with the Eastern Associated Company, which it did in 1877.[16] This outcome doubled the resolve of the Eastern Associated Company to gain control of the Turkish lines to the Persian Gulf and to bring the Indo-European Department into the cartel. Accomplishing this would give Pender's group of companies control over the most important aspects of the Ottoman Empire's national telegraph system and the only remaining telegraph lines that offered any real competition to its submarine cables in the region. Of course, this bid to monopolize the Indo-European communications market was recognized for what it was and opposed by the Central Telegraph Administration and the government of India, albeit for different reasons. For the Ottomans, it meant retaining control over a politically potent symbol of their remaining authority and aspirations to create a modern cosmopolitan nation state fully integrated into world affairs. At no other time was the gap between such ideals and reality larger than in this time, but this only made the symbolic links between communication and modernization all the more significant. For the government of India, the rationale was still to maintain an alternative route to Europe, although the need to retain a "minimally" effective Ottoman state also played a larger role in British foreign policy, and the desire to preserve Turkish control over the Constantinople to Persian Gulf lines sought to advance these goals.

In the end, the Central Telegraph Administration retained its independence, but its role shriveled as the Indo-European Department finally joined the cartel in 1878.[17] The change sensitized the department to the private company's commercial concerns and diminished, but did not eliminate, its support for the Central Telegraph Administration. The Indo-European De-

TABLE 5   The Indo-European Cartel's Impact on Ottoman Telegraph Agency's Share of
Messages, 1871–88

COMPANY	1871–72 (PERCENT)	1887–88 (PERCENT)
Eastern Associated Telegraph Co.	52.81	64.01
Indo-European Company and Indo-European Department	29.11	34.50
Ottoman Central Telegraph Administration	18.08	1.49

Source: Wardl, *Joint Purse*, 8

partment continued to offer modest amounts of technical and financial support to the Central Telegraph Administration over the following two decades and continuously pointed to its cheaper rates to encourage rate reductions by the cartel members, moves which only caused the Eastern Associated Company to bristle and fulminate in indignation.[18]

Nonetheless, these overtures could not disguise that the Indo-European Department's motives were now skewed to supporting the cartel and, specifically, the Eastern Associated Company. Whereas in the past the Department had incentives to improve the Central Telegraph Administration's lines and, indeed, it had constantly done so, after joining the cartel its aim was to send as much traffic over the Eastern Associated Company system as possible and only as much to the Turkish line as would prevent its collapse altogether. The consequences were easy to see. Within a decade and a half, the Ottoman lines were decimated, while the share of the three other firms—the Eastern Associated Company, the Indo-European Company, and the Indo-European Department—grew from a little over 80 percent to almost 99 percent, as indicated in table 5, above.

As Wardl put it, the Turkish line was never very strong, but under the new arrangements its problems were "intensified . . . as it at once became the interest of the Indian administration to *divert traffic to the lines of its partners and to handicap their competitor.*"[19] Wardl removes any ambiguity about this by stating that "the action of the Indo-European Department has thus been to practically cripple the Turkish route as a competitor."[20] In short, the British empire was now complicit in the underdevelopment of the Ottoman Empire's national communication system, whereas previously it had at least been a nominal aid to its development.

The increasingly precarious position of the Ottoman Central Telegraph Administration was recognized as early as 1880 when the Indo-European Telegraph Department made overtures to take over its management. The Eastern Associated Company renewed its efforts to do the same, offering to lease its Persian Gulf lines for $90,000 per year. Such overtures continued throughout the 1880s and into the 1890s, and on each occasion they were refused.[21] Comparatively speaking, though, things could have been worse for the Central Telegraph Administration, given the contrast between its conditions and those in Persia, where control of that country's communication system had been relinquished decades earlier. Likewise, the Ottoman Empire also fared well relative to the conditions in one of its outlying republics—Egypt.

COMMUNICATION, MODERNIZATION, AND THE TAKE-OVER OF EGYPT

Just as in the Ottoman Empire, the telegraph, linked to the growth of the press and modern approaches to journalism, had become a signature feature of attempts by Egypt to carry out a grand project of political and cultural modernization and economic development. Juan Ricardo Cole describes the relationship between communication and modernization in Egypt in these terms: "The founding of private newspapers . . . occurred simultaneously with the extreme speed of telegraph lines—new politics and political journalism grew together. By the 1860s, telegraph services allowed reception of international news through the wire services, and Ottoman and European newspapers could easily be shipped to Alexandria and taken . . . to Cairo and the interior."[22]

Several privately owned Arabic newspapers were also being published in the region by 1870 and brought with them, especially in Ottoman and Persian territories, issues of censorship and internal opposition to corrupt regimes. By 1871, as the British consul in Cairo noted, "every town or village of importance in Lower Egypt has a telegraph station." A year later, the government telegraph system carried 389,225 European and 238,521 Egyptian messages, and by mid-decade it was a major source of revenue for the government.[23]

Cole is clear that all of these things were an integral part of the broader modernization efforts that took place between 1850 and 1880 and that were based on the adoption of new technologies, the spread of literacy, and political change. Others go even further and point to the way in which these new media networks and the greater circulation of people and ideas across national borders in this region and during this time facilitated social change throughout the region. Persians, Egyptians, Turks, Indians, and others came into far greater contact with one another, directly and indirectly, as well as with European political and cultural dynamics. And as this occurred, the contours of local political cultures assumed some common features, notably a nuanced critique of globalization as imperialism, although not necessarily of globalism in general (although, to be sure, conservative factions within these regions resisted *all* changes). In fact, beyond the Ottoman Empire, Egypt adopted a constitution penned with the assistance of European legal experts during this time (1866), as did others, notably Persia, albeit after the end of the nineteenth century.[24]

But as in the Ottoman Empire as a whole, all such efforts were fueled by foreign loans and numerous economic development projects (most notably the Suez Canal), some legitimate, others woefully corrupt. Because of these obligations and entanglements, when the debt crisis of 1876 hit, Egypt's experiment in modernization was terminated and its experience with European colonialism begun. Indeed, it lurched from being on the margins of Europe's informal empire to become a formal appendage of the British Empire over the next six years. At first, there was ostensibly no intention among the Europeans to establish colonial rule in Egypt but only the goal of "regime change." The debt crisis led to the ousting of one Pasha (Khedive Ismail Pasha) and the installation of a new one (Ahmad Urabi Pasha) as part of the "reorganization" of the Egyptian state which was led by yet another International Public Debt Commission. The Egyptian telegraph system—along with the collection of customs and taxation revenues, the management of ports and railways, and even the Treasury—were taken over by the Debt Commission so that its revenues could be used to pay down foreign debts. As the Council of Foreign Bondholders nonchalantly explained, the whole idea that "eastern countries" could undertake modernization on their own was a delusion. Instead, what these failed experiments all showed, according to the Council (speaking directly in terms of Egypt), was that the European countries needed to bring about an "entire remodelling of the Administration . . . subject . . . to close and constant inspection and supervision from headquarters." As if to put a fine point on the matter, the Council went on to make the even more sweeping statement: "In the East nothing can be done without distinct authority, *and authority is in the hands of those who hold the purse strings*, and who have the power to nominate, to reward, and to punish. Without authority, the practical and full authority which the administration alone can possess, the best efforts are thwarted by the wonderful inertia of the country and other influences."[25]

But the problem was that even the regime of Ahmad Urabi Pasha installed by the Europeans, under close supervision, proved to be insufficiently compliant. Moreover, even if Urabi Pasha desired to be a more pliable puppet of the European administrators, his options were constrained by the fact that there was growing opposition to this kind of "financial imperialism" and "modernization from above" among a broad cross section of Egyptian society, and this opposition erupted in open revolt in 1882. At this point, Britain decided to take matters in hand, deposing not only Urabi Pasha but the International Public Debt Administration as well, much to France's consternation, and imposing direct colonial rule on Egypt.[26] The Eastern Associated Company played a supportive role in these events—not surprisingly, given Pender's position on the Colonial Defence Committee—temporarily disconnecting Alexandria from the global communications grid and locating the severed end of the cable on a British naval ship situated offshore so that military commanders could stay in touch with London and, as Barty-King cheerfully notes, journalists could deliver a "blow-by-blow" account of the invasion to the papers back home.[27] Such was the end of Egypt's experiment in modernization and the beginning of British rule, a status that lasted until well into the 1940s. Afterward, Egypt's telegraph system was leased to the Eastern Associated Company, in essence, a monopoly backed by the British Empire and a kind of prize of war, thus establishing yet another national telegraph system in its broader global communications network.[28]

VON STEPHAN'S BIG IDEA: A TRANS-EUROPEAN
COMMUNICATIVE SPACE AND THE PROBLEM OF
BORDERS IN THE AGE OF THE TELEGRAPH AND
CABLE COMMUNICATION; OR, WHERE IS EUROPE?

If the above events represented the macropolitics of global communication, there were others taking place at the micro level that subtly shaped relations between European and non-European countries. This could be seen in a series of initiatives taken up within the International Telegraph Union and tenaciously pursued by the German postmaster Ernst von Stephan during the last quarter of the nineteenth century. The main focus of von Stephan's initiative was to create cheaper, uniform rates for state-owned telegraph systems across Europe. The aim was to make the state-owned telegraph systems more competitive with the privately owned cable firms by adopting standardized, cheaper, and more efficient terms for the interconnection of telegraph systems across national borders. The effort also aimed to create a "trans-European communicative space" by fostering cheaper social communication across national borders within Europe. It was an ambitious proposal and reflected the liberal internationalism of the times; it was also opposed at every step of the way by the Eastern Associated Company.[29]

But besides the staunch opposition of Pender, the proposal turned on the crucial issue of how to distinguish between the European countries that would be entitled to the trans-European rate and those that were not? By convention, it was fairly easy to consider colonies as European as a way of flattering and validating those that had them, but what about those zones of informal empire typified by the Ottoman Empire and Persia? The answer to such issues turned on the structure of the Indo-European communication market, the Pender-led cartel's place in it, and the status of telegraph administrations in countries such as Turkey, Persia, and Egypt. The most important consideration was that the seemingly esoteric questions of national identity and what constituted Europe had broad economic ramifications (and contemporary parallels), insofar as defining the Ottoman Empire as European would entitle it to the cheaper trans-European rates, thereby directly threatening to divert traffic and revenue away from the Indo-European cartel. Without a cartel, in contrast, the Ottoman Empire's Telegraph Administration bid for an intra-European designation would more likely have been backed by the Indo-European Department, thereby allowing even cheaper rates than those that Anderson, Hay, and Pender already found so irksome. Of course, such a designation would happily coincide with the Ottoman's aspirations to accumulate as many symbols as possible attesting to its Europeanness, and it fit nicely alongside the constitution of 1876, cosmopolitan Ottomanism, the Tanzimat movement and Young Turks, and the complex cultural politics of identity that surrounded such issues.

But the global communication barons were not about to let all of the hard work they had put into raising rates be overturned by a multilateral institution and the politics of modernity and identity. In this regard, the cartel served them well by uniting its members in opposition to von Stephan's scheme and therefore, albeit incidentally, to the political and cultural aspirations of the Ottoman Empire.[30] This, coupled with the relentless lobbying of the British government and the refusal of the state-run British telegraph system to endorse the idea, helped block the initiative. The fact that there were tensions within German and Russian circles as well on account of the fact that communication policy experts in each country held interests on both sides of the issue—as administrators of state-run systems, but also as directors on the Indo-European Telegraph Company—further compromised support for what otherwise seemed like a good idea. In the end, the proposals were never adopted and von Stephan's death just after the end of the nineteenth century was celebrated by the private companies as the "end of a dream."[31] It was a significant blow to one of the most significant episodes in global media reform to that point.

But it is crucial to stress that the Indo-European cartel members were not motivated by a diabolical scheme to crush the dreams of cosmopolitan Ottomanism or von Stephan's vision of a trans-European communicative space. Instead, the aim of the cartel was to divert traffic away from the state-owned European telegraph systems and to the Eastern Associated Company's cables. Von Stephan's proposal sought to do the opposite and thus was destined to be resisted at every turn by the cartel members. In fact, part of the cartel's allure for the Indo-European companies was that they received the same proportion of revenues whether they carried the traffic or not. In addition, routing traffic to the cables allowed the cartel members to avoid paying European state-owned telegraph systems altogether, and this only further increased the incentives for the Indo-European companies to do just that.[32] But, by so boldly challenging the financial basis of the state-owned European telegraph systems the cartel raised their ire; the fact that it compromised the goal of affordable communication and a common trans-European communicative space compounded the effect. All added to Europe's growing irritation with Britain's premier role in global communication, to say nothing of what was felt among those in Persia, Egypt, or the Ottoman Empire.

These are crucial points in their own right and especially because the British Balfour Committee report on cable communications in 1902 would later lay the blame at the feet of the German and Russian administrations for preventing the adoption of rate reductions on the Indo-European route. However, this was pure apologetics and turned reality upside down, by failing to acknowledge how the efforts of von Stephan, the Ottoman Administration, and even the Indo-European Department to reduce rates had been opposed by the Eastern Associated Company at every single turn. The impact of this claim went well beyond just the micropolitics of this particular case as it was mobilized by the Balfour Committee to buttress a bigger argument against multilateralism. The extent to which the Balfour report turned reality on its head in the pursuit of this new unilateralist direction while claiming to adhere to well-worn policy conventions was astonishing.[33] It was a clear signal of turbulent times ahead. In short, the rejection of the German Cable Company's landing permit (chapter 2), threats to withdraw from the International Telegraph Union, and broadsides against multilateralism as a constraint on British interests all reflected Britain's turn away from the ideals of liberal internationalism to those geared more toward realpolitik, self-interest, and rivalry. Yet, even here, this was a matter of degree, rather than of kind, at least for the time being.

COMMUNICATION, AFRICA, AND THE NEW IMPERIALISM: INTERIMPERIAL RIVALRY OR COLLABORATIVE EMPIRE?

Africa was the last place on the planet to be connected to the global communications infrastructure.[34] This changed quickly between 1879 and 1887 largely because of the confluence of three factors which are briefly introduced here and explored more fully in the rest of this chapter. First, the revival of subsidies in the 1880s in Britain and other European countries. This was of particular benefit to three new affiliates in the Eastern Associated Company's corporate fold—the Eastern and South African Telegraph Company (1882), West African Telegraph Company (1886), and African Direct Telegraph Company (1886)—which received generous subsidies from Britain, France, Germany, Spain, and Portugal from the early 1880s to just after the beginning of the twentieth century to lay and operate the cables that encircled the African continent and which met their respective and collective needs during this period.

Second, the cable communication networks that encircled Africa were driven by the gold and diamond trade on the east coast and on the opposite side of the continent, not by the sudden appearance of viable markets, but because the Atlantic coastal economies of South America were once again in the midst of a "fantastic speculative carnival."[35] These two seemingly disparate regions of the world were linked by the fact that the west coast of Africa served as an intermediary point of connection in the cable systems linking South America to Europe. The renewed economic frenzy was visible in the grandiose urban development projects, funded by a seemingly endless supply of foreign capital, as property values in Buenos Aires multiplied tenfold in just three years (1886–89) and as the Argentinean government paid hundreds of thousands of Europeans, Chinese, former slaves from the United States, and the Caribbean, indeed pretty much anyone, to come and help develop the interior of the country and build the railways and the ports that would bind the new nation together and stitch it into the global economy.[36] As had been the case in the past, the economic boom in South America triggered the expansion of transatlantic cable connections between there and Europe by way of the Cape Verde Islands off the coast of Senegal, but this time finally with extensions that ran down the length of the west coast of Africa (circa 1883 to 1887).

The third factor, and yet another index of the "collaborative model of empire" that was taking shape at the time, was the convening of two international conferences by Germany in Berlin, in November 1884 and February 1885. In addition to the host, Britain, France, and the United States, ten other European countries (including Turkey) all attended to decide on the principles and lay out the rules for one of the most ambitious projects of development and civilization ever contemplated: the colonization of Africa.[37] The Berlin Agreement of 1885 exuded humanitarian and liberal justifications for the international control of Africa as it called on the participants

> to watch over the preservation of the native tribes, and to care for the improvement of the conditions of the moral and material well-being and to help in suppressing slavery, and especially the Slave Trade. They shall, without distinction of creed or nation, protect and favor all religious, scientific, or charitable institutions and undertakings created and organized for the above ends, or which aim at instructing the natives and bringing home to them the blessings of civilisation. . . . Freedom of conscience and religious toleration are expressly guaranteed to the natives, no less than to subjects and to foreigners.[38]

Amen! How could this "project," this "gift of civilisation," animated by the high ideals of liberal internationalism, be anything other than applauded for the selfless generosity and regard that was being shown for the less fortunate?[39] The fact that it was pretty much a self-portraiture—given that not one black African representative attended—assured that any of its blemishes were discreetly covered or brushed away. In the next step, the General Act laid out the rules of cooperation that required the "superpowers" to inform one another whenever they acquired new lands or protectorates in order that each had a chance "to protest against the same if there exists any grounds for their doing so."[40] All in all, the colonization of Africa was not supposed to be an act of violence, but one of civilization and salvation; it was to be conducted with civility among the superpowers and through principled talk and debate rather than bullets and bombs.

In an amazingly compressed period of two decades the scramble for Africa after 1880 saw an unknown and unmapped continent (with less than 10 percent of its land under foreign control before this period) go to one where 90 percent of it had been taken over by Britain, France, and Germany and, less so, Portugal, Spain, Italy, and Belgium. And, just as the General Act suggested they should be, the relations between the "great powers" were largely conducted in a civil manner, with a few exceptions here and there marking the zones of friction between them, while it was the African people who bore the brunt of violence that made a mockery of the ideals supposedly at the heart of what Naill Ferguson calls the "scrambling of Africa," in reference to the fact that "ten thousand African kingdoms were transformed into just forty states" in less than twenty years.[41] Pender's new cables around the continent of Africa and from there across the South Atlantic to Brazil that were laid at precisely the moment the conferences in Berlin were being brought to a close were a material and symbolic representation of the fact that globalization was in fact a system; events on one side of the world were indeed connected with those on the other.

The development of the west coast of Africa cable system in the mid-1880s will be considered later, but immediately we look at the east coast where Pender formed the Eastern and South African Telegraph Company in 1880. The new cable was built off his Indo-European network that passed along the northeast coast of Africa past Egypt and through the Red Sea before crossing over to Aden en route to India with a new link down the east coast of Africa. It began operations some time between 1880 and 1882.[42]

It was only after 1879, following the consolidation of diamond and gold mining interests in South Africa in the hands of Cecil Rhodes and the Rothschild Bank, that military needs and a humanitarian cover (a role in suppressing the slave trade) all came together to entice Pender to form this company. Cecil Rhodes, the business dynamo, imperial visionary, and British robber baron, and the Rothschilds were driving forces of these events, with the two working hand-in-glove to consolidate a near monopoly over diamond mining in South Africa. Rhodes parlayed his interests in diamond and gold mining into a spate of speculative ventures by way of his British South Africa Company and had ambitions to connect the Cape to Cairo by rail and communication networks.[43] Those ventures came to naught.

While Rhodes failed in his attempt to wire the continent, his preliminary ruminations on the issue suggested that there was ample enough economic activity concentrated in a handful of coastal cities on the east coast and southern tip of Africa, namely, Mombasa, Dar es Salaam, Delagoa Bay, Durban, and Capetown, and strong willingness within the British government as well as from Portugal and local British colonial administrators to subsidize whoever would lay a cable communications system to these areas. Under these conditions, it is not surprising that John Pender now saw the eastern coast of Africa as a natural extension to his company's existing cable network. Thus, nearly two decades after the British and European cable and telegraph systems had been brought to nearby India (1865), by 1880 the Eastern Associated Company began to extend its system down the east coast of Africa. Over the course of two years, and with total subsidies of $300,000 per year for twenty years from Britain which gave $175,000, British colonial governments in Natal and the Cape Colony ($100,000), and the Portuguese government ($25,000), the Eastern and South African Company laid a new cable from Aden, with connections at Mombasa and Dar es Salaam (in present day Kenya and Tanzania, respectively), Delagoa Bay, and farther south to a few cities along the coastline of present-day Mozambique and then to Durban, South Africa, where it was linked by a telegraph line across the southern tip of the continent to Capetown. And that is where things stayed for most of the next decade, with a couple of thousand messages traveling back and forth between Durban and London every year at the extraordinary cost of over $40 per word, a rate that largely stayed intact for the next decade until yet more subsidies brought about a reduction to just over $25 per word around 1895. However, given that most customers were probably government and colonial officials a large proportion of all the messages sent over the company's cables would have been at half-rates. But even if only 2,000 messages were sent over the line in a year, revenue would still be around $1.6 million (before subsidies) based on the conservative estimate that each message would have been around twenty words on average.[44]

In addition to connecting the east coast of Africa, or more properly, a few colonial and important mining cities there, to the global communication system for the first time, the lavish subsidies granted to the Eastern and South African Company represented a major shift in Britain's global communication policy: the revival of subsidies after they had been all but abolished twenty years earlier. In the next two decades (1880 to 1900) state subsidies financed the wiring of the African continent. For the Eastern Associated Company whose subsidiaries owned and controlled every single cable that circumnavigated the continent until 1902, this made Africa a viable business proposition. The Eastern Associated Company received nearly $7.5 million in subsidies from 1880 to 1900 from the British government and British colonies for such purposes, and another $2.5 million from the French, German, and Portuguese governments. For the Eastern Associated Company, then, the "new imperialism" was good business, even though for European empires, according to the cold calculus of business prevailing at the time, imperialism was more of an economic burden than a benefit. As a result, more missionaries than money continued to pour into Africa, and if governments and colonial administrators wanted access to the global communications infrastructure, they had to pay for it, and pay for it they did, with over half of all British cable subsidies from 1880 to 1900 going to Africa and, thus, to Pender's companies.[45]

But more than just subsidies were being revived in the scrambling of Africa. Several years after the east coast of Africa was linked to the Eastern Associated Company cables, Pender resuscitated another company that seemed to be playing a bizarre game of peek-a-boo throughout the history of global communication: the South American Cable Company. The company first appeared on the scene in 1859 in conjunction with British bids to lay cables to India and the Far East. Thereafter, it weaved in and out of obscurity, emerging in 1863 under the directorship of Arturo de Marcaortu and Ferdinand de Lesseps as part of their bid to the U.S. Congress for support of their proposal for a vast transatlantic cable system to connect Europe and all of the Americas; a few years after that as the holder of valuable concessions throughout Latin America that eventually were used by others who ultimately built the cable systems in the region; and once again in the late-1860s as the cornerstone of a bid by the French government to erect a kind of "global cable commons" around it that would hold a ninety-nine-year monopoly on transatlantic communications and subsidies from governments throughout Europe and the Americas.

While none of those efforts had ever amounted to anything, Pender resurrected the South American Cable Company again in the mid-1880s in the context of the bid to carve up Africa among the European superpowers and to lay a cable across the South Atlantic to the red-hot economies in the River Plate zone and along the Brazilian coast. The intrigue only became more mysterious in the present context and for the next two decades, as Pender and his partners both revived the company and registered it in Britain, not just as the South American Cable Company but with an odd notation on the articles of association stating that the firm would occasionally be operated as the

Spanish National Submarine Telegraph Company.[46] And so it was even by the standards of Pender's corporate empire an elaborate shell game. We will reflect in the pages ahead on why this might have been. For the moment, however, it is crucial to set out clearly the origins, ownership, and control of the company en route to establishing our claim that, between 1880 and 1900, most European governments in Africa were quite content to rely on a communications system that was owned and controlled entirely by one British-based cable conglomerate: the Eastern Associated Company.

One thing is for sure, this shell game has confused subsequent historians of the global media ever since. Even the best among them such as Daniel Headrick assume that just because the alternative name was the Spanish National Submarine Telegraph Company, that it was a communications arm of the Spanish state. It was not. Why Headrick suggests this is understandable and fits with the well-worn groove of media historians to slot everything into a framework of interimperial rivalry, and here Headrick and others rise to the occasion without fail. Headrick argues that the French favored the "Spanish company" precisely because it *was not* British and that this had become crucial in the context of growing rivalry in the scramble for Africa. At the same time, Headrick points to the British Colonial Office steering its own subsidies and support away from the company for precisely the same reason and in favor of a "more British" company—the African Direct Company (also Pender owned). As Headrick puts it, the Treasury opted against the Spanish National Telegraph Company because it was too closely allied with the Spanish government "to suit the Colonial Office's sense of . . . national security" and because the Colonial Office "would not hear of official messages passing through French or Spanish territory."[47]

Such conjecture is plausible, but the fact that the South America Company and the Spanish National Submarine Telegraph Company were commonly owned by the Eastern Associated Company is made clear in the Memoranda and Articles of Association filed in London. The document not only registered the company in Britain but explicitly states that it would sometimes use the title of Spanish National Submarine Telegraph Company. The Spanish National Telegraph Company was recognized by Charles Bright, the leading authority on cable communications in Britain after the beginning of the twentieth century, as having been British from the outset and its status as a member of the Pender Group of companies formally registered in 1885.[48] In addition, the only company to lay a cable across the South Atlantic at this time to Brazil was the Brazilian Submarine Telegraph Company, also owned by

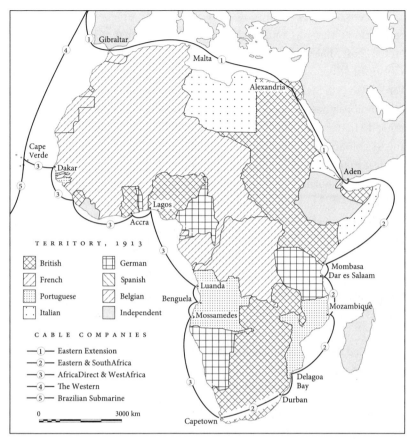

MAP 8 Imperial carve-up of Africa, cable routes, and main companies, circa 1900

Pender, which did so in 1885. As soon as the Brazilian Submarine Telegraph Company laid its cables between Senegal and Brazil, it and the Eastern Associated Company formed two new companies to bring the cables down the west coast of Africa from Senegal in the north to Luanda in Angola: the African Direct Company and the West African Company.[49] Thereafter they connected to another cable laid by the Eastern and South African Company running northeast from Capetown, thereby encircling the continent. Altogether, the west coast project took four years to complete, beginning in 1883 with the initial cables laid by the Spanish National Submarine Telegraph Company between Cadiz and the Canary islands and then two years later and just in time to connect with the Brazilian Submarine Telegraph Company's new cable from Brazil to St. Louis and Dakar.

This was nearly the same route followed a decade earlier when the Brazilian Submarine Telegraph Company laid its first cables from the Cape Verde Islands off the coast of Senegal to Brazil. But this time the French government paid a subsidy of $65,000 per year for twenty years to bring the cable from the Cape Verde islands to Senegal rather than continuing to bypass Africa altogether, as had previously been the case. The German government paid another $33,500 to bring it a bit farther down the west coast of Africa, and the British government and its colonies in Sierra Leone and Nigeria contributed a further $98,000 to connect major British colonial cities along the west coast of Africa between Senegal in the north and Capetown in the south. Thus, between 1883 and 1887, and with nearly $200,000 in annual subsidies for twenty years from three of the world's most powerful governments (and another $300,000 on the east coast of the continent from Britain and Portugal chipping in $25,000), Pender resuscitated the South American Cable Company/Spanish National Submarine Telegraph Company; created two new subsidiaries—the African Direct Telegraph Company and the West African Telegraph Company; and laid a vast electronic harness around the perimeter of the continent. And when everything was said and done, all of the various subsidiaries, and here Headrick is in agreement, were folded together and tucked neatly behind the corporate veil of the Eastern Associated Company.[50]

CONCLUDING REFLECTIONS ON COMMUNICATION AND EMPIRE

This was an electronic empire in its starkest form and there was probably never anything like it either before or since. Pender was not just meeting the military communication and security needs of the British Empire but of a kind of global empire. It is no doubt true that there were clashes between rival imperial interests, such as the so-often repeated references to the Fashoda incident of 1899, where British and French soldiers struggled over who got to use the telegraphs first, or mounting Anglo-German tensions during the Boer War (1899–1902). There was also a long list of instances when French or German dependence on British-owned cables meant that Colonial and Foreign Office officials in London knew before Paris or Berlin what was happening in some key French or German colony or sphere of influence. As Naill Ferguson notes, Britain and France clashed repeatedly over Egypt, Nigeria, and Uganda, but these were more akin to "bizarre showdowns" and "surreal contretemps" than full-scale military battles.[51]

A bit more perspective is added to these exaggerated examples of imperial rivalry when we realize that such clashes were nothing compared to the application of new military technology—especially the machine gun—to the massacre of Africans in a matter of hours. Thus, while the Zulus could send British soldiers home in coffins by the score at Isandlwana in 1879, never again would this occur. As Hiram Maxim's new machine gun was test marketed on the plains of Africa (long before the trenches of World War I),[52] the mercenary and professional soldiers of France, Germany, Britain, and Portugal took their instructions from afar by way of cables that were owned, controlled, and operated by the Eastern Associated Company. This, of course, represented British hegemony in the truest sense of the term, but it was also an indication of the extent to which globalization, communication, and empire were bundled together as a package.

This is only confirmed by other examples that are usually offered as evidence of the overwhelming influence of interimperial rivalry, in particular as France and Germany went about laying a series of new cables in southeast Asia and the Far East around the beginning of the twentieth century. But these examples, like the Fashoda incident, do little to provide such evidence. France and Germany (the latter collaborating with Holland) did build regional networks—the former between present-day Vietnam and Shanghai and the latter between Indonesia and Qingdao—but they still relied on the Eastern Associated Company for the long-haul transportation of all their diplomatic, military, and commercial communications back to Paris, Amsterdam, and Berlin.

Where there were complaints—as indeed there were—there was also recognition that "he who paid the piper called the tune," and while all the European superpowers continued to pay subsidies to the Pender regime during the new imperialism era (1880–1900), Britain's subsidies outweighed all of the others combined. Moreover, Pender's position on the British Colonial Defence Committee indicated clearly where his deepest loyalties were. But just in case this was not understood, the British government put it in writing: it had priority access to Pender's cables over all others.[53] Thus, in the last instance, the company was British and was required to behave as such when the chips were down, as they most certainly would be in World War I.

So, we shall leave it at that, but not before rounding off one more point that was introduced above but left hanging, namely, that the exclusive dependence of empire, at least insofar as Africa was concerned, on the Eastern Associated Company did not go unchallenged or was completely without problems. There is no doubt that by the beginning of the twentieth century there was disharmony in the global communication system. One such incidence, for example, was the small unremarked on piece of evidence in one of the tables to the Balfour Report showing that France had stopped paying its subsidies to the West African Telegraph Company in 1897.[54] Why was this? We really don't know for certain, but maybe it was one more sign of mounting tensions. Remember, the French reports had just come out in favor of more autonomy for France's global communication needs, and furthermore, the Fashoda incident (1899) was just around the corner.

But even if this were an index of mounting tensions, several things followed quickly to reveal that these were problems to manage rather than to destroy the world for. First, these tensions occurred far after the time when most historians identify as the moment in which rivalry becomes the driver of global communication. Second, the fact that France stopped paying its subsidies to the West African Telegraph Companies may have been an index of growing dissatisfaction, but the French and German governments continued to rely on the firm's network in Africa until just after the end of the nineteenth century. Third, at that point (which was also coming close to the expiration of the original twenty-year agreements), the French government gave up paying long-term rent for the cable between Cadiz and Senegal and purchased the financially troubled South Atlantic Cable Company and the West African Company from Pender outright in 1902. If dissatisfaction were at the root of this development, it was not obvious, and the exchange was made according to the standard business practices of the time: by forming a cartel, not just along the west coast of Africa, but for both companies' cables from Senegal, on the one side of the South Atlantic to Brazil on the other. The German company also established agreements with the French-owned South American Cable Company at this time and tightened its ties to the company after laying its own cable to Brazil in 1910.

By the beginning of the twentieth century, the initial development phase of the global communication industry was coming to an end. As it did so, the expiration of concessions and early subsidy deals and the emergence of new wireless communication technologies opened up new vistas and led to a reshuffling of the deck. But as time passed, the gap between the ideals of the age and reality widened, and as it did collaboration among empires and competition in the marketplace were eclipsed by war and carnage.

Electronic Kingdom and Wired Cities in the
"Age of Disorder": The Struggle for Control of China's
National and Global Communication Capabilities,
1870–1901

> Being born into the world of disorder and having witnessed enough misery
> of the people, I was pondering about the way to salvage the world. After
> serious searching, I found out that the only way to ensure peace is to create
> a cosmopolitan community, namely Da Tong.—KANG YOUWEI,
> *Da Tong Shu*

> Undiscouraged by repeated failures, the advocates of progress have contin-
> ued their efforts to move the Chinese mind and to arouse this old empire
> from its inertia.—BENJAMIN AVERY to U.S. Secretary of State
> Hamilton Fish, December 20, 1874

COMMUNICATION AND THE POLITICS OF DEVELOPMENT

European and American interests did not seek to take over and create a new
nation-state from scratch in China but to remodel the existing structures of
political organization and adjust them to what they perceived to be the realities of a modern world economy and nation-states. Reformers and radicals
within China sought to do the same, while a bulwark of conservative manda-
rins sought to resist *all* change unless it helped to preserve the outmoded way
of life represented by the Qing Dynasty (1644–1911). For the foreigners, and
Chinese modernizers and radicals as well, the key question was whether or not
change and the creation of a "new China" could be best achieved by working
through the ossified channels of the Qing Dynasty in Beijing or cultivated at
the edges of the faltering empire, by supporting the more open and commercially dynamic city-states arrayed along the southern coastal regions of China,
from Hong Kong to Fujian and Shanghai. That is, the politics of global
communication in China were entirely entangled in the politics of the nation,
a politics that would determine the life or death of an "old empire" and the

8   "The Open Door that China Needs," from the *Brooklyn Eagle*, reproduced in the *Literary Digest*,
July 14, 1900. It links American conceptions of modernization and its open door policy to
the perceived need to sweep out Chinese "anachronisms" by the "international community"
at the time of the Boxer Rebellion. The chasm between the "ideals" of this civilizational discourse
and the brutality of the intervention of the "Allied Powers" during the Boxer Rebellion is neatly
ignored.

prospects for a new nation-state in the celestial kingdom. In their bid, the
foreign powers oscillated between two models of empire and foreign policy regimes: the conventional spheres of influence approach and the internationalization of control model. In the nooks and crannies offered by these
cleavages, Chinese reformers and radicals offered their own diagnoses of the
old kingdom, their critiques of imperialism, and their prescriptions for a
new China.

The internationalization of control model evolved in stages, first in the
context that we cover in the following pages, as one that had yet to be formally
named but which was unequivocally present, particularly in the field of communication. This was followed later at the beginning of the twentieth century
by the United States' formal announcement of its open door policy, which
tried to capture this idea of a China equally open to all, rather than one that
was carved up into mutually exclusive spheres of influence. Although con-
trasts between the two methods were blurry, there were substantive differences, and this was revealed by the constant changes in the constellation of
interests arrayed around both models that took place and the cleavages that

opened up regularly not only between the imperialist powers, but also between them and the business interests they claimed to represent. Because the city-states were seen as more dynamic and simply more eager to open their doors to Western businesses and influences, business tended to support the city-states and spheres of influence approach, while those with a longer-term view believed that only a stronger central state could ensure the long-term viability of China as a market and nation. The internationalization of control model was formalized with the creation of the international banking consortium in 1909, used as a template for the Treaty of Versailles negotiations, and constituted a major part of the United States's global communication policy in general after World War I and with respect to China in particular.[1] As an aside, all of these factors indicate that far from encountering an impenetrable wall of Chinese resistance, as Erick Baark and Daniel Headrick suggest, the main factor shaping the evolution of the telegraph and cable communication networks in China was the ongoing struggle between Beijing and the southern city-states over the future of the country and its relationship to the rest of the world.[2]

PRINCE GONG SWITCHES FROM PREVENTING TO CAUTIOUSLY PERMITTING "NEW MEDIA" TO ENTER THE HEAVENLY KINGDOM

The conservative Qing dynasty obstinately blocked each and every overture of British, Russian, and U.S. foreign ministers to introduce telegraph and cable communications in China during the 1860s, all the while promising each that should one be permitted to introduce these new means of communication in China, they all would. Things began to change in the 1870s, initially in a very cumbersome way with the granting of the first concession to the Eastern Extension Australasia and China Telegraph Company (a Pender-controlled company) in 1870 and then by the mid-1870s in an all-out struggle between Beijing and the southern city-states over which would control economic development and foreign relations in China.

Beijing's first tentative steps toward a more permissive approach was clear in the approval by the Zongli Yamen—the secretary of state for foreign affairs, a post created at the behest of European powers in the 1860s—of a request by the British Minister Thomas Wade to bring submarine cables to commercial cities and foreign treaty ports in southern China. The approval marked a decisive break with the 1860s. Just as important, however, that the agreement was hardly satisfactory was revealed by the impossible conditions attached to it. The most notable constraint was the stipulation that the cables could not connect directly to the cities they served but were required to terminate at a buoy, ship, or rock off the coast and messages brought ashore manually.

That the arrangement was arrived at grudgingly was also evident in the fact that it was silent about just how many cities could be connected to the global communications infrastructure; Prince Gong assumed that one (Shanghai) would suffice, while the companies and foreign diplomatic communities assumed that this was a question to be settled by the private companies themselves. And even here the fact that the agreement was riven with problems could be barely concealed. For one, the Zongli Yamen assumed that the British minister had acquired the concession on behalf of Pender's Eastern Extension Telegraph Company but quickly learned that the rights had been transferred to C. F. Tietgen and Henrik G. Erichsen's Danish-based Great Northern Telegraph Company. The Zongli Yamen was probably also surprised by the fact that both companies, much as they had done in South America (as described in chapter 2), had struck an agreement even before the first cables were laid to divide the Chinese and Far Eastern communications markets between themselves. According to the agreement, everything north of Shanghai, and including Korea, Japan, and the east coast of Russia, would go to the Great Northern Company, while Hong Kong and everything south of there was reserved for the Eastern Extension Company. The most lucrative Chinese market—the southern coastal cities between Shanghai in the north and Hong Kong in the south—were to be shared between the two companies. The next assumption, at least for Prince Gong, was that the concession contained at least an implicit recognition that the Zongli Yamen and Wai Jiao Bu (Ministry of Foreign Affairs) were China's main authorities on matters of global communication.

Most of these assumptions quickly proved to be misguided. Reflecting Tietgen and Erichsen's tendency to operate in a high-handed manner, the Great Northern Company laid its first cables to Shanghai in late 1870 and immediately smuggled them ashore under the cover of darkness.[3] Two years later, and just after its cable had been extended from Shanghai to Hong Kong, the company opened another cable to Xiamen. But more than just obliterating the conditions of the company's license, the cables were quickly connected to a clandestine urban network of telegraph lines in both cities and between them and nearby suburbs, mostly ports and suburban enclaves where foreigners took up residence in the treaty ports along the coast of China and important inland centers of commerce. This was not necessarily a novel practice, and even railways had been built by stealth between Shanghai and the nearby port at Wusong without official sanction. But in the field of communication, the development by stealth approach took on a new magnitude as Tietgen and Erichsen authorized construction on at least three lines—between Shanghai and Wusong, between Fuzhou and Mawei (Pagoda Anchorage—20 kilometers) and from Fuzhou to Xiamen (approximately 350 kilometers)—between 1871 and 1874, all without approval from Beijing.[4] C. F. Tietgen, the Danish banker, industrialist, and one of the most significant players in the global communication business at this time, saw these urban telegraph networks not just as feeders for his extensive system of cables but as the cornerstones of China's national communication system. As Tietgen and company officials in China saw things, this national communication network would contain a mix of coastal cables and overland telegraph lines from Fuzhou, Xiamen, and Shanghai farther south to Guangzhou, Quanzhou, and Hong Kong, and to Tianjin and Beijing in the north, as well as inland to Nanjing and Hankou along the Yangzi River.[5]

In this venture of wiring the celestial kingdom, the Great Northern Company was, at least before 1875, the darling of the international community, with the British transferring the original Chinese concessions to the company; the French allowing its first cable to be smuggled into their military quarters at Wusong; and Britain, Russia, France, and the United States throwing their diplomatic weight behind the company. The company's most ardent supporter from the outset was the United States, whose official position was that international cooperation in such matters was essential to the development of the global communications system.[6] This bulwark of support was so crucial because it was universally acknowledged that the Great Northern Telegraph Company's activities completely violated the agreement signed by Wade and Prince Gong. As correspondence from Benjamin Avery, the U.S. consular official in Beijing, noted, the "shore cables were laid without distinct concessions or explicit permission from the government."[7] Even J. Hegerman Lindencrone, of the Danish legation at Beijing, confided as much with respect to all of the company's operations in a letter to U.S. Secretary of State Hamilton Fish, in which he candidly acknowledged that the company had not received "official authorization to lay cables along the coasts of China, and to establish lines and stations in the two aforesaid Chinese ports [Shanghai and Fuzhou]," although "the local authorities have . . . tacitly recognized that the operations of the company are authorized."[8]

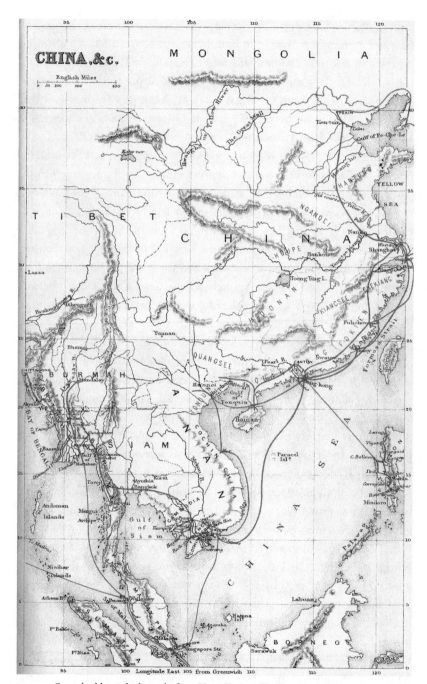

MAP 9   Coastal cables and telegraphs from Hong Kong to Tianjin, circa 1874

But to read the foreign correspondence of the time, the Great Northern Company's actions were not condoned on account of a giant conspiracy among rapacious imperialist powers, but rather as part of an earnest commitment to the "international development of China," to "arouse this old empire from its inertia," as the second epigraph at the beginning of this chapter stated.[9] And in Lindencrone's letter to Fish was another clue as to the preferred approach: a focus on local officials to gain at least their tacit approval of the company's actions. This was part of an effort to absorb China into the global communication system, an approach that relied on the fait accompli; the cultivation of local support among politicians and merchants, both foreign and Chinese; and the hope that all the legal niceties could be ironed out with Beijing when the right time came. The most active in this regard were the British, American, and Danish consular offices in the key city-states of southern China, although Russian officials were never far offstage either. And in particular, all recognized the leading role of the American consular official at Fuzhou, Warren Delano Jr. (who was, incidentally, the maternal grandfather of future U.S. President Franklin Delano Roosevelt and a distant cousin of President Theodore Roosevelt), in cultivating support among local officials in Fuzhou.[10]

CITY-STATES, WEAK STATES, AND GLOBAL COMMUNICATION

At first, Delano's cavalier attitude toward the niceties of international relations appeared to play well. In fact, it was Delano who obtained the first permission from any authority to construct telegraph lines in China, and this from local officials in Fuzhou, the capital of the southern Chinese province of Fujian. The agreement retroactively sanctioned telegraph lines that had already been constructed or were in the process of being built. It also gave the Great Northern Company a monopoly over communication along the routes served—Fuzhou to Mawei and Xiamen—for thirty years and precluded any level of government from constructing competing lines for the duration of the agreement, although in a minor nod of deference to Beijing, it gave the government the option of taking the lines over at any time, at fair market value. Additionally, and as an incredible mark of either great trust between Delano, the company, and local officials, or just plain gullibility or duplicitous behavior on everyone's part, the company was obligated to maintain "state secrets" contained in messages sent by the Chinese government over its lines and, with

9　The first graduating class of the Telegraph School at Fuzhou, 1876. The school was opened by the Chinese government in 1870 under supervision of the Great Northern Telegraph Company. Reproduced by permission of GN Great Nordic and author Dr. Kurt Jacobsen from GN Great Nordic's 125th anniversary publication, *From Dots and Dashes to Tele and Datacommunications*, Copenhagen, 1994.

a nod to the modernizers now gaining in ascendancy, the company was required to "instruct Chinese lads . . . in the art of telegraphy free of charge."[11] Last, the telegraph network was to have two lines, one reserved for commercial uses and the other for government purposes, although both were operated by the Great Northern Company.

But reflecting on this agreement for even a moment leads to several observations. First, in its outline, the agreement looked very much like the "Persian," or "weak state," model of communication development that we have seen before, with its promises of "training" and the possibility of transferring control over the system to national authorities at some unspecified future date. The dependence of the Chinese state on a foreign company for service—both within China and to the rest of the world—and for the protection of state secrets pointed in the same direction. The weak state model of development could also be seen in the way in which the company and foreign diplomats took advantage of cleavages in the domestic political system, in this case playing local officials, such as Li Henian, off the central government in Beijing and largely relying on the former to push their interests before the latter.[12] But in the end, the key question remained: Did the agreement really embody any kind of authorization at all?

The international diplomatic community liked to think so, but even the tenor of their correspondence expressed trepidation. And if there was trepidation among those who sought to benefit most, Prince Gong and others in Beijing made it crystal clear that such agreements were not worth the paper they were written on. Even before the Delano agreements in Fuzhou, the Zongli Yamen protested to British and American consular officials about the urban telegraph networks and unauthorized shore landings in Shanghai. Apparently perturbed by some rather callous and gloating newspaper articles on the subject, probably in the British-owned newspaper *Shenbao* that was operating in Shanghai at this time, the Zongli Yamen began a campaign against the unauthorized Shanghai to Wusong network beginning in October 1873 that lasted for over a year. In face-to-face meetings with British and American consular officials, the Zongli Yamen stated that it wanted the Wusong to Shanghai line removed. British and American officials concurred with the Zongli Yamen's assessment, but then they adopted obstructionist policies of their own. As part of this policy, all of the foreign consular offices were contacted and advised to do nothing in response to the Zongli Yamen's protests. Subsequent follow-up letters from the prince met with further delay.[13]

In the meantime, the central government registered its opposition immediately to the Fuzhou to Xiamen scheme brokered by Warren Delano with Li Henian on behalf of the Great Northern Telegraph Company. As the Zongli Yamen and Wai Jiao Bu saw it, the agreement was worthless. In their eagerness to be linked to the global communication infrastructure, the city-states were acting way beyond their sphere of authority, signing agreements that lent legitimacy to projects not sanctioned by Beijing and which constrained Beijing's options by way of their exclusive concessions. City-states did not make foreign policy and conduct international relations, Beijing did, and in this instance the deals struck between Delano, the Great Northern Company, and officials in Fuzhou were trampling all over the central government's turf. The British and American consuls who heard the protests this time took them a bit more seriously and asked Prince Gong to wait until a Danish mission now en route to China was present to discuss the matters further.[14] From the Zongli Yamen's perspective, the first half decade of telegraphs and cables in China vindicated its original position. They were a constant source of irritation, the thin edge of the wedge as foreigners relentlessly pushed to expand their control over China's national communication infrastructure.

THREE PATHS TO A REVOLUTION: IMPERIAL CONQUEST, MODERNIZATION, AND THE NEW CHINA MOVEMENT

More than just a few telegraph lines and submarine cables were at the core of Prince Gong's struggle to get a hold on these developments. Like the hard-not-to-notice railway chugging along between Shanghai and Wusong, the Western powers' policy of build first, ask later was an affront to Chinese sovereignty. For the Western powers, the country would either be prodded into modernizing on its own accord or taken over directly, that is, as a failed state put under international tutelage or the imperial power of one country, likely Britain. But within China itself, the prospects were seen dialectically, both as a threat of Western domination and a stimulus to those seeking to transform China.

For the "technocratic modernizers" such as Li Hongzhang, Shen Baochen, Sheng Xuanhuai, Ding Richang, to name just a few, the task was to absorb the material forces of the "new world"—capitalism, new administrative styles for businesses and the state, and technology—into China, while guiding them through the maintenance of Chinese values, an older version of contemporary debates around Asian values. But there were others far more radical in their inclinations, at least for a time, such as Kang Youwei, the source of this chapter's opening epigraph, and Tan Sitong and Liang Qichao, for whom the just-mentioned reformers represented but the bland religion of modernization. For them, nothing less than a complete reconstruction of a new China, sans the stifling constraints of traditional feudal Chinese values, would do.

In short, the foot-in-the-door strategy of the global communication companies reverberated far beyond just a question of cables and wires to become an integral part of a far-reaching process of total political, social, and cultural transformation, or at least the potential for such things. And nowhere was this more evident than in the treaty ports, or what we have been calling city-states, where the arrival of cable communication networks and telegraphs paved the way for a new era in Chinese journalism and press history. Within a short period after their arrival, news sources and styles that came to define the history of the Chinese press were ushered in, as Reuters made Shanghai the base of its Far Eastern operations in 1871, and as numerous foreign-owned newspapers were started in major Chinese cities.[15]

10 Caricature of Li Hongzhang (often spelled Li Hung Chang) as sketched for *Vanity Fair*, August 13, 1896. Perhaps the most influential of the Chinese modernizers, he became special envoy of the emperor to the West and a leading businessman whose legacy continues to influence current generations of Chinese technocrats. *Vanity Fair* called him, patronizingly, "a man of liberal views (for China) even favouring railways."

These papers, while operating mainly in the treaty ports, could neither be restricted to just the foreigners for whom they were mostly directed, nor prevented from spilling over into the adjacent areas within and beyond these city-states. And as they spread, they had a dramatic influence on Chinese politics, culture, and social transformations, in essence creating urban media-scapes that weaved in and out of Chinese society in unintended and indeterminate ways. In the process, questions of press censorship were foisted upon the Chinese state. Previous practice within China and a recent crackdown in Japan on the basis of that country's new heavy-handed press censorship law (1875)—of which the Chinese authorities were well aware—suggested an obvious course of action, that is, if the authorities were as bent on preventing change as on shaping it, as is so often assumed. But in the main and formative years, no such crackdowns on the foreign press were conducted, and the press was largely regulated by the principles of extraterritoriality and the application of foreign, mainly British law, that governed the expatriate communities in these states within a state. The foreign-run press did not limit their target audiences to foreign merchants and diplomats, but consciously addressed themselves to Chinese merchants, statesman, and *zhi zhe* (intellectuals). They also republished articles from the official Chinese press, most notably from the *Jingbao*, the official gazette of the Qing Court in Beijing. And via this publication, republication, and circulation of their news beyond the expatriate communities living in the international settlements, the foreign media weighed heavily in the political affairs of the day.

That they did so was amply shown by the numerous editorials and discussion pieces carried by, for example, *Shenbao* on the development-by-stealth approach to railways and telegraphs, and the character and desirability, or otherwise, of foreign loans. While the Chinese press historian Robert Wagner minimizes the extent to which these papers were the "mouthpiece" of the "foreign establishment," and accentuates their even-handedness to an extent which we feel is naïve and unsupportable, there is no doubt that their influence, at least among elites—foreign and Chinese—was great. Their journalistic practices and style fomented both an explosive growth in the indigenous Chinese press as well as an intense movement to carve out what the radical "new China" intellectuals and journalist practitioners (most notably Kang Youwei and Liang Qichao) called the "new journalism," albeit in the 1890s rather than in the 1870s that have been our focus up to this stage. But, the point is, the connections between telegraphs, cables, urban media, new journalism and, indeed, the whole idea of a "new China" were bound together. Prince Gong, even if he could not precisely put his finger on it, knew this, and thus sought to gain control over the "new media," not to suppress it, but to steer it between the rocky shoals of imperial conquest on the one side and the revolution promoted by Kang, Tan, Liang, and similar activists, on the other.

FROM CITY-STATES TO THE NATION-STATE: THE POLITICS OF MODERNIZATION AND THE INTERNATIONALIZATION OF CONTROL OVER CHINA'S COMMUNICATION CAPABILITIES

The idea that the central government had no intention of suppressing the new media but rather was now embarking on a more enthusiastic approach became clear from 1875 to 1885. The fact that matters were being put on a different footing was readily apparent by 1874. At this time, the Zongli Yamen had gone further than just declaring the Fuzhou agreements with the Great Northern Company worthless; it had instructed local officials to buy the lines immediately for $150,000. Now it was the Great Northern Company that was in the awkward position of having unauthorized telegraph lines and a generous offer in hand. The company spurned the Zongli Yamen's offer and continued to plow ahead with the completion of the Fuzhou to Xiamen line, likely bolstered in its convictions by the continuation of unanimous support among international diplomats as much as by renewed assurances from local officials that they would smooth over political relations in Beijing, so long as the company agreed that the government could buy out the system at any time in the future.[16]

The Great Northern Company was no doubt reassured by the fact that the much touted high-level Danish mission arrived in Beijing in 1874 to take up the issues surrounding the company's coastal cables and telegraph lines. Moreover, a flurry of correspondence and the adoption of a protocol assured the Danish mission that it had the support of the entire foreign diplomatic community in its negotiations with the Zongli Yamen. But even with this unanimous endorsement, the basis for a new approach was being laid. Indeed, the very presence of the Danish delegation signaled that the old policy of proceeding without official authorization and through the city-states was inadequate and that, ultimately, the effective use of the new technology and cultivation of a viable communication market relied upon official recognition.[17]

One apparently insurmountable problem with the weak state model of communication development was that the Great Northern Company's cables and telegraph lines were constantly interrupted because there was no adequate security for them. This was underscored by the fact that after spurning Beijing's offer to take over its domestic lines, the Great Northern Company's Fuzhou to Xiamen lines were dismantled by local villagers, ostensibly on the grounds that they violated ancestral graves and the feng shui in the areas they passed. While some wondered if the central government had not assisted in the "manufacturing of dissent" to show the company the folly of trying to proceed unilaterally, this was never proven and the company's own responses demonstrated even more clearly the perils of trying to take network security into its own hands. In conjunction with local officials, company officials burnt several houses to the ground and those held responsible were beaten, whipped, or tied to telegraph poles alongside the road with the heavy weight of cangue around their necks—a traditional form of Chinese discipline. Not only did the company's rampage of retribution enrage the local population, it displeased the Chinese government. Even the foreign diplomatic community could barely conceal its contempt for the company's actions.[18]

Beyond the disruption of feng shui, there were also pirates, thieves, and boat anchors to contend with. Prince Gong had always told foreign companies and diplomats they would face these threats, but now pirates and thieves really were stealing sections of the company's cables and selling them on the black market. And where such miscreants did not destroy the company's communication system intentionally, Chinese junks plying the coastline unintentionally dragged the cables up with their anchors.[19] In short, if the political situation was not already complicated enough, cultural conditions threatened to ruin a viable communication market in China altogether. Even the company had become "convinced that its future in China would be assured . . . if it could succeed in obtaining effective official support."[20] In fact, this was the raison d'être of the aforementioned Danish delegation's visit to Beijing in 1874. The American consular official in Beijing, Benjamin Avery, expressed the same belief in a letter to the U.S. Secretary of State: "The importance of . . . telegraphs as essential adjuncts of commerce is more and more apparent, and the policy of insisting upon official permission for their safe enjoyment is sure to be strongly urged."[21] As a result of these changes in perspective, the status of the Zongli Yamen was transformed seemingly overnight from an obstacle to progress to becoming indispensable to it.

Official authorization would accomplish several objectives in one fell swoop: first, it would legalize the coastal cables and telegraph systems; second, it would offer better network security; third, it would provide a salve for troubled international relations; and fourth, it would signal to the public that the new technology now had the mandate of heaven, as Beijing, not Shanghai or Fuzhou, was where that resided. Even Chinese peasants knew that. Together, all of these factors demonstrated that new media technologies and the realities of global markets, diplomacy, and so on could not simply be imposed on a nation, but had to be deftly inserted into existing cultures while also trying to leverage change in the directions sought. Working through the national government in China would help to cultivate support among the local population and, in so doing, undercut culture as a barrier to "globalization."[22] The cultivation of an effective nation-state was not only of the utmost importance to avoid yet another imperialist adventure but touted as wholly consistent with the foreign policy objectives of the international community. The United States explicitly made this a cornerstone of its foreign policy objectives.

Even though the United States had no direct interest in the matters, it thought that the importance of cables and telegraphic communication to commerce and its own foreign policy aims meant that it should play an active role in promoting them. Such support was supplied in copious amounts and indeed it was the American Warren Delano Jr.—consular official in Fuzhou and active in the "China trade"—who had obtained permission from local officials for the Danish-based Great Northern Company's operations in southern China. In a moment of self-congratulations, Delano expressed the American position well, stating that "my action in connection with the telegraph projects has . . . been in the general interest of commerce, civilization, and progress. . . . I think we may justly claim that American influence has largely contributed to the disinterested support it has given to initial telegraph effort, *no matter by what nationality made, in the interest, solely, of commerce, civilization and progress.*"[23]

Of course, such idealism could merely be cover for the fact that while the United States did not yet have any direct interests in the Chinese communication market, it had long coveted a position in China and was therefore motivated to "keep the door open" for any American companies that aimed to enter China in the future. Thus, far from pursuing idealist ends, perhaps American diplomacy was merely serving the needs of the American global communication players, notably Cyrus Field and others closely allied with Western Union and the Anglo-American Telegraph Company who regularly emerged in the U.S. Congress or the press with some scheme or another to lay a cable across the Pacific en route to China. Of course, such interests no doubt weighed heavily in the mix of considerations, but it would be a mistake to ignore the much "thicker" firmament behind the approach to China of the United States and certain other countries.

The U.S. approach represented the ideals of its global communication policy announced in 1869 and an ascendant liberal internationalism. In fact the same architect of the 1869 policy—Secretary of State Hamilton Fish in the administration of President Grant—was responsible for American foreign policy in China. In many ways, China was the testing ground for both the American global communication policy and a broader orientation, not just in the United States, but among all of the great powers, toward what can best be described as a world order organized around the principles of "liberal internationalism," not in the John Stuart Mill kind of way but more along the lines pursued in our own time toward globalization. The record of foreign correspondence exudes such an orientation. Even the "shore cables" were seen as "cosmopolitan,"[24] a characterization which surged to the fore in China around the mid-1870s to promote the need for their adequate protection. As Benjamin Avery argued, "as the foreign communities . . . in China grow in numbers and enlarge their trade, it becomes more and more necessary to them that they should possess the same facilities for quick communication as are enjoyed by mercantile communities elsewhere with which they have business relations."[25] Secretary of State Hamilton Fish concurred, observing that new communications technologies had already "caused . . . a re-adjustment of the laws of trade and finance [in] America and Europe, and unless the mercantile communities which have pioneered the way to a better civilization in China can accommodate their business fully to this re-adjustment by full enjoyment of the same advantage, they must lose ground; and what they lose will react injuriously upon China."[26] In short, the collaborative development of China's communication capabilities was vital to China's development *and* to global trade. As such, the *official* policy of the United States was to promote their development and to do so in concert with the other great powers.

The collaborative approach was in full view in 1874 as the high-ranking Danish mission arrived in Beijing to discuss the state of the Great Northern Company's coastal cables and telegraph lines with the Zongli Yamen. Upon arriving in Beijing, the delegation immediately contacted the American and other embassies to see if they were still behind both the company and the delegation as it prepared to meet with Chinese authorities. The answer was unanimous: yes. Although the company had already begun to squander some of the deep well of goodwill that existed among foreign diplomats and merchants, cooperation was the order of the day. In late 1874 the United States, Britain, France, Denmark, and Germany signed a protocol calling on the Chinese government to protect the coastal cables. Among themselves, although not in view of the Chinese, they also noted that any agreement by the Zongli Yamen to offer better protection for the coastal cables would also go a long way toward achieving "quasi legal status [for] what has been already accomplished in the way of telegraph-building."[27] Hamilton Fish characterized the approach as follows:

> The Government of the United States has heretofore invited consideration to . . . an international convention for the protection of all cable-lines, as well as to encourage the future construction of new ones, and is still desirous to see this project carried out. . . . Considering the peculiar circumstances of foreign representation and residence in China, and the identity of in a large measure of all foreign interests here, we would . . . be acting wisely to cooperate . . . , not with a view to obtaining any exceptional advantages for any country or scheme, but to provide for the general protection and encouragement of all cables in Chinese waters."[28]

In the general scheme of things, impediments to the development of the telegraph were viewed not only as impediments to China's progress but to foreign trade and global diplomacy. And here, such views fit into an even larger foreign policy framework. Again, the United States made these points clear in the correspondence among its own officials as well as with other countries, including the Chinese government. Indeed, this record lays out the vision so well that it is useful to consult it again to further elaborate the ideas being sketched here:

> The peaceful advancement of China, and to the extension of foreign trade, of the introduction of railway and telegraph lines. The interest in these means of progress is world-wide and appeals to the humanitarian as well as the political economist. *While the* United States *justly disclaims any right . . . to dictate to China how . . . she shall advance,* that disclaimer surely was not meant to stop us from advising or asking that forward steps be taken when practicable and when demanded by the common welfare. If we rest on the assertion that China will be left to advance in her own time and way, . . . we are simply allies of the native apathy and inertia which oppose progress now as they have opposed it heretofore.[29]

This was liberal empire. In short, it was realpolitik, but with a conscience and an internationalist twist. The right to intervene was, notably, not denied, but asserted. But the difference between naked imperialism and the above approach was that soft power—advice, the stimulation of "progressive change," the refusal to acquiesce to inertia—and humanitarian ideals would be the preferred agents of change.

Within this context one thing is absolutely clear: the foreign powers were no longer bypassing the Zongli Yamen and the central government, but actively relying on them. Several things had emerged to harmonize the interests of the government in Beijing and most of the Western countries. The first decisive factor was the Japanese invasion of the Chinese island of Taiwan in October 1874. The Japanese invaded the island ostensibly on the grounds of protecting its citizens from "native savages" who had recently killed three Japanese sailors and, according to Japan, were living beyond Beijing's control. It was a thinly veiled announcement of Japan's imperialist ambitions, although on this occasion such aims were subdued through British diplomatic intervention and China's agreement to pay compensation to the families of the slain sailors. During the crisis, the Chinese had scrambled to lay a cable to the island as part of their military effort, but the rapid de-escalation of events placed the effort on hold. However, as a result of these events, the Chinese government's perception of cables as an instrument of imperialism was modified and adjusted to the view that new communication technologies were, in fact, essential tools of national security and a modern nation-state.[30]

Beyond the clumsy brutality of the Great Northern Company's attempts to provide network security and the Taiwan crisis, Beijing's shift from the cautiously permissive approach of the early 1870s to one of promoting telegraph and cable networks was revealed by four other instances: a remarkably favorable response to the international community's call for better network security, renewed offers by China to take over the clandestine urban telegraph systems already in place, the rising influence of the *ziqiang yundong* (self-strengthening movement), and the creation of the Imperial China Telegraph Administration in 1881. As the American consulate remarked to the Danish mission in Beijing: "The central government has now very friendly sentiments toward telegraphs."[31]

With these changes in Chinese politics, the international community now redirected all of its attention to the reformist elements in Beijing. Among other things, Warren Delano, who was always lavishly praised for his pioneering role in bringing telegraphs to southern China but also seen as a bit of a loose cannon, was put on a shorter leash. As evidence of this and the bigger changes taking place, Delano's call to support the southern city-states as they tried to further the Great Northern Company's Fuzhou to Xiamen line and to "assert their right to manage the affairs of their province independently of Shen" fell on deaf ears.[32] Shen Baochen was one of the formative forces behind the ziqiang yundong movement and the fact that Delano had singled him out showed just how important Shen had become and also that Delano had picked the wrong adversary at the wrong time.

The Zongli Yamen announced two elements of China's new communication policy in 1875. First, it sent instructions to local authorities to provide the best protection they could for the coastal cables and existing telegraph lines, albeit without offering any guarantees as to the ultimate effectiveness of these efforts or by any kind of formal recognition. And to show the central government's resolve, a crackdown on cable piracy and telegraph thievery was launched. As Avery stated in a letter to Fish, "The instructions of the Zongli Yamen have been construed to mean that the central government is not unfriendly to telegraph enterprise."[33] Other diplomatic correspondence expressed the point even more forcefully, calling the new approach "an advance upon anything the yamen have heretofore said on the subject of telegraphs, for they offer nothing in hostility or by way of objection, but announce that they have given orders . . . to the local officials to see that the cables are protected."[34]

But China's new communication policy was built on more than providing security for foreign-owned networks. In fact, the Chinese government aimed to create its own national telegraph system. While of course, this pitted the central government against the Great Northern Telegraph Company, the central government had a surprising degree of support from the international diplomatic community. After all what was really at issue was the sacrifice of the company's clandestine network of overland telegraph wires—which *every-body* knew had no basis whatsoever in any concession ever granted by the *central* government—to the goal of putting the coastal cables, which were really the core concern anyway, on a more secure footing. And here, the Chinese had no intention of moving in on those areas, a move that would have been opposed by the international community given its desire to maintain at least one network that was "independent of Chinese surveillance."[35] Within this context, then, the Great Northern Company's domestic telegraph lines were the sacrificial lamb to China's modernization and national ambitions. And even then, the idea of sacrificing the company's ancillary interests was made all the easier by its squandering of a great deal of support that had been there for it from the outset. The company's recent decision to raise prices for its services; an uneasy sense, especially in British circles, that it was too closely aligned with Russian interests; and its bull in a china shop approach to the Xiamen-Mawei-Fuzhou lines only widened the breach between the Danish company and the rest of the international community even further.[36]

Thus, when the Zongli Yamen renewed its instructions in 1875 to take over the company's lines in Fujian it did so with considerable support from foreign governments, notwithstanding the company's protests. Delano's advice to bypass Shen Baochen and the central government's new initiative fell on deaf ears. In fact, far from heeding Delano's advice or yielding to some of the foreign press operating in the southern city-states who continued to support the company, Avery wrote a letter to Prince Gong, commending him for the adoption of a communication policy that, in his words, "was of marked significance in the history of China. . . . I shall . . . congratulate the Yamen on their new departure, and to hope they will authorize a northern line at an early date."[37]

The Great Northern Company, of course, did not acquiesce easily to this diminished role, but persistent work by Li Hongzhang, Shen Baochen, and Ding Richang between 1875 and 1876 finally brought about the changes sought. Throughout, the government consistently offered roughly $150,000 for the southern Fujian lines. While there were minor changes in the Chinese position, the Danish company tried to stall the inevitable by excluding the short Fuzhou to Mawei line from the deal and by insisting that the Chinese government compensate the company for "lost profits." Even a formerly supportive local official involved in the negotiations, Li Henian, found the company's "greed unbelievable, and even unheard of . . . among foreigners."[38] But, by 1876, the company concluded an agreement with Ding Richang that finally resulted in the southern telegraph lines being transferred to China. Over the next four years, the Fuzhou to Xiamen line was completed under the guidance of Shen Baochen, new telegraph lines erected in Taiwan by Chinese engineers, and agreements signed with Li Hongzhang for new lines to Tianjin and Beijing.[39]

Li, Shen, and Ding were among the most important modernizers and nationalists of the time, and they loomed large in Chinese politics during the last third of the nineteenth century. Befitting their position, Shen Baochen and Li Hongzhang also held two of the most important posts in the Chinese government, as the commissioners of trade for the southern ports (Fujian) and the northern ports (Tianjin), respectively.[40] Although Shen had previously opposed the telegraph and cable communication on the grounds that they were tools of Western domination, and Li was reprimanded in the 1860s for supporting unauthorized telegraph networks in Shanghai, by the mid-1870s both played a key role in developing a national telegraph system. Li Hongzhang also spearheaded, with the help of his close associate Sheng Xuan-huai, the creation of the Imperial China Telegraph Administration in 1881. Yet, at the same time, it became clear that there was a huge irony at the core of China's new national telegraph system and policy: a large and *officially sanctioned* role for the Great Northern Company.

MODERNIZERS, THE NATION-STATE, AND THE GREAT NORTHERN TELEGRAPH COMPANY'S MONOPOLY

In 1882, Russell and Company, an American firm doing business in Shanghai, applied to the Zongli Yamen for permission to create a network of coastal cables from Shanghai to Fuzhou and Hong Kong to compete with the Great Northern Company. The proposal was promoted as an international cable consortium that would involve not just Russell and Company, but other foreign and Chinese merchants as well. It also had the strong backing from the new American ambassador to Beijing, J. Russell Young, as well as German, British, and French consular officials. What the Zongli Yamen probably did not know was that Russell and Company had been running a clandestine urban telegraph system in Shanghai since 1875.[41] In and of itself, there was nothing unusual about urban telegraph networks, and in major world cities such as New York, London, Paris, Buenos Aires, they were commonplace. Shanghai was a major world city, and thus the fact that Russell and Company provided specialized commercial and financial news services to merchants, traders, and those who played the stock market was not untypical. In China, however, there had never been any mention of Russell and Company's Shanghai network, at least in any official correspondence. But here it was, in 1882, applying to Prince Gong to approve its plan to challenge the Great Northern Company's monopoly on coastal communication. The Zongli Yamen's response astounded the company as well as consular officials. The application could not be approved because this would violate the Great Northern Company's *officially authorized* monopoly on international communication services in China.

Specifically, Prince Gong informed J. Russell Young that an agreement had been signed by Li Hongzhang and the Great Northern Company in 1881 that prevented him from approving the application. Apparently, the company had not sat idly by as diplomatic and commercial support for its operations in China slipped, but cut a deal with Li Hongzhang that officially recognized its cables and their termination in Shanghai and Xiamen and which gave the company a twenty-year monopoly over global communications in China. The American ambassador was livid. For one thing, Young noted, the Zongli Yamen had promised American companies since the 1860s that they would be able to enter China once permission had been granted to others. But while Young was critical of the Zongli Yamen and other government officials, his most trenchant criticisms were reserved for the Great Northern Company. Thus, in Young's words:

> There can be no reasonable doubt but that the Danish company profited by the inexperience of Chinese officials to lead the Chinese Government into covenants which no western Government would allow, covenants the nature of which must in time become apparent to the Chinese themselves, and *probably be regarded to our detriment as another evidence of the unjust and grasping spirit of the western world in dealing with the Oriental people. . . .* By this contract the Government grants the Danish company for twenty years a monopoly of all the submarine cables already landed in China, during which time the government engages itself not to allow any cable whatever unless with the consent of the company. During this period the Government will not permit the construction of any land lines that may be in opposition to the submarine cables. Thus a concession for a submarine line is so worded as to put the whole land telegraph system of this vast Empire at the mercy

of a private corporation. . . . Practically, the whole question of the future of telegraphs in China, so important to the welfare and development of the Empire, is surrendered by the Government for twenty years.[42]

There is little more to be said on the nature and the criticisms of the agreement than that noted by Young in his letter to Prince Gong. But to be fair, one has to see in Young's criticism the fact that it too was self-serving. His criticisms skirted over the fact that the United States and Britain, most notably, had also ignored such arrangements in the past, their supposed commitment to free trade in cables, notwithstanding. Prince Gong was quick to point out such facts in support of his contention that there was nothing exceptional about the arrangements between China and the Great Northern Company. But in a small bid to mollify Young, the Prince pledged to deal with the problem of frequent service disruptions and to more tightly regulate the company's rates to avoid the increases that the company had regularly imposed on the merchant and diplomatic communities.[43]

While, it is pointless to go any further into the details upon which this aspect of the debate turned, it is useful to explain some of the other reasons why the Chinese government would strike a deal along the lines it did, especially so soon after seeming to have disentangled itself from the clasp of the Danish company. For one, Prince Gong, Li Hongzhang, and Shen Baochen—or at least the official posts they represented, namely, the Zongli Yamen and the Northern and Southern Trade Commissions—received free telegraph privileges, although at the time of the agreement this only amounted to a paltry sum of $33,000 per year. Such an amount could hardly justify a twenty-year monopoly, and Young was quick to point out just how cheaply the Chinese seemed to have sold out their sovereignty. The far more important factors were related to how the agreement fit into and reinforced the modernization project being pursued at this time, even though that was hard to tell at first glance. From China's perspective, the deal was so valuable because the most important foreign communications company in China, and one that had been supported throughout its existence by the *entire* international community, officially recognized Beijing as the sole authority in such affairs. The agreement appeared to provide China official recognition by everybody, a validation coveted by the central government as much as the Great Northern Company sought recognition of its cables. Little did the government seem to realize that the company was fast becoming the bête noire of the international community and this deal would only reinforce such views.

More vital than affirming Beijing's exclusive authority in global communication policy issues, the deal recognized the Imperial China Telegraph Administration's monopoly over the *national* communication system. In fact, the Great Northern Company and Imperial China Telegraph Administration had formed a cartel, with the former agreeing not to compete with the latter in return for the same guarantees with respect to its own monopoly over global communication. For the Chinese government, the recognition of its monopoly over the national telegraph system was crucial, given that the company had traditionally been the arch-opponent of Beijing and, in particular, of key modernizers such as Shen Baochen and Li Hongzhang. Thus, the agreement was probably seen as closing a difficult chapter in adjusting Chinese politics and culture to the new realities of global communication.

In the first of a series of agreements setting out these new arrangements, Li Hongzhang signed a contract with the Great Northern Company on December 20, 1880, for the latter to build a new Shanghai to Tianjin telegraph line and to transfer it to the Imperial China Telegraph Administration upon completion. The Chinese Telegraph Administration would operate the new line once it was finished, but the system would use Danish engineers and technology and be built by the Danish Company. The interests of the company were obvious as well. It was not to be shut out from the Chinese market so much as made the supplier of choice to the national administration. It was a valuable and tangible benefit. And as with other such agreements that had been signed, for example, with the Persian and Ottoman governments, the Great Northern Company agreed to open telegraph schools in China, the most important being in Shanghai, to help train Chinese telegraph operators and engineers. Such aspects undoubtedly reflected the Chinese perspective that the Great Northern Company, far from exploiting China, was being used as a tool in China's own modernization project.[44]

This, of course, was not how others saw the arrangements. Among its other most distinguishing features, the agreement was a brazen bid to usurp control over the entire Chinese global communications market, and this put the Danish company on a collision course with the Eastern Extension Company, the Asian affiliate of the world's most powerful cable company: the Eastern Associated Company. More than just that, it was an extraordinary attempt to repudiate the joint purse agreement that the Great Northern had established with Pender in 1870 for the division of the Chinese communication market. In fact, the agreement signed between the Danish company and the Chinese government explicitly called for *all* messages from China to the rest of the world to go by way of the Great Northern Company's northern route through Russia and Europe rather than by the southern route, where the company had obligations under a previous agreement with the Eastern Extension Company for the distribution of messages and revenues.[45] Not only was it ill-advised to take on this global communications colossus, the idea that all communications north of Hong Kong would go by way of Russia was intolerable to the United States, Britain, and European countries.

THE ONE-SYSTEM APPROACH: THE EASTERN EXTENSION AUSTRALIA AND CHINA TELEGRAPH COMPANY REASSERTS CONTROL OVER THE CHINESE AND ASIAN COMMUNICATION MARKETS

The Eastern Extension Company's response to this act of betrayal came swiftly and forcefully. Within two months of the secret deal becoming known, the Great Northern Company was forced to capitulate through a series of new agreements. Although couched in diplomatic language, the basic thrust of the agreements was clear: that whatever benefits the company had gained in its deal with Beijing were to be shared with the Eastern Extension Company. Moreover, the new agreement, signed on January 12, 1883, and followed by several others over the next few years, not only reinstated the status quo ante but expanded the original joint purse by

Reaffirming official recognition for the Great Northern Company's existing network of coastal cables and obtaining the same for a new Hong Kong to Shanghai cable laid by the Eastern Extension Company in 1883;

Making it absolutely clear that the Chinese market for international cable services was to be shared *equally* between the two companies;

Setting out the ground rules that would guide both companies' coordinated approach to the development of the rest of the Asian communications market (the single system approach);

Carving out a position for the new Imperial China Telegraph Company in its relationship with the two companies;

Spelling out their positions with respect to new media technologies (i.e., the telephone); and

Explicitly setting out the chain of command between all three companies and the steps to be followed whenever new concessions were desired or obtained: full disclosure, approval by joint committees of the companies, review by the council of foreign ministers in China, and then submission to the Zongli Yamen and to the Northern Trade Commissioner (Li Hongzhang) and the Southern Trade Commissioner (Shen Baochen) for final approval.[46]

All these elements combined, the Eastern Extension Company informed the others, constituted the new communication policy for China and Asia. No discussion, just sign. The joint heads of instruction summarizing the goals signed a few years later stated that the agreements were designed "to establish good feeling and harmonious working between the . . . services."[47] While the phrase may have helped the others to save face, the agreements, less diplomatically put, put the Great Northern Company on a short leash and ensured that insofar that anybody called the shots with respect to the global communication services market in China and Asia, it was the Eastern Extension Company. In their entirety, these agreements, periodically revised and renewed, set the communication policy regime for China and the rest of Asia for the next two decades.

As both a symbolic gesture and as a way of giving the new arrangements an undeniable physical presence, the Eastern Extension Company laid a new cable between Hong Kong and Shanghai in 1883.[48] Now, there was no question about relying on the goodwill and integrity of the Great Northern Company. Pender's company had its own tangible stake in the control over the two main electronic gateways to China: direct cable connections and offices in Hong Kong and Shanghai. Moreover, while the agreements spoke in terms of collaboration and respective spheres of influence, their tenor overall suggested that the Asian communication system was being put, first and foremost, under the general planning authority of the Eastern Extension Company. The most important agreement, namely, the first one signed in January 1883, acknowledged that both companies had developed spheres of influence outside of China and that these needed to be protected as much from the competitive threat posed by others as from the two companies themselves. The Great Northern Company had been providing services to Japan from Russia and between Nagasaki and Shanghai since 1871, and the Eastern Extension Company had been doing the same with respect to Australasia (1876) and the Philippines since 1880.[49]

At the time of these efforts, both companies were planning additional cables in the region and the agreement spelled out the terms that would be followed as they did so. First, the agreement preserved the status quo positions of the companies with respect to Japan, Hong Kong, and the Philippines. Second, it recognized that the Great Northern Company had recently gained concessions to develop a new regional network between Korea, Japan, and China, while the Eastern Extension Company possessed the same with respect to a new cable between Hong Kong and to the Portuguese colony off

11 The Eastern Extension Company and Great Northern Company Office in Hong Kong, an example of corporate cooperation of Western cable interests in China and Hong Kong. The counters of the Great Northern and Eastern Extension companies sit side by side in their shared offices on the Hong Kong waterfront, 1922. The headings "Via Eastern" and "Via Northern" are just visible above the counters. Reproduced by permission of GN Great Nordic and author Dr. Kurt Jacobsen from GN Great Nordic's 125th anniversary publication, *From Dots and Dashes to Tele and Datacommunications*, Copenhagen, 1994.

the southern coast of China, Macao. The new agreement required that all of these deals be put on the table for review by both companies and that both had the option of joining the other to lay, and own, these new cables together.[50]

This was the single-system approach, and its logic begins to pervade the official corporate documentation of the companies at this time. It was, in short, an expression of the idea that communication markets would be regulated and developed collaboratively by corporate interests insofar as possible. The scramble to put Humpty Dumpty back together again after the Great Northern Company had so cavalierly toppled it exuded the rationality of regulated markets that governed global capitalism in the late nineteenth century. The events under discussion represented the clearest application of the new economic orthodoxy yet with respect to China and the Far East. Collaboration and cartels, not competition and free trade, ruled.

Beyond just regulating regional communication markets, a second set of agreements extended to the regulation of new media (i.e., the telephone) by both companies. Prior to becoming aware of the Great Northern Company's backroom deals with China, the Eastern Extension Company had signed an agreement with the company that gave it authority to obtain concessions in China and Japan for a new telephone company that Pender's group of companies had recently established, the Oriental Telephone Company.[51] The basic thrust of the agreements was to introduce this new means of communication into the major urban centers of China and Japan—Fuzhou, Xiamen, Shanghai, Hong Kong, Nagasaki, Tokyo, etc.—in a way that complemented rather than competed with the companies' commitments to the old media of telegraphy and submarine cables. While that was the plan in 1882, another agreement two years later required the greedy, double-dealing Great Northern Company to renew its vows to develop telephony in China and Japan in a way that complemented rather than competed with either of the company's other operations. This meant, in its most basic form, that the telephone would be developed as an urban communication system and one that would rely either on the companies' cables or, in China, the long-distance lines of the Imperial China Telegraph Administration. The same would apply in Japan where the Japanese state-owned telegraph system would provide the long-distance connections and in the city-state of Singapore, where the Oriental Telephone Company set up its first operations in 1882.

The reconstitution of the joint purse not only carved out and reaffirmed relations between the Danish and British companies, but also spelled out the roles of the Imperial China Telegraph Administration and even the various political officials involved. In many ways, the agreement was probably seen as a good thing by China. While Baark suggests that Li Hongzhang and others opposed the cable laid by the Eastern Extension Company in 1883, they also had goals of their own—which Baark presents with exceptional clarity—that were facilitated by the new agreement. First and foremost, the new agreements recognized China's monopoly over the national telegraph system. Second, the new cable was laid in 1883 only after the principles set out in the original 1870 concessions were formally accepted. Last, and most important, agreements were made that not only protected the Imperial China Telegraph Administration's national monopoly but also allowed it to establish direct connections with Hong Kong and to open its own offices there. In addition, the Imperial China Telegraph Administration was admitted as a full-scale member in the cartel in 1887, making it party to the deals that would set the rates, distribute the revenue, and impede others that were seeking to compete for a slice of China's communication market.[52] Throughout this period, agreements were revised and new ones struck that extended the Danish and British companies' monopoly over global communication services in China, first to 1903, then to 1910, and ultimately to 1930.[53]

COMMUNICATION AND THE CRISIS OF EMPIRE(S)

The agreements were remarkable for all of the above reasons and for the fact that just behind the veneer of establishing normal relations between the private companies as well as the Chinese government and the Imperial China Telegraph Administration lay glimpses of the fact that conditions in China were sliding from bad to worse. The first clue was the steady expansion of the foreign companies' exclusive rights over China's electronic connections to the rest of the world. While, as we have seen, exclusive concessions were in fact quite normal in zones of informal empire, the fact of the matter was that such rights had always been complicated by diplomatic commitments that made it difficult to explicitly assert their existence. Their explicit statement, at least in the agreements of the 1880s and 1890s, did clarify things and, by the standards of the time, their limitation to fifteen years was quite reasonable. However, the fact that exclusive concessions were constantly revised and prolonged—finally, until 1930, as just noted—revealed that the Chinese government was being forced to yield more and more to the companies. And this was for a variety of reasons that were only incidentally linked with the agreements. The most important of these factors were related to the acquisition of foreign loans to help finance the national telegraph system and for payments that were extracted from the Chinese government as a form of retribution and punishment for the anti-imperialist uprising in the Boxer Rebellion of 1900. At this point in time, even the pretext that the foreign powers were there for the international development of China was abandoned in the face of a full-scale anti-imperialist uprising consisting of a mixture of backward-looking conservatives, modernizers, and new China revolutionaries. It was the most violent anti-imperialist uprising in China to date. In response, a 20,000-strong "international expeditionary force" led by Germany but including soldiers from the United States, France, Germany, Japan, and Britain, with a large element of the latter force being Indian, was quickly assembled to crush the uprising.[54]

The Eastern Extension Company rose to the task by hastily laying another cable between Shanghai in the south to Tianjin, Beijing, and other cities along the way deemed vital to squelching the uprising. While the cable served the collective security needs of foreign powers in China, it was ordered by the

British government and paid for by China. Yet, while nominally owned by China, the cable was controlled by the Eastern Extension Company until 1925. It was another example of cooperative imperialism in action, and the Chinese paid dearly for it, both in terms of paying for the cable that coordinated the military assault on them, the lives lost, and the dreams of a modern China forsaken. As one British banker observed, "This is a war of extermination. No prisoners are being made, everything living is being killed. It is the only way of dealing with these people horrible though it may seem."[55]

The Politics of Global Media Reform I, 1870–1905: The Early Movements against Private Cable Monopolies

> Monopolies as being opposed to the public interest are discouraged by the law, and obnoxious to public opinion. But the cable monopoly is the creation of very able men, the ablest of whom is unquestionably Sir John Pender, Chairman of the Eastern Telegraph Company and various American Cable Companies; and it is so fenced about by contracts, subsidies, reserve funds and capital, that it is difficult to strike a fair blow at it.—JOHN HENNIKER HEATON, "Postal and Telegraphic Reforms," *Contemporary Review*, March 1891

At various points in the past 150 years, demands for the right and opportunity to access specific services or institutions have pervaded the political landscapes of many countries. In Britain, during the late nineteenth century and early twentieth, the dominating issue was the right to the franchise, first for men and later women. Concurrently, the right to education in the form of mass public schooling was a significant driving force in many countries, and in the twentieth century this became tied into the concept of "equality of educational opportunity," especially after 1945. Contemporary political trends and the veneration of "self-regulating markets" have undermined some of the primacy of this concept of rights and access today, but the spread of new media technologies, notably the Internet, has revived calls for "universal access" and "communicative democracy," even on a global scale.

There is a tendency, however, to believe that movements aimed to facilitate access to information and communication technologies and services are a relatively recent phenomenon. They are not. As a case in point, cheap, fast, and frequent postal communications, both nationally and globally, spurred a long campaign in Britain and its empire which culminated in the imperial penny post in 1898. And then, such movements encompassed subsequent technologies of communication: witness the slow transformation of the telephone from an instrument of business to a popular means of communication, followed by the cheapening of long-distance phone rates, concurrent with the rise of the Internet, that has occurred during the past two decades. Indeed, Tom Standage draws some fascinating parallels between the Internet and nineteenth-century electronic technology in *The Victorian Internet*, stating that "the telegraph unleashed the greatest revolution in communications since the development of the printing press."[1]

While there is little doubt that the impact of the telegraph was as great in its day as the influence of the Internet is now, there is a fundamental difference, namely, that the telegraph, and even less so transcontinental cable communications, never became a means of mass communication in its own right. According to many of the "media reformers" covered in this chapter, the stunted development of long-distance social communication in the late nineteenth and early twentieth centuries was due mainly to the fact that powerful cable monopolies charged high rates and focused solely on "premier customers." The companies' sense of entitlement seems also to have been a factor. How else, for example, can one judge the response of Sir James Anderson, managing director of the Eastern Associated Company as well as the Anglo-American Company, to the complaint, in 1879, of a British General Post Office official that cable tariffs to India were too high? "We really represent the capital and organisation which unite Great Britain with the external world, and upon every principle, commercial and political, may fairly claim not only the support of your Department, but the amplest assurance that none of our interests shall be adversely affected for the sake of any new idea which is not of vital character in respect of international telegraphy."[2]

Anderson's response, redolent with a sense of self-importance, infuriated the reformers. Their response was to criticize the cartels, their exclusive privileges and their rates, and, for some, to argue against the very principle of private enterprise in international cable communications. Admittedly, their gains were limited, but, in Britain, their efforts did culminate in 1902 in the first state-owned public service–oriented cable, the imperial Pacific Cable. Beyond the British Empire and transatlantic economies, there were also various reasons for criticism of cable companies: sometimes the main issue was one of revenue loss from public telegraph systems as in some South American countries, undue restrictions on international communications as in states such as Japan, which had its own growing imperial ambitions, or as in China and much of the Middle-East, governments' wariness of the political and cultural ramifications of communication systems developed without regard to their own ambitions and view of their place in the world. As that wariness grew, the cable business found critics in plenty throughout the world.

CUSTOMERS AND USES OF THE
INTERNATIONAL CABLE SYSTEM

The domestic telegraph, with its links to the international cable system, was certainly considered by many countries to play a vital role in their domestic communications, and hence in many countries became a state-owned medium. As such, it was frequently priced below the cost of administration and long-term maintenance. Thus, for example, the British General Post Office telegraph fell into major deficits between the 1880s and 1914 after successful campaigns by the press and other groups for lower charges on inland telegrams. Yet, as Kieve observes, even though greater facilities and lower rates resulted in a vast increase in use between 1870 and 1914, "It was still very much the ancillary of the penny post . . . [and] probably not even being used much in business by small tradesmen."[3] Nonetheless, in 1902, about 90 million telegraph messages were sent in Britain, the equivalent of 2.2 for each person in the population, a ratio far higher than in the United States and Canada, where the equivalent telegram-to-population ratios were about 0.9 and 0.7, respectively. Looking at these circumstances, there was a strong belief among American supporters of state-run telegraphy, such as Professor Frank Parsons, head of the U.S. National Public Ownership League, that the fact that the U.S. telegraph system was privately owned and run for a handsome profit rather than for the needs of the people prevented it from being a popular communication medium.[4] Thus, in 1899, Parsons showed that, on average, telegram charges were double those in Europe and the numbers of telegraph offices proportionately far fewer. But the fact is that while he held up European countries, notably Britain, in favorable contrast, the demand for telegraphy actually fell off dramatically in Britain after 1911, as the telephone finally made some impact as a means of inland social communication. In Canada and the United States, perhaps partly due to the limited use of telegraphy, the telephone was actually far more popular than in Europe. However, while it might have offered tough competition to the telegraph, on neither side of the Atlantic did it rapidly become available as a household item.

The bottom line was, however, that a communication medium which, unlike the phone, required a visit to the nearest telegraph office, was still not competitive with the cheap and efficient postal services maintained in Britain or indeed Canada. Beyond that, the fact that the practice of most

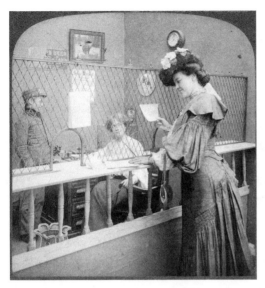

12 The telegraph in popular culture. Originally for use with a bioscope and distributed internationally, this image portrays the use of the telegraph in popular Western culture, circa 1904. The technology, like that of the telephone in figure 13, is shown as "a way to get the news quickly," though both telegraph and telephone were actually too expensive for many people at this time. In the photograph, a sign on the wall behind the counter reads "Western Union Telegraph and Cable Company." Image produced by Underwood and Underwood, publishers, New York, 1904.

13 The telephone in popular culture. In this photograph, the telephone is shown as a popular household instrument, though the attractive lady, in line with early sales pitches to North American customers of the service, is clearly well off. Image produced by Underwood and Underwood, publishers, New York, 1904.

European states was to encourage telegraph use, whereas Western Union, in the early 1900s, focused on the very profitable business of leasing out wires to stock brokers, traders, and press associations.[5] While this made good business sense, it reinforced the popular image of an uncaring and grasping monopoly. In more ways than one, the long-distance cables were the domestic telegraph writ large. Writing as late as 1918, British engineer and policy analyst Charles Bright noted that efforts should be made to popularize the cable, since "at present, the average person regards the cost of 'cabling' as prohibitive . . . [and that] at the moment, quite a large number of educated people are even igno-rant as to what course to pursue if obliged to despatch a cablegram."[6]

To be fair, there were some instances where cabling prices underwent substantial declines. And there were also cases where the high rates of capital-ization and technical risks actually did justify high rates, while still bringing the cable companies slim, if any profits, as was most notably the case for the Western and Brazilian Company and the West Coast of America Company in South America (see chapter 2), but in other cases such claims of financial necessity were dubious. And the best example of that was probably the first two decades of experience in the transatlantic communications market, where arguments from capitalization and technical risk were trotted out to justify the immensely high initial rates of $10 per word, but which fell first to $5 per word and then finally to $2.50 within two years (1866–68). Overall, after the early plunge in rates, they stayed remarkably stable for the next two decades. Subsequently, regular rates, under the press of competition mounted by the Commercial Cable Company, plummeted again to 25¢ a word by 1888 and remained at that level, by agreement of the Atlantic cable cartels, until 1923.[7] Customers in Asia were even less fortunate, where the per-word rate sky-rocketed from its initial level of $12.50 for twenty words in 1872 to $40 for the same number of words in 1875, a rate that remained until just before the end of the century before then falling by about 30 percent and another 15 percent by 1905. In early 1902, the word rate from London to India also finally fell nearly 25 percent after having been stuck at $1 since at least 1890, and from London to most Australian states it was 75¢, a major drop from the earlier rate of $2.37 a word in 1890 but still a heavy cost to merchants and public alike.[8] More information on ordinary cable rates from London to various countries can be seen in table 6.

There were some significant drops in tariffs between 1890 and 1902. Yet, in several key areas—the transatlantic route and to China and Japan, for exam-ple—rates had been remarkably impervious to change, while in others, most

TABLE 6   Ordinary Cable Rates (per word) from Britain to Selected Countries, 1890–1902

COUNTRY	1890 (US$)	1895 (US$)	1902 (US$)
United States	0.25	0.25	0.25
Canada	0.25	0.25	0.25
Australia	2.37	1.23	0.75
Japan (via Great Northern Co.)	2.00	2.00	1.94
(via Eastern Extension Co.)	2.66	1.94	
Argentina	1.75	1.12	0.98
Brazil	1.50	1.00	0.75
Chile	2.21	1.54	1.43
China	1.77	1.75	1.87
India	1.00	1.00	1.00 (0.72)
Nigeria	2.41	2.41	1.70
East Africa	1.93	1.25	0.75
Jamaica	1.45	0.75	0.75
Trinidad	3.20	2.60	1.27

Source: Britain, Inter-Dept. Comm. Report, 1902, 36–39

notably in the case where it cost more to send a message from London to Calcutta in India than either to Sydney or Shanghai, the cost of service bore no relationship to distance. The Caribbean rates had for a long time stayed among the highest in the empire (and also for the non-British colonies in the region), until dropping a third in 1895 after the advent of competition, and by half again in 1902.

Not surprisingly, these tariffs imposed heavy burdens on the cable companies' customers. From the outset, the news agencies and foreign correspondents of major newspapers paid dearly to be first with the international news, although much of it—features, syndicated material, and the like—was not so timely and could, in later decades, be sent by cheaper "deferred rate" categories. According to one "insider account" of the late-nineteenth-century American news business,[9] cable rates were "the most costly convenience in newsdom" and the U.S. news services often came close to the brink of bankrupting themselves in their competition with one another. The advent of much cheaper "press rates" in the 1870s and their use all over the world lightened the burden somewhat for the news agencies. However, even though the rates were one-third to one-half the cost of "ordinary rates" the prohibition against the press's use of "code" minimized their effectiveness. Private codes were almost always used, both for purposes of security and cost reduction, by commercial companies, traders, and governments, and the fact that the press could not use them in the standard press rate category was a significant drawback. Yet, even when press rates did drop in the 1880s and 1890s newspapers and the agencies usually just upped the volume of international news that they produced and distributed instead of cutting back on their cable bills.

The cost of news was still high enough, however, that the global news agencies cut back in other ways. One such way was to share facilities, as Havas and Reuters did with respect to the former's foreign bureaus in South America (1869–1876). They also had quasi-formal agreements that set out a rough division of labor between them, with Havas focusing on selling its news service to the South American press and developing what Rhoda Desbordes calls a "a new flourishing branch of service known as the 'private' dispatches service." Desbordes goes on to describe the private dispatch as "a more intimate, personal and rapid mail than the one offered by the current postal system. Despite the fact that commercial telegrams were still very successful, 'privates' contained chiefly financial or commercial information as well."[10]

These private services were Havas's most important source of revenue in South America and, consequently, most of its agents there were not journalists at all but commercial representatives who focused on soliciting clients for the privates and selling the agency's news service to the South American press. Consequently, most of the production of news in South America for export to Europe was done by Reuters's foreign correspondents. Given this rough division of labor, and the fact that the cable systems through which South American news messages and private dispatches were sent all terminated in London, Havas and Reuters added another layer to their mutual interdependence when they jointly opened the "Bureau Amérique Sud" at Reuters's London-based headquarters in 1874, an arrangement which lasted until 1916 when Havas moved to New York in the midst of the war. These arrangements did not eliminate all sources of friction between the two, of course, but considered alongside the "ring circle" agreements they do show that Reuters and Havas were willing to extensively share resources and markets in their bid to better manage the economics of global news production and distribution.[11]

In weaker markets, the issues were more pronounced. Thus, of the nine joint Havas-Reuters news bureaus opened in South America in the mid-1870s, seven were in the commercially vibrant cities of Brazil and the River Plate, and only two were located on the west coast. Indeed, only a few newspapers, merchants, and public officials in Chile or Peru used the cable system during the 1870s and 1880s, and Peruvians often wondered whether they were connected to the outside world at all.[12] Rates dropped in the 1890s after the joint purse between the West Coast of America Company and James Scrymser's Central and South American Company collapsed, but not greatly. In fact, the cable companies were not all that interested in more customers. As Scrymser told a U.S. Senate committee in 1900, four hundred customers, the vast majority of whom were Europeans, accounted for 90 percent of his firm's Latin American business.[13] And while the press and news agencies were vital sources of income, they should not be overrated as such. This was because most areas of the world were allocated to just one of the major global news agencies by the ring circle agreements, and in those cases where the operations of the big four—Reuters, Havas, Wolff, and Associated Press—did overlap they typically shared facilities to keep a tight rein on the costs of news gathering and dissemination. Until competition among news agencies, both inside and outside the cartel, increased after the first decade of the twentieth century, this meant that their role as a source of income for the cable companies, although great, was subject to the very strict limits imposed by the economics of the global news business.[14] Nonetheless, even in 1913 John Denison-Pender, the son of the Eastern Associated Company's founder and now among the firm's managing directors, identified Reuters and Lloyd's of London as the company's two most valuable clients. But Denison-Pender also made another crucial point, namely, that "the large merchant would . . . prefer to pay the higher rates and not bring in the small merchants."[15] That is, high rates served the cable company and its premium clients equally well. So, in addition to enriching the cable company's coffers, high rates helped to deter competition in their best customers' markets and thus the incentive for cheap rates was unlikely to come from that quarter.

To be sure, the cable companies' position was not without justification, and one of the key factors that did have a significant impact on rates was the initial limited carrying capacity of the cables and the fact that customer demand was highest during business hours. This was especially important in the transatlantic economies where, with London at five hours ahead of New York, an 8 A.M. start in New York meant that there was about a four-hour time span, 8 A.M. to noon in New York and 1 P.M. to 5 P.M. in London, during which to send a limited number of urgent messages from priority customers and get a reply. Even after the introduction of "duplex technology" in 1879 (which doubled capacity to around twenty-four words per minute) and the threefold to fourfold increase again ushered in by the advent of new "fat cables" in 1894, there really was a limited amount of time and capacity available for people in London and New York to conduct business. Or, as Denison-Pender stated, the companies were justified in trying to keep peak periods free for their premium customers instead of filling their systems to capacity at "unremunerative rates."[16] However, Denison-Pender's argument ignored the fact that there was a great deal of "dead-time" and that, at the very least, the international cable system should be available at cheaper rates during those times so as to spread the benefits of long-distance communication. In this, his attitude reflected a key fact of global communications in the late nineteenth century and early twentieth: that it was a luxury to be used only for the most urgent business, international journalism, and matters of state rather than as a social necessity.

CABLE REFORMERS AND IMPROVING THE MEANS
OF SOCIAL COMMUNICATION

While many saw a straight line between new telegraph and cable communication technology on one hand and progress and modernization on the other, not everybody was naively optimistic about their impact on world affairs. Thus, at the very beginning of the transatlantic electronic age in 1866, the *New York Times* showed a remarkable clarity of editorial opinion as it reviewed reactions to cable communications: "We see no reason for the excessive jubilations of some writers, viz, that these international telegraph lines are going to introduce us, by a short cut, to the millennium. We believe that . . . we shall have just as much international bickering and bluster after they are built as we have now. Human nature is not going to be changed by any material improvement of this sort. How much has the direct line of telegraph between London and Berlin improved the tone of discussion between England and Prussia?"[17]

While the *New York Times* saw little reason for the need to speculate on "Utopian achievements," it was quite prepared to wax on at length about the "vast stimulus" that cable communications would bring to trade and industrial enterprise. Over the next two years, the paper followed this initial thread with further editorials focused on the high costs of cabling and the problem of foreign-controlled cable monopolies; thus "To fix rates which will prohibit the use of the line except at an exorbitant rate, is at once, it seems to us, to detract, from its usefulness to no small degree."[18] And more ominously, "The absolute monopoly of telegraphic communication with this Continent, involves a power more vast and terrible than has ever been enjoyed by any nation in the world . . . The French Government is interested, as we are, in preventing an absolute monopoly of the Atlantic for telegraphic purposes by England or any other single power."[19] That was a rather simplistic view of the matter which glossed over how the global communication business was already becoming deeply multinational in character, but it did register the fact that the issues of rates, monopoly, and foreign control were already at the center of the politics of global communications and had been put there by one of its main customers: the press.

But besides the criticism sometimes heaped upon the cable companies by the press, who else was involved in trying to reform global communication? There were, of course, many such individuals and agencies, some of whom we have seen in previous chapters, especially in South America where the politics of national and global communication were always intertwined and intense as well as in China, Persia, the Ottoman Empire, and Japan where similar dynamics were close to the surface. But in this chapter our focus is more on those who were active in the transatlantic economies and the British Dominions, although near the end of this, and the chapter following, we also focus on events in China and, related, the views held by modernizers and imperialists alike in Japan regarding the role of long-distance cable communications. With this focus in mind, the history of the early cable reformers is one of individuals who showed a great deal of organizational talent, immense energy, dogged persistence and, in one case—Scots-Canadian engineer, Sandford Fleming— outstanding creative verve, in pursuing goals which attracted around them many supporters. After 1880 and until World War I Fleming was always on the scene, but others soon joined him: notably, from the 1880s, John Henniker Heaton, long-term Conservative Member of Parliament for Canterbury in Britain, and, for some twelve years, 1899–1911, Edward Sassoon, a fellow Conservative member of the British Parliament. Then, there was the British engineer Charles Bright, son of the famous cable engineer, Charles T. Bright, who, as previously noted, stood out as Britain's top "communication policy expert" and as the author of many accounts of the contemporary communications business, its technology, and military significance. In the United States, Captain (later Major General) George Squier of the Army Signal Corps figured prominently in debates over state-ownership and the Pacific Cable after that nation's rise as an imperial power (1898) and just prior to World War I. In Europe, the German postmaster and delegate to the International Telegraph Union, Ernst W. H. von Stephan, also stood out as a persistent and prescient cable critic, with his perceptive ideas about how cheaper trans-European rates could facilitate European commercial, political, and social integration.

In a detailed history of international communications in 1912, the London

14 Sir Sandford Fleming, circa 1906. During his
long life (1827–1915) this Scots-Canadian engineer showed a remarkable capacity to promote a diversity of ambitious projects. Many,
like advocacy of Universal Standard Time
and the Pacific Cable, were brought to fruition. Reproduced by permission of Queen's University Archives, Kingston, Canada.

Times credited Sandford Fleming with being, since his first public musings on
the subject in 1879, the real visionary behind transpacific cable communication and a steady advocate "from that time onward [who] never wearied of
pressing the need for its completion on the statesmen, journalists and businessmen of the Empire."[20] The tenacity with which Fleming pursued these
ideas was one of his outstanding features; it was also one of his flaws. Typically
Fleming's response to frequent setbacks was to write yet another pamphlet, in
which he would lay out the nature of whatever problem had seized his and the
public's imagination, chastise the powers-that-be responsible for the continuation of the problem, and then offer his preferred solution, with Universal
Standard Time and the Pacific Cable being the two most well known.

This was an age of conceptual innovation and from the mid-1880s on,
Fleming began to correspond with Henniker Heaton, the originator of the
imperial penny post, internationalist, and another leading member of the
"global media reform" group.[21] Henniker Heaton was born in Britain, but had
spent much time in his early years working for newspapers in New South
Wales, and indeed represented that state at many international events, including the International Telegraphic Conference of 1885, where he successfully
advocated reduced cable rates between Britain and the Australian states. This
proved to be the genesis of his subsequent career as a British Member of
Parliament, the tireless critic of cable monopolies, and advocate for cheaper
and more accessible communications. His first campaign, in the teeth of
General Post Office opposition, was for the extension of imperial penny post-

age to the rest of the empire. After this was achieved in 1898 with the influential support of the Canadian Postmaster General William Mulock, Henniker Heaton turned to focus on cable monopolies and the Eastern Associated Company in particular. Known jokingly in Parliament as "the member for Australia," he knew well the role of high cable rates in cutting British immigrants off from speedy communication with home; consequently, he built on his success with imperial penny postage to incorporate the radical proposal for global "penny-a-word cables."

Heaton's prolonged attack on cable monopolies was carried out in the British House of Commons and in numerous articles and pamphlets. Backed in later years by Sassoon, who had acquired some critical views of the Britain to India cable links while in business in the subcontinent,[22] these two could always expect Canada and New Zealand to be notably supportive of their efforts. Canada's position as a crossroads of empire made it a natural starting point for these cable reformers and visionaries to pursue what were, at the time, audacious projects to lay cables across the Pacific to Asia and on to New Zealand, Australia, Hong Kong, and other locations throughout the Far East. New Zealand's position, not at the crossroads but at the periphery of empire, meant that it would benefit immensely from such schemes, and partly on that account New Zealanders were among the strongest proponents of the Pacific Cable.

Taken together then, the organizers of the "global media reform" movement knew each other and were acutely aware of the fact that there was much public resentment to build upon. There were, to be sure, some key differences on matters of principle among them. For instance, Charles Bright typically stood outside the popular reform movement because, although often critical of the cable companies, his analysis of the cable business did not lead him to its wholesale condemnation. There is also no doubt that Captain Squier was well versed in the politics of the global media reform movement and approved of many of the principles on which it was based. However, like others in certain U.S. military circles, he took a more nationalistic view and this became more pronounced as time wore on,[23] especially between 1914 with the start of World War I in Europe and 1920. Of the four most prominent cable reformers at this time—Fleming, Sassoon, Bright, and Heaton—it was Heaton who went furthest toward adopting an internationalist, rather than imperialist, stance. Indeed, Heaton's position on cable reform went well beyond the mental boundaries of empire as he began moving in internationalist circles, meeting the well-known American literary star, anti-imperialist, and internationalist, Mark Twain. Twain's well-known quip of the time, "it's easier to stay out than

15 John Henniker Heaton and Mark Twain, 1907. Sir John Henniker Heaton was included among
a group of notables who posed with Mark Twain during his visit to the British House of Commons, July 2, 1907. Mark Twain is in the foreground holding a cigar, and Henniker Heaton
immediately behind on his right side, balding with a white beard. Both were known as for
their anti-imperialist views and as peace activists. Reproduced by permission of the National Portrait
Gallery, London.

to get out"[24] in response to the rise of the American empire, was likely a
sentiment that Heaton found quite compelling. Over time, Heaton pushed
harder and harder to break apart his call for state ownership and cheap cable
rates, on one hand, from the needs of the British military and empire, on the
other. Thus, in 1902, standing before the most important British Cable Inquiry of its time, Heaton took a solidly anti-imperialist position when he
stated "it is carrying Jingoism to an enormous extent to ask for an all-British
cable. It is cutting our own throats; [and again] . . . It is sheer madness unless
we are always in antagonism. I earnestly wish the Committee to disassociate
trading, social and commercial from war telegraphs."[25]

MAIN MOTIVES FOR GLOBAL COMMUNICATION REFORM

Unlike Heaton, Sandford Fleming was more attached to the idea and the
ideals of the British Empire. And also in contrast to Heaton, Fleming realized
(even though he often sounded utopian in outlook) that the popular All Red

Line project for world-circling cables touching solely on "British" imperial
territories could only be advanced if it furthered the strategic and military
needs of the British Empire. Indeed, a great deal of Fleming's efforts rode on
the coattails of a revival in the 1880s in Britain of a security conscious discourse on the military and strategic needs of the British Empire and, after
having gone into abeyance for over two decades after the 1859 Red Sea to India
fiasco, the reinstatement of state subsidies as inducements for the private
companies to bring their cables to the far-flung but otherwise commercially
weak regions of the British Empire, notably in Africa and the Caribbean.[26]
Thus, in this sense, there can be no mistake that military and imperial considerations were among the main drivers behind the revival of subsidies and the
thrust for state-owned cables, and it was this reality that caused Heaton to
launch his broadside against the militarization of communication. However,
while Fleming refused to follow Heaton down the anti-imperialist and internationalist path, he was not excessively militaristic in his views. While recognizing the place of military considerations, and even that his own agenda
depended on meeting such needs, Fleming continued to place the stress in his
advocacy on the role of improved cable communications as a kind of "cultural
glue" by which to bind together the citizens of the British Empire. As he put it
"no people so greatly needed speedy communications as the British, and yet,
owing to the prohibitory rates at present charged, scarcely one in a million of
our fellow citizens ever uses the existing means of telegraphy between Great
Britain, Canada, Australia, New Zealand, South Africa or India."[27]

Fleming's views were perhaps expressed best in a 1907 address by the
president and council of the Ottawa Board of Trade on the status of the Pacific
Cable. The address, written by Fleming, exudes his belief that evolutionary
social progress and improved communication advanced hand in hand. And as
markers of each, he pointed to the progress in national communications in
Britain which commenced with the advent of inland penny postage in 1840,
followed by the private development of the telegraph during the next three
decades and their eventual take-over by the British state in 1869–70 because
"the public interests demanded that they should be," and finally, to the adoption of the imperial penny post in 1898, which he referred to glowingly as "a
great onward movement in . . . civilization, and in the development of wider
national sympathy and sentiment."[28] The next step along this arc of communication and progress, in Fleming's eyes, was the development of the All
Red world-circling state-owned system of cables at dramatically reduced rates.

That task, however, Fleming knew full well, would be vehemently opposed every step of the way by the private cable industry and unlikely to be enthusiastically supported by either the tight-fisted Treasury or the General Post Office which clung steadily to the commitments that it had made to private industry on the eve of nationalization in 1870 never to compete with them in global markets.[29] For his part, Henniker Heaton sought to parlay his successes with the imperial penny post into the realm of global cable communications. His early proposals for cheap global communications were relatively modest —for example, he suggested a 12¢ rate to India and a 24¢ rate to Australia in 1891[30]—but by 1902, he was tentatively advocating penny-a-word (2¢) rates before the Balfour committee. The proposal brought a condemnatory response from the committee and scathing comments from Bright, and the penny-a-word cable rate proposal was later to meet polite demurral as impractical even from Sassoon and Fleming. Still, they all wanted cheaper rates and, with the exception of Charles Bright, believed that nationalization, or possibly internationalization, of the cable industry was the only route to these ends. Indeed, later in life, Sassoon went so far as to propose the radical idea of internationalizing control over the cable business, a position very agreeable with Heaton's own internationalism and to supportive trends elsewhere.

Notably, in their push to nationalize the British cable industry, or alternatively for competition between state-run cables and the private companies, the British and empire reformers were in line with trends in the United States, France, Germany, and Japan at the end of the nineteenth century where support for state intervention in cable communications was on the rise. In the United States, for example, Captain Squier told a House of Representatives committee in 1900 which was considering the construction of a Pacific cable that the existing high rates excluded all but the most urgent traffic, with the effect that the long cables were actually idle for much of each twenty-four hours. In his words, "And why should any cable, binding a country to her colonies, spanning an ocean and connecting two continents together, ever be permitted to be idle? Is it because these countries have nothing to say to each other? The enormous volume of the present mails effectually answers this. The world's cable plant could be duplicated many times and kept full to overflowing if the cable rates were sufficiently reduced."[31]

Squier was supported by General A. W. Greely, chief signal officer of the army, who responded to cable magnate James Scrymser's claim that private enterprise could run a Pacific cable more cheaply with the caustic comment, "Possibly it can but it means an inferior cable. What are these gentlemen in the cable business for? They are in it to make money. They say that they can run it cheaper than the Government. Very probably they can, but that means grinding the faces of American labor; it means hiring a man for less than his services are worth."[32]

Such a radical observation was rarely to be heard, even from Henniker Heaton, among the British and colonial critics. The words of Squier and Greely were also important, however, insofar that they reflected the rise of the United States as an imperial power after 1898, and in the circles in which they moved it was only right that the United States should have an imperial communication system befitting its new-found status. Yet, as Scrymser put it before the committee in 1900, cables served the needs of business, and indeed a very select few, and mostly foreign at that, so why make American taxpayers support such a clientele? The obvious answer was that it was the high rates that made cable communications a luxury rather than a necessity and, furthermore, that the goal of state ownership was precisely to change these conditions.

THE PURSUIT OF A DREAM: THE PACIFIC CABLE
AND ALL RED LINE

The one constant lacuna in the late-nineteenth-century politics of global communications was the absence of any cable from the west coast of North America to Asia. Equally prominent was the number of plans put forward to change that state of affairs. Indeed, as Robert Boyce has observed, not a year went by in the 1870s without someone proposing to connect North America and Asia by cable.[33] Sandford Fleming moved to the foreground of these initiatives by the end of that decade, with his advocacy of a Pacific Cable from the west coast of Canada to Japan, China, Hong Kong, Australia, and New Zealand. In two instances while working as an engineer with the Canadian Pacific Railway and Telegraph Company, once in 1880 and again in 1891, Fleming, with the assistance of the Privy Council of Canada, appealed to the British for surveys of the Pacific Ocean and asked that ongoing negotiations between the British and Japanese governments be expanded to include discussions of a Pacific cable.[34]

A few years after his first proposal, Fleming joined several well-connected Canadians, British, and Australians to form the Pacific Telegraph Company in 1886. The company presented itself as a public service organization that would foster trade in the Asia-Pacific region, meet the strategic needs of the British Empire, and improve people's access to long-distance communication by offering more affordable rates. In addition to having $10 million of its own capitalization, the company requested an annual subsidy of $375,000 for twenty-five years, half to be paid by Britain and the other half split between Canada, the Australian colonies, and New Zealand. However, leery of such a long-term commitment of public funds and confronted with a similar proposal apparently backed by well-heeled Australians, the Canadian government vacillated and both projects fell through.[35] Attempts to gain support from the British government similarly came to naught, but perhaps most decisive was that both James Anderson of the Anglo-American Company and John Pender of the Eastern Associated Company resolutely opposed both of these initiatives. As Pender told Anderson, "The American traffic with Asia and the Far East is more important than I anticipated and no doubt a cable from Canada to China with a reduced tariff would do the existing companies a good deal of harm and it is just possible that there is some way of finding the money with a subsidy, for example, necessary to make this line."[36]

Perhaps Anderson and Pender were able to rest content once both proposals fell through, but the fact that they had thought to stand in the way of one of Fleming's ambitions was just the kind of challenge he relished. Thus, buoyed by corporate intransigence as well as lethargy and obstruction from key elements within the British imperial bureaucracy, Fleming brought out his arsenal of arguments against the cartels and in favor of a Pacific cable which, given the support it was receiving in certain quarters, was soon transformed into the more ambitious idea of the "the All Red Line." Fleming became obsessed with the Eastern Extension Company's obstructionist tactics and its monopoly over cables to Australia and New Zealand, chastising its high profits, unreliable service, and, most of all, the fact that it had maintained rates to these areas at the astronomical rate of $2.33 a word for twenty years. And all this, he lamented, subsidized by the Australian colonies to the tune of between $3.3 million and $3.8 million during the same period.[37] Yet, under just the threat of competition, the Eastern Extension Company dropped its rates to $1.00 in 1891, and went so far as to adopt a penny-a-word rate for the news material sent between the Australian and New Zealand press. That latter change saw the news flows within the Australasian press system blossom from just under a quarter of a million words in 1891 to nearly 400,000 words four years later, with the cable company's revenues dropping by half. The cut in regular rates also caused a 150 percent rise in the volume of "ordinary messages." Clearly, then, the Pacific Cable was already having an effect before it was even laid, and this only galvanized support for it and, for

Fleming, just reinforced his belief about the importance of cheap rates to the citizens of empire. He did not see the gains as reflecting so much the partial culmination of an enduring social struggle over the means of communication, but rather as supporting his more technologically determinist belief that the reformers were merely releasing the potentials of new communication technology from the prison of hostile business and political interests. Hence his pregnant observation that

> [technological] conditions have been continually changing and the process of growth and development goes on. True, change has been met with resistance from individuals and companies and classes, but resist it who may, the law of development follows its steady course.... The telegraph practically annihilates space ... to bring the British people, so widely sundered geographically, within the same neighborhood telegraphically.[38]

Yet, alongside this remarkably powerful critique of how the potentials of new media were stunted by class interests, note again Fleming's focus on "the British people," which is why, unlike say Heaton, we see him more as a "progressive imperialist" and a modernizer. And indeed, as the Pacific Cable came closer to fruition, it faced opposition in the British House of Commons (not least from Irish parliamentarians) that it was indeed an imperialist venture into which the government had been inveigled by Canada and Australia, who would be the main beneficiaries. For its part, the British government certainly agreed that the Australian colonies, Canada, and New Zealand would be the main beneficiaries, and that accordingly a substantial part of the cost of the cable would have to fall to them. Perhaps this is why, as Heaton remarked in 1896, "all I know is that Canada is doing the lion's share of the work."[39]

The fact is that Canada *was* doing the lion's share of the work. Several conferences in 1887 kept the idea of an imperial public service communications network alive, with Fleming out front, as usual, heaping scorn on John Pender's interventions about how a Pacific cable was technically impracticable because of the "unsurveyed depths" of the Pacific.[40] By this time, as Pender knew full well, the technology, while still difficult and demanding, was well established. But it was not the technological issues that were the most important anyway. That could be seen as the Pacific Cable idea continued to languish for much of the 1880s and early 1890s, not for lack of technology but political support, with its prospects only really brightening considerably at the "mini" Colonial Conference held in Ottawa in 1894. The Ottawa Conference resulted in the first practical steps toward the "global media reformers" dream: a Naval survey of the Pacific, a published call for competitive tenders, and, finally, detailed financial studies of the cost and feasibility of the Pacific Cable. Concrete proposals were also made to extend the cable from Australia to South Africa, and less well thought out ideas bandied about on extensions to Japan, Hong Kong, the Philippines, Hawaii, and the west coast of the United States. Three years later at the full-blown Colonial Conference in London, visible opposition had melted and in its place now stood a rather solid phalanx of support for the project by the British Colonial Office, Canada, New Zealand, and among several, though not all, of the Australian colonies. Nonetheless, there was still some reticence to the idea in certain quarters. Typical was the criticism of the British Secretary of the Post Office John Lamb in 1896 that while the Pacific Cable would "cause a perfect revolution in communication . . . it would be a matter of extreme difficulty, I think without precedent, for the English government itself to . . . constitute itself a competitor with an existing commercial enterprise carried on by Citizens of the British Empire. There would be very serious questions raised, and it would be possibly extended to other forms of British enterprise."[41]

Obviously, achieving the Pacific Cable required more than just the realization of a dream and tenacious work by its backers; its success demanded a fundamental transformation in the policy, organization, and ideology of Britain's approach to global and imperial communications. Since, even at this time, the cable systems controlled by British interests, which for all intents and purposes were the interests of the Eastern Associated Company, still constituted the central node in the worldwide system that had been built up over the past forty years, embracing change on the magnitude implied by the Pacific Cable was no small matter. Indeed, because of these conditions, as we will see, the struggle to realize the Pacific Cable was far from over.

But in the meantime, and as agreed, immediately after the Ottawa Conference of 1894, the Canadian government published advertisements in leading British and Canadian newspapers calling for cable tenders and stipulated that the cable had to be ready to use by 1898. While this seemed to keep the momentum alive, there were still several continuing sources of delay. Among these, as intimated above, was the fact that not all of the Australian colonies were on board. Federation among the Australian colonies would only take place in 1901, and in the meantime it was clear that not all the future constituents were seeing eye-to-eye on the Pacific Cable. On the one side, Queensland, Victoria, and, with more diffidence, New South Wales supported the scheme from the outset. On the other, however, Western Australia, South Australia, and Tasmania did not. The first three were closest to the projected Pacific Cable's terminus and the fact that it would land in Queensland also carried some weight. There was also varying levels of support from the commercial and newspaper groups in each of these states, with Queensland particularly keen to enhance its access to the ever-expanding global communication infrastructure of the late nineteenth century. Given state differences in population and wealth, the Pacific Cable's future economic viability, as well as its imperial role, rested primarily on the support of the governments, commercial classes, and press of the two most populous states, Victoria and New South Wales. This support it would need, since the three hold-out states— Western Australia, South Australia, and Tasmania—saw the Pacific Cable either as irrelevant to them, or, in the case of South Australia, as more of a threat than an opportunity. Such views, in turn, were fostered by the Eastern Extension Company's rear guard but powerful campaign between 1890 and 1902 to place itself in an impregnable position in the Australian market. One cornerstone of those efforts, as we have seen, were dramatic cuts in the price for "ordinary" and "press" rates with the business groups of Australia as well as the Australasian Press Association as major beneficiaries. And the fact that South Australia had spent substantial money on a telegraph line to Port Darwin in order to connect with the Eastern Extension Company's cables also helps to account for its tendency to see the Pacific Cable as a threat to its interests.[42]

The very substantial rate reductions were a crucial factor in the politics of global communication being waged at this time in Australasia: on one side, Pacific Cable supporters saw them as proof of the benefits that competition would bring, and on the other side, elements of the population leaning toward private enterprise were quite satisfied with the Eastern Extension Company's newfound willingness to work more closely with state governments. And if that humbled corporate attitude was evident in the new cheaper rates, which did indeed cut deeply into the company's revenues, it was even more conspicuously so in the Eastern Extension Company's unusual gesture of laying, without government subsidy, a cable from the Cape of Good Hope in South Africa to Perth in Western Australia in 1902. That move offered both more efficient cable service and indeed looked like a very generous contribution to the goal of the All Red Line. Going yet further, the company pressed the state governments to allow it to open offices and deal directly with customers in Perth, Adelaide, Melbourne, and Sydney, rather than, as was then the case both in the Australian colonies and more generally (the United States, Canada, and Britain, excepted), acquiring its business through the colonial post offices.[43] The governments of Western Australia and South Australia accepted these proposals immediately, since they had already made it clear that they had no intention of participating in the Pacific Cable scheme. The government of Victoria rejected the proposals and New South Wales was initially uncertain, but by early 1900 New South Wales, followed soon by Victoria, permitted the Eastern Extension Company to open offices in return for cheaper rates. After federation, the Australian Commonwealth felt legally obliged to take over these agreements for a period of ten years, a fact that undoubtedly jeopardized the economic prospects of the Pacific Cable.

While there is little doubt that local benefits were already beginning to flow into the Australasian communication and media systems, there is equally no doubt that the Eastern Extension Company directors believed that they were being badly treated by their critics. As hints of an opposition cable began to surface, as early as 1887 John Pender complained that Australians did not understand the complexities of the global communication business and strongly defended his company against charges of monopoly: "We believe that we have now, not a monopoly. But such a widespread system that it is now impossible for private enterprise to compete with us, because, in the first place, we can do the work cheaper than they can, and in the second place, there is no necessity for any opposition line, because we can do a great many more times the work than is required of us at present."[44] Pender followed this up with a sarcastic bite at an early statement from Henniker Heaton in the *Pall Mall Gazette* that a one shilling (25¢) tariff from Britain to Australia would pay well: "The gentleman who wrote that article, and is posing now as a great apostle of telegraph reform . . . [that proposal] I thought scarcely worth taking notice of, but still when a man writes who is ignorant of the subject, people who do not understand the question are apt to believe what he says. We who know the whole of the circumstances know how absurd the proposition is."[45]

In addition to the Eastern Extension Company the reformers also faced practical as well as verbal obstruction put in their way by the British General Post Office. This could be seen especially clearly when, right at the beginning of a visit to the Eastern Australian states by the Canadian trade minister Mackenzie Bowell and Fleming in late 1893, designed to drum up support for the Pacific Cable, the Post Office circulated a scathing report on the prospects for the Pacific Cable. Adding insult to injury, right in the midst of their visit, the Colonial Office approved the renewal of the Eastern Extension Company's monopolistic concession in Hong Kong for a further twenty-five years, a move that weakened any prospects that the Pacific Cable might have had for extensions into Asia. And for good measure, the British Admiralty petulantly continued to refuse to survey the Pacific Ocean and, for reasons of its own, threw more cold water on the scheme by pronouncing the Pacific Cable to be technically difficult, militarily insignificant, and an unjust interference with the private business interests of the Eastern Extension Company.[46] Thus, with opposition to the company in several of the Australian colonies silenced by a new cable and cheaper rates, and an apparent long-term phalanx of opposition from within the British state, the prospects for the Pacific Cable in the mid-1890s looked dim, indeed.

BREAKTHROUGH: THE PACIFIC CABLE AND THE REFORMERS' DREAM REALIZED

As usual, Fleming remained undeterred by the opposition, and immediately after visiting Australia traveled to England to collaborate with the Canadian High Commissioner Sir Charles Tupper to organize the 1894 colonial conference. There were also other changes afoot that likewise signaled that the Pacific Cable project was not dead but rather ensnared in the politics of global and imperial communication, the outcome of which would determine its fate.

One area where a residue of hope was kept alive, somewhat surprisingly, was in the pages of the business press, where there was a discernibly more open attitude in the 1890s than earlier toward the complex issues at stake. *The Economist*, for example, already a doyen of the business press in Britain and abroad, published several articles critical of the Eastern Extension Company and other members of cable cartels, charging that they were "watering" their stock to hide excess profits and to make unreasonable rates appear respectable.[47] That charge, in the past, had been the preserve of the "global media reformers" as well as of the critics of big business and the trusts in the United States such as Frank Parsons, to whom we referred earlier in this chapter. The fact that these ideas had now migrated into the pages of the mainstream business press was a critical index of just how acceptable the "critique of global communication" had become. William Hay, now marquis of Tweeddale, expressed hurt corporate feelings in 1896 to the effect that "the chief arguments adduced in favor of the companies have been more or less overlooked."[48] Actually, whereas it had been nearly impossible for critics to get a hearing in the press as late as 1890, the advocates of the Pacific Cable and defenders of the Eastern Extension Company subsequently skirmished in *The Times* (London) on several occasions, notably just before the Ottawa Colonial Conference of 1894, and again prior to the opening of the Pacific Cable in 1902.[49]

The changes were also reflected deep in the corridors of power in the British Empire. This was especially noticeable after the election of a new Conservative government in 1895 and Joseph Chamberlain's consequent appointment as secretary of state for the colonies. Chamberlain took a broad view of government's role in the provision of public services and cited cable communications alongside life assurance, banking, and the distribution of parcels as examples of such services and areas where competition by the government with private enterprise could be beneficial. The new imperialism and its progressive underpinnings also allowed him to cast the global media, both in terms of network infrastructures and the British and imperial news that flowed over them, as a kind of "cultural glue," similar to the ideas long pronounced by Fleming, and as agencies for the promotion of greater empire consolidation, and even federation. In short order, Chamberlain made it clear that he saw no insurmountable difficulties standing in the way of the Pacific Cable.[50]

Chamberlain convened a Pacific Cable Committee in 1896 to study and report on the feasibility of the project as well as the model of ownership and financing which should be adopted. On each of their seven measures—technical practicality, route, cost, revenues, ownership, management, and contract details—the committee supported the Pacific Cable. In fact, so great had the tide turned in favor of the cable that the committee's deliberations showed elements of naiveté and excessive optimism. Thus, it accepted uncritically Fleming's wildly sanguine projections of vast increases in the volume of messages that would flow from reduced rates, the assumption that the Pacific Cable would immediately gain 40 to 50 percent of all traffic, and, finally, that it would be profitable from the outset.[51]

Optimism aside, severe official tensions surrounding the committee's report indicated that the twenty-year impasse over the Pacific Cable had yet to be surmounted. The most obvious indication that something was amiss was that the publication of the report was delayed over two years "for some occult reason," to quote a bewildered marquis of Tweeddale.[52] Although almost everyone, including the cable companies, were "kept in the dark,"[53] the delay was due to the fact that "the Treasury and Chancellor of the Exchequer opposed . . . the [Pacific] cable root and branch."[54] The Treasury's efforts to suppress the report were not surprising, given the notoriously parsimonious habits displayed by both it and the Foreign Office. The fact that the Eastern Associated Company could play on the importance of its contribution to

British imperial strategy over the years, and not least as the race for colonies heated up, was also not lost on those British government officials who were quite happy to have its representatives as formal members of the Colonial Defense Committee. Thus, in the eyes of the Treasury, the company's contributions deserved special consideration, as the following statement from a Treasury official illustrates: "These companies represent a real British interest, and one entitled to great consideration from the Imperial Government. . . . In times of emergency they have always been ready to render the Government any services which were in their power . . . It is therefore a matter of no small importance that the Government should continue to maintain friendly relations with them. But this could hardly be expected if the Government were to take an active part in the establishment of a cable in direct competition with the Eastern Company's system."[55]

It helped, of course, that the Eastern Associated Company consistently retained a remarkable combination of influential aristocrats, businessmen, politicians, and talented engineers on its various boards of directors. The fact that its tentacles reached into the "power elites" in regions where its affiliates were strong was also of considerable importance, especially in Britain's informal empire in South America. By comparison, the reformers, especially colonials, were definitely "outside the club." Inside, Arthur Balfour, later a director of Eastern's South American subsidiary, the Western Telegraph Company, was First Lord of the Treasury during the negotiations which followed the Pacific Cable committee report. The 1902 Inter-Departmental Committee which bears his name also pointed in its report to the tangible benefits the company received in return for its loyal service: roughly 80 percent of the $15 million in subsidies paid by the imperial and colonial governments before 1900 went to the Eastern Associated Company, with the biggest recipient in the group being its African subsidiaries (approximately 54 percent of the total) and the Eastern Extension Company, which served Asian markets beyond India (approximately 25 percent of the total). Several smaller Caribbean companies that were only loosely connected to the Pender group, if at all, received the balance. For some British politicians, such amounts seemed quite enough, without adding on the cost of state-owned cables, as, for example, the Chancellor of the Exchequer, Sir Michael Hicks Beach who complained, in 1902, that the fetish for the All Red Cable revealed a growing tendency for strategic interests to be defined so expansively that there was no end in sight to the demands potentially placed on the Treasury.[56]

While the Treasury worried about the impact of a Pacific cable on public

16 Lord Balfour, Michael Hicks Beach, and Joseph Chamberlain. This sketch by Sir Francis Car-
    ruthers Gould, circa 1902, shows (from left to right) Arthur Balfour, Michael Hicks Beach, and
    Joseph Chamberlain sitting on the government front bench in the British House of Commons.
    The contrasting attitudes toward state-owned cables between Balfour and Chamberlain were
    symptomatic of a wider split within the reigning Conservative Party which ultimately lost them
    an election. It might also, by a flight of imagination, account for Balfour's dour expression.
    Reproduced by permission of the National Portrait Gallery, London.

finances, both it and the Post Office also feared that a direct link between Asia
and North America could bend world communication in favor of the United
States. Secretary of the Post Office John Lamb aired these concerns explicitly
in his evidence to the Pacific Cable Committee in 1896. According to Lamb:

> At present if you look at a map of the world you will find that practically all the
> telegraphs of the world are centred on England; but under this arrangement another
> centre, a rival centre, would be established on the west coast of America . . . I hold the
> opinion that it is important to this country in every way that its position as the centre
> of telegraph communication should be maintained. Under present arrangements
> the British merchant should get his information first, and he ought to get it cheaper
> than the American merchant . . . It seems to be that the mother country is here asked
> to contribute to a scheme which will damage her position of telegraphic communi-
> cation and which will at the same time facilitate her rivals in trade[57]

However, in face of an announcement from President McKinley, in 1899, that
the United States should no longer rely on foreign cables, specifically those of

the Eastern Extension Company, to meet the needs of its own growing imperium in the Asia-Pacific region, Lamb's objection to the All Red Pacific Cable soon began to look almost unpatriotic.

Matters were advanced considerably once Joseph Chamberlain brought the Treasury around to recognizing that the Eastern Extension Company had a "practical monopoly," which is exactly the interpretation to be given to Pender's 1887 defense, noted earlier, of his corporate empire against the charge of monopoly control. In the end, it was this position that prevailed, and the Pacific Cable was laid from the west coast of Canada to Australia and New Zealand, with service opening to the public on December 7, 1902. As a testament to the vital role of Sandford Fleming in bringing it about, he had the honor of sending the first message around the world.

The Pacific Cable was a joint financial arrangement of Australia, Britain, New Zealand, and Canada. Australia and New Zealand each paid subsidies of three-ninths and one-ninth, respectively, of the capital costs and sinking fund, and Canada and Britain shared the remaining five-ninths, to an annual cost of about $385,000, and an additional $175,000 was added to a reserve fund for ultimate cable replacement. The arrangement was truly unique, never having been tried before and never would again. Some benefits immediately followed its initiation, most notably greater correspondence between Australasia, North America, and Britain. This was no doubt due to the fact that the Pacific Cable itself charged a reduced rate of 75¢ per word on ordinary messages, down from the $1.18 which had prevailed before the Eastern Extension reduction to $1.00 in 1902, a reduction which that company was, however, bound to follow. But despite the reformers' celebrations, this price reduction, combined with the threat but not the actuality of further price competition, was the only significant benefit, prior to World War I, which the Pacific Cable provided. This point is followed up in the next chapter.

THE OTHER PACIFIC CABLE: CORPORATE IDENTITY, CORPORATE COLLUSION, AND ACCESS TO ASIA

We have noted previously that the race to link Europe to Asia by telegraph and cable resulted in the Danish-based Great Northern Company and Eastern Extension Company forming a cartel before their cables were even opened for business. And as we have also shown, most governments and businesses willingly relied on their systems, regular criticism of high rates and poor service notwithstanding. But as the imperial and commercial interests of the Netherlands, Germany, Japan, and the United States began to swell at the beginning of the twentieth century, particularly in Asia, each began to chart a more independent course and, as part of that course, encouraged the development of cable systems under the control of their own nationals. In 1903, Germany and Holland established the German-Dutch Telegraph Company, a relatively small regional cable system that met both countries' expanding imperial and business interests in the Chinese province of Shan Dong and Indonesia, respectively, while still relying on the British and Danish companies' networks for their long-haul connections to Europe. Three years later Japan, with rapidly enlarging imperial designs of its own, also began laying its own cables and quickly signed deals with the Eastern Extension Company and Great Northern Company which spelled out its share of revenues and role in the Asian cable communication system. Thus, while the rising tide of nationalism and imperialism was unmistakable, the long-standing private structures of control were still firmly in place, regulating access to foreign markets and managing relations among corporate and state actors alike. And so it was that Japan joined the cartel in 1906, just as had the new German-Dutch Telegraph Company two years before that.[58]

Similar dynamics were in play with respect to the role of the United States in the region. Thus, as noted, President McKinley announced in 1899 that it was no longer acceptable to rely on foreign cables in accessing its Asian Pacific colonies or the vast potential markets of Japan and China, a remark that came amid a flurry of activity in Congress, with at least thirty-five bills to promote an American Pacific Cable being laid before Congress between 1896 and 1902.[59] In many ways, this was the American version of the All Red Pacific Cable campaigns waged concurrently in Britain. Indeed, both campaigns seemed to be mutually reinforcing. Similar substantive issues were covered as well, with U.S. participants to the debates focusing on whether their prospective cable should be state owned or subsidized or left solely to the private sector. Supporters of state ownership, especially within the U.S. Navy, spoke of the strategic military advantages and reasonable rates that would follow from it. In contrast, those opposed—including many, but not all, of the American cable firms—warned of the difficulties of linking a state system to the private lines of the Eastern Extension Company, of the government's lack of experience with handling such a venture, and the danger that a state-owned cable could not land, for reasons of sovereignty, in China and Japan.[60] Ultimately, despite a remarkable volume of congressional paper and the launching of several potentially competitive companies, it was all "whistling down the wind." John Mackay's recently formed Commercial Pacific Cable Company, a subsidiary of his Commercial Cable Company and the Postal Telegraph Company, which was, in turn, Western Union's main rival in the United States, proceeded unilaterally to enter into an agreement with the Telegraph Construction and Maintenance Company to manufacture and lay a cable between San Francisco and Honolulu and Guam, with proposed future extensions to the Philippines, China, and Japan.

The Commercial Pacific Company's vice president, George Ward, told a 1902 U.S. congressional inquiry that his company desired no subsidies and that the new venture was set up entirely "for the benefit of the U.S. and to give a reasonable rate."[61] In opposing the Mackay interests, James Scrymser—who, besides being doyen of American cable interests in Latin America, was president of the Pacific Cable Company of New York, a rival company set up with the purpose of achieving a subsidized cable to Asia—told the committee that he had been trying to negotiate with the foreign cable companies for years to gain access to Asian markets, but found the task impossible. In the end, Scrymser came out in favor of a state-owned Pacific Cable rather than face the prospect of the Commercial Company getting into Asian markets first and thereby preempting his own plans. Western Union felt much the same way, claimed Scrymser, and even more so, since a unified Commercial Cable Company system with operations spanning the Pacific, the United States, and the Atlantic would pose a grave threat to Western Union business. Given the Commercial Cable Company's run of luck in the transatlantic economies between 1888 and 1900, where it had assembled a vast network of alliances with several British, French, and German firms (see chapter 2), the threat was real enough since it could determine whether that company or Western Union was to be the major American global communications powerhouse. And for that reason, anything else was to be preferred: even the last resort of a state-owned Pacific Cable.

The fact that so much was at stake meant that the committee proceedings were clouded with recrimination and suspicion from the beginning. Indeed, the element of suspicion hung heavily from the outset due to the strongly held belief, at least within the industry, that whichever company ultimately gained access to the Asian communication markets, it would only be able to do so after coming to terms with the Eastern Extension Company and Great Northern Company, given the ironclad monopolistic concessions that these companies held, singly and jointly, in the Philippines, Hong Kong, China, Korea, and Japan. In the context of mounting nationalism and imperial rivalry, the specific concern was that access to Asia would be gained only at the expense of relinquishing control to foreign interests, namely, the aforementioned British- and Danish-based companies. And as the proceedings progressed, each of the two main contenders—the Pacific Cable Company of New York, backed not just by James Scrymser but also J. P. Morgan and Western Union, versus the Commercial Pacific Cable Company, backed by John Mackay and George Ward—sought to outdo the other by wrapping themselves in the American flag and extolling their corporate virtues and national purity: their ability to serve the needs of the American empire and their willingness to help prize open Asian markets for U.S. commerce.[62] But it was George Ward, a British citizen, in particular who faced tough questions from the committee chair- man, William Hepburn, who asked pointedly whether Mackay did not hold the majority of shares of the Commercial Pacific Company "in the interests of the . . . English stockholders."[63] His denied the charge, but Hepburn's sus- picions were actually dead on the mark.

In actuality, the majority of the directors and shares in the newly laid "U.S." Pacific Cable were held by the Eastern Associated Company (50 percent) and the Great Northern Company (25 percent), with the balance falling to the Commercial Cable Company.[64] So much for George Ward's candied words about the primacy of "American interests"! The threat of competition had been neutralized with the aid of American entrepreneurs. The underlying deal was not exposed until nearly two decades later, and indeed all but one copy of the agreement were destroyed, and that copy was kept in a box with six locks and deposited at a Safe Depository in London in 1905 with instructions that the box was never to be opened, except by a court, without the written orders of each company.[65] Though worthy of a Sherlock Holmes scenario, such caution was deemed necessary in light of the scheming which it kept hidden. Thus, an especially cynical Eastern Extension Company memo observed, in 1905, that "it is believed . . . that the Commercial Pacific Company is earning less than 1 percent dividend; consequently much sympathy has been felt for the company, and a readiness shewn by the various Government Departments [in the United States] to assist it in overcoming its difficulties in Midway, Guam, etc."[66] The memo then advised that if it was known how well the company was actually doing, there would be a danger of competition, de- mands for cheaper rates, and even nationalization, the three biggest fears of the cable companies. Hence arrangements were made for the Eastern Exten- sion Company to pay the Commercial Pacific Company investors the differ- ence between direct earnings and a percentage of the joint purse accruing to

MAP 10   The Asian cable cartel—network routes, circa 1906

the company so that the facade of low revenues could be maintained. In 1912, *The Times* was to comment on the secrecy of the British cable cartels, and no better example could be found than this.[67]

The Commercial Pacific Company's cable reached the Philippines in 1904 and, after the firm reached agreement with other cartel members in 1906, new cables were laid from the Philippines to China and Japan, the former to Shanghai and the latter to the Bonin islands east of the Japanese mainland. In turn, Japan laid its own cable to Bonin from Yokohama in the same year, thereby providing a direct cable route from the United States to Japan.[68] Thus, between 1903 and 1906, the Asian cable communication system had undergone its largest transformation since the 1870s. Amid the rising nationalism and imperialism that were propelling these changes, one thing that did not change was the British and Danish companies' impenetrable monopolistic rights. And this they continued to use to cartelize Asian communication markets, although whereas in the last quarter of the nineteenth century there were only three ventures directly involved, now there were six: the Eastern Extension, Great Northern, Imperial Chinese Telegraph Administration, Imperial Japanese Telegraph Administration, the German-Dutch Company, and the Commercial Pacific Company. Hence the six locks, and accompanying keys, on the London strong box.

CABLE CARTELS AND IMPERIAL JAPAN

Given Japan's growing role in Asia it is interesting to briefly explore how that country dealt with the European and American cable companies. And through that exploration, one thing is clear: Japan had learned much following its initial concession with the Great Northern Company in 1870. The initial reaction of European diplomats and corporations to Japanese society and culture was similar to the view taken of China, and typified both by the unequal treaties of 1858, and the patronizing attitude of such long-term officials as the British consul, Sir Henry Parkes, who expressed the belief that the Japanese were incapable of running an efficient postal or telegraph service. Such an attitude was swiftly changed when, during the Meiji reforms, the Japanese government developed an efficient penny-post service and, in the early 1870s, established telegraph communications between Tokyo and Yokohama and Nagasaki and Yokohama, specifically in order to forestall moves by the Great Northern Company which, having obtained its cable concession, wanted to move inland, as it had done in China.[69]

The Japanese, by contrast, had no intention of allowing such foreign control, so although the members of the Iwakura fact-finding mission, which was sent to Europe and America in 1872–73 partly to seek revision of the unequal treaties of 1858, marveled at the amount of business conducted by the London telegraph office, the Japanese government's response was characteristic, namely, to use British engineers to help develop the rapidly expanding telegraph system, and to lay cables which, for example, linked the northern island of Hokkaido with the mainland, but also to focus heavily on training its own specialists. Thus, a school for telegraph operators was established, some students sent overseas for training as engineers, and advanced courses in cable and telegraph technology were offered by the Imperial College of Engineering beginning in the 1870s, initially with help from British teaching staff.[70] However, unlike the case in China, foreigners were useful tools, paid well but not extravagantly, and not allowed into positions of real authority.

One should not, however, appear too sanguine about Japan's ability to maintain its sovereignty in the face of the foreign cable companies. A published British commercial intelligence report of 1908 observed that the long concession held by the Great Northern Company in Japan was not likely to be easily renewed, since "at that time [of the original agreement] Japanese statesmen were not as well versed in the burdens and restrictions connected with the monopolistic possibilities of a cable company as they are now."[71] Such lessons had been learned early on, for example, during Sino-Japanese rivalry in Korea in the early 1880s when Japan was anxious to lay a cable across the Korean strait to connect with its settlement at Pusan. However, since the Japanese government lacked the technology and finances to carry out the task on its own, it asked the Great Northern Company to undertake the task in 1884 in return for a twenty-year extension of its monopoly over Japan's international communications. After asserting its own twenty-five year monopoly over the Pusan telegraph connection, Japan achieved the important ability to send Japanese-language telegrams outside the Japanese islands, a vital element in its growing presence in Korea which was ultimately annexed in 1910. But, as Daqing Yang notes, the outcome of the deal with the Great Northern Company was "continued foreign dominance of Japan's overseas communication for many more decades, with serious economic and strategic consequences."[72] Indeed, because of these consequences, and for reasons of national security and prestige, the Japanese government decided to build its own cable system. The result, notes Yang, was Japan's rise as a real "player in submarine telegraph communications,"[73] albeit one which still had constantly to play a tough game of diplomatic chess with the Great Northern Company.

Thus, by the 1890s, Japan was becoming an increasingly powerful and rapidly modernizing country which could not be treated with the diplomatic contempt often shown by the international community to successive Chinese governments, hence the abandonment of extraterritoriality by the European powers in certain Japanese treaty ports in 1899 in return for an end to Japanese government prohibitions which had prevented foreigners from living or owning property outside these ports. Despite nationalist fears that the repeal would allow "foreign insects" to culturally poison the country,[74] the general official Japanese attitude in the early twentieth century was that the Europeans and Americans were imperialists to be used as needed, and this included the United States which, in 1899, had brought its first proposal to Japan for an "Open Door" allowing all powers equal access to China.[75] This proposal was looked on favorably by the Japanese in light of their movement into the northeast area of China known as Manchuria, following the Russo-Japanese war of 1904–5. The Japanese government also looked favorably on moves by the Commercial Pacific Company which resulted in direct cable connections with the United States in 1906. But, far more sensitively, Japan had laid a cable to Lushun (Port Arthur) in Manchuria, captured from the Russians, and asserted control over the Manchurian end of a telegraph line which reached well into Chinese territory. This gave Japan direct telegraph connections with China outside of the control of the Great Northern Company and its exclusive concession, as well as a more direct route to Europe over the telegraph systems of Russia and central Europe, and at roughly half the cost of current rates paid to the cartel.[76]

The complex negotiations which followed and some of their consequences are picked up again in chapters 6 and 9. However, the point for now is that Japan was by this time (circa 1900–06) taking a far bolder approach to the European and American cable companies and to meeting its own rapidly expanding industrial, commercial, and imperial needs in Asia. Japan had much experience with monopolies, having built its economic and technological revolution around a series of capitalist monopolies, or zaibatsu. Amid all the calls for cable reform during this period, and the exploitation of foreign concessions in many countries by the cable companies, the Japanese succeeded in preventing their worst excesses in their country while embarking on imperial ventures in Korea and China. Japan, with its increasingly educated workforce and substantial newspaper-reading public, was a force to be reckoned with, increasingly to trade with, and its national policy on telegraphs and cables ultimately made it more of a partner than an exploited concession-granter in the global communication business.[77]

THE END OF A STAGE: THE BALFOUR REPORT
AND A PAUSE IN REFORM

In its major report on the international telegraph business in 1912, the London *Times* noted that the history of cable communications since 1902 had fallen into two phases. The termination of the first phase, following the laying of the imperial Pacific Cable, witnessed a "certain falling of interest in the subject, coupled with a considerable hardening, at any rate in British imperial circles, of the feeling against state ownership."[78] It also coincided with the earlier-mentioned publication of the two reports on cable communications in

1902 by the Balfour committee inquiry on cable communications, which was chaired by the First Lord of the Treasury, Arthur Balfour, who was shortly to become prime minister. The reports focused a good deal on the military dimensions of cable communications, but they are first and foremost a treasure trove of insights into contemporary political attitudes towards the cable communications business. And in that regard, their most important consequence lay in their vindication of the privately owned British cable industry and the rates which they charged the cable-using public.

Although many reformers had the opportunity to present briefs to the committee, their arguments were rather cavalierly dismissed by committee members who consisted, besides Balfour, of a small number of aristocrats and military men. The *First Report* did recommend that the complex cartel governing communications between Europe and India be revised, since the German and Russian stakes in the Indo-European cartel gave them, in the committee's view, too much influence over British communications policy and tariffs. However, these recommendations were more a sign of the fractures in existing nineteenth-century patterns of cooperation than a commitment to improving long-distance communication. Ultimately, the *Second Report* assumed an ideological commitment to laissez-faire notions which leaned heavily toward the interests of the companies. Thus, early on, this report noted that "there is no widespread feeling of dissatisfaction with the present state of British cable enterprise, and that such dissatisfaction as exists is of somewhat vague and unpractical character."[79] The committee also announced that it was strongly opposed to the general purchase or subsidization of private cables by the state and accepted the private companies' view that cables catered to a "special class," the upshot being that any attempt to lower rates so as to broaden that class would be inappropriate.

Indeed, the committee support for cheap rates was thin indeed and its treatment of the proponents of this movement, notably Henniker Heaton, was almost brutal. The committee was undoubtedly correct that his suggestion for an international penny-a-word rate was impractical, but it revealed its disdain for the proponents of "cheap rates" when it claimed that "the advocate of the scheme [Henniker Heaton] broke down altogether when cross-examined on the point."[80] This egregious comment suggested that truly conservative souls found Heaton's constant publicity for cheap rates and trenchant criticisms of the cable monopolies to be highly irritating. On top of this, the committee's observation that the importance of all-British routes was overrated (a point where they were actually in agreement with Heaton), also cast the nascent Pacific Cable and other such moves that the reformers might have had in mind in a negative light. Yet the Pacific Cable was a fact, and one element of the next chapter is an analysis of its shaky progress during the following twelve years as well as its links to a broader set of efforts designed to transform the global media system more generally.

6

The Politics of Global Media Reform II, 1906–16: Rivalry and Managed Competition in the Age of Empire(s) and Social Reform

> The Tyrant who has sundered
> colonies from Fatherland,
> sons and daughters from mothers,
> more than doth the mighty sea,
> the mountains and the rivers.
> Who has brought the voices of love
> and of the law to silence.
> Be silent, passer-by!
>
> —J. HENNIKER HEATON,
> "Epitaph to a Cable King," in
> "The Imperial Conference and
> Imperial Communications," 1911

The changes covered in this chapter on the global media system between 1906 and 1916 were decisive and shaped the evolution of global communications for the next quarter of a century. Never before were the prospects for competition more apparent and, at least partially, realized. Nor had the interdependence between the networks of communication (the network industries) and news organizations (the content industries) that relied upon them been tighter. Illuminating this interdependence, Walter Lippmann observed later that among the main determinants of foreign news coverage—the allocation of resources, the routines and ethics of professional journalists, reliance on official sources, and so forth—was the high cost of communications. As Lippmann put it, "The real censorship on the wires is the cost of transmission. This *in itself* is enough to limit any expansive competition or any significant independence."[1]

Lippmann's comments highlight this chapter's focus on the role of news organizations, alongside Fleming, Heaton, Sassoon, and others, in the revival of the reform movement and their push for cheaper transatlantic tariffs and a

British state-owned Atlantic cable during the period from 1906 to 1916. The chapter also explores some of the ties between the global media reform movement and the broader social politics of the times. In particular, we connect our work with what Daniel Rodgers called "the new social politics [which] from its first stirrings in the 1890's . . . was to emerge as a powerful political force by the 1910's, with representatives in every capital in the North Atlantic world."[2] According to Rodgers, the "new politics" associated with the rise of regulated markets and the early stirrings of the welfare state shook every political system within the Atlantic economy in the first two decades of the twentieth century. We also draw on Michael Sklar's analysis of capitalist restructuring in the United States during the emergence of the progressive era (1890–1920).[3] Together, the idea of a new social politics and the progressive era in the United States helps us to understand how the reform movement and shifts in ownership and control, notably over the transatlantic cable system, fit into the broader dynamics of corporate consolidation, administrative rationalization, and government regulation which defined the times.

And while these processes were strongest in the North Atlantic region, they also penetrated South America as a source of conflict and of ideas and ideals about modernization, the shape of revolutionary politics, and the language of democracy. This factor obliges us, in the second half of the chapter, to step back in time somewhat to the 1890s and to address the formation of a wider transatlantic economy, as defined by O'Rourke and Williamson,[4] and the role of cable communications systems and the news media, therein. While the main South American economies (Brazil, Argentina, Uruguay, and Chile) more often than not offered large and stable markets, during the early 1890s they were convulsed by economic and political crises. During economic booms, the buildup of communications systems peaked; conversely, when such booms collapsed, those same efforts ground to a halt. In addition, then, as now, there was nothing more attractive than coups and crises to gain the attention of the global news media. Combined with the global news agencies' desire for new markets and the emergence of wealthy South American newspapers, these factors raised South America's status in the global news system.

Our last focus is on the intensifying seeds of conflict that could be seen during the years covered in this chapter. This was embodied in the main maritime nations' stress on the naval arms race, the security of communications in time of war, and the consolidation of the scramble for colonies, especially among three new imperial powers: the United States, Germany, and Japan. The vast expansion of Japan's industrial and imperial interests in Asia, and China in particular, are the focus of the last section of this chapter. And those events, we will see, were bound up not only with Japan's rise as an imperial and commercial power, but also with growing concerns in the United States about the format of its own foreign policy, not least with respect to communications companies, in Asia generally and China particularly.

THE IMPERIAL PACIFIC CABLE—PUBLIC SERVICE COMMUNICATIONS NETWORK FOR THE EMPIRE

The emergence of the imperial Pacific Cable in 1902 carried with it the hopes of the progressive imperialists who had pushed long and hard for its realization. Yet, from the outset, it was also clear that such hopes had been wildly optimistic and, furthermore, that it would be some time yet before the "public service" remit that had been attached to the cable would be realized. An early indication of hard reality occurred with the appointment of the first director, Sir Spencer Walpole (1902–07) to the cable's governing board. Walpole, given his laissez-faire credentials, seemed an odd choice for the chief executive officer of a state-run organization since from the outset he saw the Pacific Cable system more as a supplement to the existing operations of the Eastern Extension Company than a true competitor.

Added to the conservative administration of Walpole was the disappointment felt by the reformers when the bright future forecast for the Pacific Cable had to be dimmed. As a 1911 report by the Pacific Cable Board noted, it was obtaining less than two-fifths of the total traffic anticipated by the Pacific Cable Committee in 1896.[5] As table 7 shows, the volume of words transmitted by the Pacific Cable rose dramatically between its first full year of business and 1910–11. However, between Australasia and Britain between 1903 and 1908, the Eastern Extension Company continued to dominate the "ordinary" cable category, while receiving between 70 percent and 75 percent of all press cables sent to Britain. For the Pacific Cable, in contrast, government messages remained steady at about 9 percent of the total, and press messages were very slim pickings until 1909–10. The table also underscores the board's overwhelming reliance on "ordinary" messages, which were mostly commercial. Only the Australasian governments committed themselves to using the imperial public service communication system extensively. They sent between 83 percent and 95 percent of their messages "via Pacific," and logically so, since they partly owned the system.[6]

The main outcome of such limited business was that, up to World War I,

TABLE 7   Numbers and Percentages of Messages to and from Australasia by Pacific Cable,
           Selected Years, 1902–17

YEAR	ORDINARY (%)	GOVERNMENT (%)	PRESS (%)	DEFERRED ORDINARY (%)	DEFERRED PRESS (%)	OTHER REDUCED (%)	TOTAL NUMBER OF WORDS (000s)
1902–3	88.6	10.6	0.8	—	—	—	228
1906–7	89.5	8.9	1.6	—	—	—	1,129
1910–11	65.7	9.0	25.2	—	—	—	1,849
1913–14	42.6	6.7	5.2	18.0	16.9	10.5	3,118
1916–17	8.9	15.9	6.2	30.5	1.9	36.6	8,769

Sources: Britain, Pacific Cable Act, 1927, 10, unnumbered table
Note: Covers paid messages between Australia and New Zealand and Britain, Canada, the United States, Europe,
and other unstated destinations.

the Pacific Cable Board ran annual deficits between $235,000 and $435,000 and, as such, was a burden on the treasuries of its sponsors: Australia, Britain, Canada, and New Zealand. The Canadian government alone paid just under $1 million toward covering the cable's costs between 1902–14, though they did benefit from a reduction on ordinary messages of almost 50 percent.[7]

Several factors contributed to the imperial Pacific Cable's problems.[8] Two factors, however, seem particularly important: first, the success of the Eastern Extension Company in keeping its Australian customers and, second, the poor performance of the Canadian Pacific Telegraph Company as the intermediary North American link between Europe and Australasia. As for the first point, one of the Eastern Extension Company's largest customers was the press, and the fact that it had a long-standing alliance with Reuters and the Australian Press Association was especially important since it ensured that all Reuters' news flowed over the Eastern Extension Company's wires between London and Australia. In fact, the 1909 Australian Report of the Select Committee on Press Cable Services found that the alliance constituted a de facto news monopoly, a factor that only confirmed the suspicion among Pacific Cable backers that Reuters and the Australian press were discriminating against its services in favor of the Eastern Extension Company. Parallel conditions applied in New Zealand, largely because the New Zealand press depended on the Australian Press Association.[9]

The report of the committee led to a loosening of the Australian news cartel in 1910. The result was that the volume of news material went from under 2 percent of all messages carried by the Pacific Cable in 1906–07 to over one-quarter immediately after the findings of the Australian news cartel inquiry were published (1910–11), though there was a subsequent move to cheaper "deferred" press messages. Yet, this was not the end of the Pacific Cable's problems. And one matter that stood out in this regard was the fact that, as the Canadian commercial agent in Melbourne, D. W. Ross, noted in 1909, the Eastern Extension Company "had not been slow to . . . to deprecate the nationalisation of cables—pointing out that such enterprises cannot be carried out to perfection under State control."[10] This deprecation made the problems of the trans-Canada connection very sensitive. Thus, as Ross earlier informed the Canadian deputy minister of trade and commerce, the trans-Canadian transmission of "via Pacific" messages was "the weakest chain in the link [sic]—and I regret to say a very weak chain. . . ."[11] As he explained further, the Canadian Pacific Company's telegraph operators were not trained to deal with the codes and ciphered messages which comprised nearly all of the Pacific Cable business. Ross concluded by noting that in the keen competition for traffic, merchants and traders "generally cannot sacrifice business advantage for sentiment or loyalty to a cause."[12]

The criticisms of the Canadian Pacific Company also helped to fuel a wider movement to nationalize the Canadian telegraphs. And unsurprisingly, Sandford Fleming led the way, spearheading a drive by the Canadian Press Association as early as 1901 to promote state-owned telegraphs and the establishment of a state-owned Atlantic cable, with rates in both cases being reduced to actual transmission costs.[13] This was not to be. Indeed the London *Times* claimed that the resignation in 1905 of the Canadian Postmaster William Mulock, who had led an investigation of telegraphs and telephones in Canada and was sympathetic to the nationalization of such facilities and to a state-owned Atlantic cable, was an outcome of Prime Minister Wilfrid Laurier's adhesion "to the old Liberal prejudice against State regulation."[14] Nonetheless, facing intense scrutiny and now the prospects of government regulation, which was accepted as a result of the inquiry, a compromise was worked out in 1910 between the Canadian Pacific Company and the Pacific Cable Board. Under the terms of the new agreement, the Cable Board leased a wire from the company that ran between the cable's landing place at Bamfield, on Vancouver Island, and Montreal, and which would be entirely operated by Pacific Cable Board staff.

Resolving these problems in part or whole did not resolve others. Notable was the Cable Board's reliance on the Commercial Cable Company and the British General Post Office on one side and the Australian post office on the other for collecting and receiving its messages. And as Ross pointed out, the

Commercial Company seemed to treat the Pacific Cable traffic somewhat as a sideline and indeed suggested that the company might be deliberately mutilating its messages as retaliation against Canadian government support for a state-owned Atlantic cable. This lackadaisical approach contrasted sharply with the dynamism of an experienced and none-too-scrupulous Eastern Extension Company which had its own offices to deal directly with customers, controlled all of its route, and was well known to its clientele.[15] The early lack of information on "via Pacific" in British post offices was particularly galling, as was the lack of any office in London. Indeed, a Canadian postal official noted in 1908 that "in the matter of canvassing, the fact that the Eastern Company controls every part of its route is an enormous advantage. Many little privileges that they are able to give customers such as franking, and occasional short messages on personal matters between London and Australia, are quite impossible for the Board as the CPR and American companies would refuse transmission unless paid for."[16]

THE ATLANTIC CABLE CAMPAIGN, 1906–12

Despite all the criticisms raised before and after the inauguration of the Pacific Cable, the Balfour Committee report complacently concluded that the British- and American-owned Atlantic companies provided an efficient service and that "no complaints against them have been lain before us."[17] In marked contrast, however, a Canadian official described the companies' rates as "fictitiously enhanced" and as an impediment to making the benefits of long-distance communication available to the public.[18] And it was this view, rather than that of the Balfour Committee, which helped resurrect the reform issue and to launch the campaign for changes in the Atlantic cable market. And as one indicator of the extent to which the issues resonated in the public mind, the doyen of British social and political commentary, *Punch* magazine, published an editorial cartoon in 1909 entitled "In the Coils"[19] (figure 17). The cables spelling out "cable rates" were the serpent strangling John Bull in the guise of Laocoon and his colonial sons.

News Organizations and the Campaign for State-Owned Atlantic Cables

To further underline the inextricable ties between the fortunes of the Pacific and Atlantic cables, we should note that, in 1912, to take a year in point, the cost of sending a transatlantic message was 25¢ a word and had been so since

IN THE COILS.

17 "In the Coils," by Linley Sambourne in *Punch*, June 16, 1909. The suffocating coils spell out "cable rates," a neat illustration of the concerns on the effects of high rates expressed at the Imperial Press Conference in that year. Reproduced by permission of Punch, Ltd.

1888. The press rate was lower at 12¢. This meant that well over half of the cost of a 20¢ per word press cable sent between England and Australia "via Pacific" went to the Commercial Company for carriage across the Atlantic. In other words, the business of the Pacific Cable, and certainly much of its potential press business, depended on the Atlantic rates. For Canada there was both disquiet over the transatlantic route press rates and an early hope among some officials that a busy Pacific Cable could be used as a lever to obtain rate reductions from the Commercial Company.

While agreements between private cable companies and newspaper cartels could obviously limit the distribution of news material, the main problem for the press, as ever, was the high cable prices. And, as such, the press associations of Britain, Canada, and the United States were in the forefront of the reform campaign. The issues were absolutely crucial and in many ways became entangled in the cultural politics of the British Empire of the time. That this was so could be seen, for example, in 1910, as Canadian Postmaster

General Rodolphe Lemieux complained that the press business was being "carried out in a crippling and unsatisfactory way," with the consequence that the Canadian press was being forced to rely on American news services, since even the wealthiest papers could not meet the expense of transmitting and receiving news at existing rates.[20] In New Zealand such sentiments were even more pronounced. In 1908 only one-third of the country's newspapers used the cables, and the other two-thirds got a rehash of British news which had already been strained through the hands of the Australian Press Association. And this was the reason why, as one New Zealand paper remarked, "We hear so much of Australian non-entities who happen to be visiting London and whose doings are chronicled for the benefit of Toowomba or Kalgoorlie."[21]

This merging of the critique of the global media with the cultural politics of the British Empire reached its apogee at an Imperial Press Conference held in London in 1909. Described by the chairman of the Empire Press Union to which it gave birth as "an epoch-making event," the conference was actually a link between the press and the numerous proposals for imperial federation which marked this period.[22] Attended by British press barons and forty-two editors from the "white dominions" as well as India and Ceylon, the top item on the conference agenda was a resolution passed in favor of state-owned cables, with a Canadian delegate tabling an even broader measure covering "state-owned communication across the Atlantic," a formulation designed to bring wireless into the mélange of concerns being addressed. The resolution was shelved, but it highlighted the fact that Marconi's invention was seen as a promising alternative to cables between North America and Britain. Most telling, conference delegates showed a healthy suspicion of the British government and its apparent complicity with the cable companies.

While the American press largely ignored the Imperial Press Conference, it did not ignore the problem of high cable rates in general and Atlantic rates specifically. In particular, the *New York Times* regularly drew on the trenchant criticism of the existing state of affairs launched by Henniker Heaton. Although Henniker Heaton had been an unyielding critic of the global cable business for decades, he was at the top of his game as a gadfly and publicist during the years of the Atlantic cable campaign, publishing a remarkable stream of detailed pamphlets and articles, including some in U.S. magazines with punchy titles such as "How to Smash the Cable Ring."[23] By now, Henniker Heaton was deeply into his utopian stage, advocating a penny-a-word international rate as a basis for world federation and international peace, and this captured the imaginations of news editors in many countries. In the best tradition of "Atlantic Crossings," Heaton's criticisms had been featured approvingly in the *New York Times* as far back as 1900 and gained more attention as time passed. Thus, in November 1908, a *New York Times* editorial noted that "it is a severe arraignment of the cable companies which the leader of the rate reform movement has made. . . . Their ownership, he holds, is in the hands of combinations, and their utilization is amongst millionaires rather than amongst the millions."[24]

The *New York Times* was even more disturbed by Heaton's charge that the cables were being kept idle for long periods of time in order to maintain high rates, concluding with the following acerbic comment: "So far as the American cables are concerned, commerce is practically throttled."[25] If this doyen of the American press was any indication, the United States was being drawn into the politics of global media reform in a far more direct fashion.

THE TRANSFER OF CONTROL, PART I: THE TRANS-ATLANTIC CABLE SYSTEM AND THE GLOBALIZATION OF NATIONAL REGULATORY POWER, 1908–12

If this chapter is seen to draw heavily upon Canadian participation in the Atlantic cable campaign, this is not merely an outcome of the authors' backgrounds. The role of Canadians in the establishment of the imperial Pacific Cable; the importance of the telegraph route through Canada to the Atlantic cable terminals; the hope that the Pacific Cable could be used as a lever to oblige the Atlantic companies to reduce their rates; and the desire for improved social communication over long distances all these stressed Canada's role as a fulcrum in global electronic communications. This centrality helps to explain the ardent nature of the Canadian support in favor of cheaper Atlantic cable rates and for a state-owned Atlantic cable joining Britain and Canada. However, while Sandford Fleming was not greatly slowed by advancing years in the flow of his increasingly eccentric ideas, it was evident, by 1908, that his mantle had fallen upon Postmaster General Lemieux, whose passionate support for cable reform made him a major spokesman for the media reform movement. In turn, he reflected the Canadian Post Office Department's long-standing support for such ideals, and it is tempting to explain this support by the central focus that postal and telegraph communications were seen to play in the nation's process of development, a public service focus which, unlike the revenue-generating concerns of the British Post Office, led successive fed-

eral governments to accept a long succession of departmental deficits alongside a remarkably advanced public communications system.[26]

Canadian postal officials' critiques of cable capitalism could be scathing
and also stood as an early example of the strong role of media criticism in
Canadian thought about communication, as the following quote illustrates:
"Every civilized state sets a limit to the time with which an inventor enjoys his
exclusive privileges. The capitalist, by the power of his wealth, is able to set
these limitations at defiance, and long after any patent rights have expired, has
been able to keep the field to himself by crushing out rivalry. The history of
the cable companies affords as complete an instance as can be found of success
in the maintenance of a monopoly."[27]

At the previous Colonial Conference in 1907, cable monopolies had been
attacked through a successful resolution by the delegate from South Africa
which favored limiting cable landing licenses to a maximum of twenty years.
In addition, if a cable company had received a subsidy, it was further resolved
that company receipts beyond a certain maximum gross revenue should be
used both to pay down the subsidy and also to reduce cable rates.[28] Clearly
then, a sense of the need for drastic action was in the air, and hence the
action taken by Lemieux in the 1909–10 parliamentary session fit the tenor of
the times. On this occasion, he continued his efforts by shepherding an act
through Parliament, passed in May 1910, which gave the Board of Railway
Commissioners the authority to regulate Atlantic cable companies landing in
Canada. The companies were given four months, with a possible extension to
a year, to file and obtain approval for their "tariffs and tolls." The catch was,
however, that the act would only "come into force upon similar provision
being made by the proper authority in the United Kingdom, and upon procla-
mation of the Governor in Council."[29] The British Parliament had a year to
pass equivalent legislation or the Canadian act would lapse. And so it turned
out to be. The British government did not take the reciprocal action, and the
act was effectively annulled.

Yet, for all that, the pressure for the British government to do something to
appease the demands for Atlantic cable reform could not be ignored: there
were the resolutions of the 1907 Imperial Conference, those of the Imperial
Press Conference (1909), and, finally, the Canadian Act of Parliament (1910).
But perhaps the most impressive was a 1908 Mansion House (London) meeting of cable reformers and their supporters; present were not only Lemieux,
Henniker Heaton, and Sassoon, but also three hundred leading businessmen
of the City of London, with the Lord Mayor in the chair. Edward Sassoon was

sufficiently enthused to tell the assembled group that it was "calculated to make history," and a correspondent for the *New York Times* in London hinted that a state cable charging just 4¢ a word was imminent.[30] Despite the optimism created at the Mansion House, the prospect of a state-owned Atlantic cable was, as usual, opposed by the British Post Office on a number of grounds: notably, that a state-owned cable offering service at cost would be swamped with business, that two cables would be needed in case one failed, and that the Post Office had that agreement with the Anglo-American Company to give the latter all messages not marked for a particular route. It was further argued that backing a state-owned cable was inappropriate at a time when wireless was emerging as a potential rival long-distance communications medium.[31] Nonetheless, the British government could not easily remain passive, and so by 1910, the press on both sides of the Atlantic was reporting upon negotiations between the General Post Office and Western Union which the London *Times* later described as having "been shrouded in mystery."[32]

THE TRANSFER OF CONTROL, PART II: WESTERN UNION AND CONSOLIDATION OF THE TRANSATLANTIC CABLE SYSTEM

At this point, Western Union had just merged with AT&T to create a massive communications colossus, the merger having been initiated primarily by AT&T and influenced by evidence that Jay Gould and Thomas Eckert (Gould's tool as Western Union president) had milked that company for profits "while its offices became shabby and uninviting, and [Eckert's] policies bred poor service and public disfavor."[33] By contrast, the president of AT&T, Theodore N. Vail, was a popular figure. In the words of one communications scholar, "Vail's policies contrasted vividly with the jungle morality of Western Union in the Gould era and Eckert's incompetence. Building an outstanding public relations department, he cultivated the confidence of the public . . . [and] newspapermen liked Vail because of his availability, friendliness and frankness."[34]

With Western Union a very powerful element in the Atlantic cable system and Vail now at the helm, its negotiations with the British authorities would be conducted with a finesse absent in the Gilded Age. Yet, as events proceeded, the vice president of the Commercial Company, George Ward, worried that the secret talks were really part of a bid "to combine all the English [Atlantic] companies with Western Union in opposition to the Commercial Cable Company."[35] And as the Commercial Company worried about a business coup, the

Come under de old Umbrella,
Come along, pickaninnies do;
Hark to Uncle Sam a-singing,
"There's room for all of you."
From the *Express* (London).

18 "Under the Trust Umbrella," originally from the *Express* (London) and reproduced in the *Literary Digest*, July 13, 1901. It expresses the common British view of American government collusion in the dominance of the economy by corporate trusts. British authorities, however, rarely queried the monopolies of their own corporate concerns, especially those working within Britain's imperial sphere of influence.

British communication policy expert Charles Bright warned about "the possibility of our transatlantic cable system becoming a huge American trust."[36] If consummated, the outcome of the discussions between Western Union and the Post Office would indeed have placed the British Atlantic cable companies under lease in the former's corporate hands, the broader outcome being that all fifteen of the Atlantic cables would be primarily U.S. controlled, with Western Union controlling eight, and the alliance led by the Commercial Company the other seven. Yet, besides these concerns, the key question remains: why had Vail taken the path which he did, and on what understanding did the British government let him?

Seen through the eyes of Vail, Western Union's actions represented a move toward a grand vision of corporate rationalization and coordination. Yet, seen from another angle, such moves were a more sophisticated version of the tactics used by Jay Gould in the past when he had threatened to cut off the Anglo-American Company from his North American–wide telegraph network. Indeed, during the negotiations, the British government and directors of the Anglo-American Company were "encouraged" by Western Union to lease their system to it by the threat that the American company would withdraw from the pooling agreements which it had made with them three decades earlier in 1882. Now, however, Western Union declared that it needed to walk cautiously in the face of the Sherman Antitrust Act, and therefore had to consider the agreements as illegal and to terminate them as of March 1, 1911. To make its point, Western Union began to withhold some of its business from the British transatlantic cable companies, the Anglo-American and Direct United States, resulting in revenue losses.[37] As Ivan Coggeshall, assistant vice president of Western Union in the 1930s, later observed, his company, "spurred on by whatever prompting the Commercial could bring from behind the scenes," had been caught in the crosshairs of President Roosevelt's "trust-busting" efforts.[38] He continued by noting that "the Attorney General of the U.S. in the early part of 1910 let it be known that the Western Union would have to get out of the pool voluntarily or be forced out by operation of the Sherman Act. Vail thus was balked in his desire to amalgamate the companies both at home and abroad into a world-wide telephone, telegraph and cable system."[39]

While the evidence supporting such claims on the impact of the Sherman Act is thin there were significant shifts in the political and legal context at the time that were leading to the consolidation of greater state regulatory powers on both sides of the Atlantic. As Martin Sklar notes in his seminal study of American corporate capitalism in the progressive era (1890–1920), the early phases of this trend had been marked by a strong tendency for the U.S. courts to interpret the Sherman Antitrust Act as condemning "monopoly" and industrial combines per se. That trend, however, encountered serious criticism during the first decade of the twentieth century under both the Roosevelt and Taft administrations on the grounds that there were "good trusts" and "bad trusts." By 1907, Theodore Roosevelt stated clearly that "reasonable business restraints" were a normal feature of modern capitalism and should be accepted as such so long as strict regulation was maintained.[40] Major American corporate leaders quite happily agreed, and at the forefront of such trends was none other than Theodore N. Vail. In a statement which would not be out of place in more contemporary merger crazes, Vail justified his company's lease of the British-owned cables on the grounds that only by bringing eight cables under unified control could Western Union "put into effect the economies and improvements contemplated," that is, act as a good trust should.[41]

For Vail, the deal ultimately struck with the British government in 1911 was a stroke of brilliance that simultaneously gave Western Union complete control over "a real system of competitive cables, free of the stigma, in the United States, of foreign domination. [While] at the same time, British nationalistic pride was sufficiently assuaged by their retention of actual ownership of the cables."[42] All this may have been true, but the London *Times* was skeptical. The *Times* did not deny the threat of antitrust, but suggested that other businesses preferred to wait for the decision of the courts in such matters and then act. However, Western Union "decided to interpret the law for itself, and having pronounced an adverse judgement, preceded to give instant effect to this decision by dissolving the pool."[43] Meanwhile, Western Union lawyers deemed the offer of a ninety-nine-year lease of the British cables not to violate antitrust law, but, should it be otherwise, the company promised to restore the "good connections" which had existed prior to 1910.

As we have already seen, however, Charles Bright, for one, was not assuaged, grumbling: "Choosing the favourable moment [Western Union] simply placed a knife at the throat of its allies, and the latter being unable to exist without the landlines of Western Union, were compelled to bow to its demands."[44] Yet, the fact that the deal evoked no obvious negative official government reaction indicated that the "knife at the throat" was not all on one side. Indeed, as part of the agreement with Western Union, British Postmaster General Lord Samuel created the principle "of government control . . . over the rates charged by the Western Union group of cables, and that the same principle would be applied to the other group when its licences fall to be renewed."[45] This new element of rate regulation, a variation on Canada's attempt to extend regulation to international communication services, however, turned out to be a paper tiger. As a senior British postal official noted in 1918, it was a control that could only be used to a very limited extent to deal with unreasonable rates or unnecessary proposals by the cable companies, and even then it was subject to appeal and the general laissez-faire strictures against state intervention set out in the Balfour report. Indeed, even though Western Union and the Commercial Cable Company's licenses had expired in 1911 and 1915, respectively, and years of negotiations between the companies and the British government had taken place, the licenses had not been formally renewed because of a disagreement over the proposal to insert in them a "control of rates" clause.[46]

CHEAP RATES, SOCIAL COMMUNICATION AND MISSED OPPORTUNITIES

So, if the British government did not get its "control of rates" clause, what benefit for press and public—for the "millions" as the *New York Times* had put it—was gained as the transatlantic cables passed into American corporate hands? The main gain was Western Union's agreement to the establishment of cheaper deferred cable rates, both for press and regular cable, as well as weekend and night cable letters. All of these new categories of messages were to be sent during the "quiet times" when the Atlantic cables would otherwise have been largely idle and were explicitly designed to improve people's access to long-distance communications. Rates in these categories were 50 to 75 percent off the regular per word Atlantic rate, plain language letters at 12.5¢ a word rather than the normal 25¢; deferred night letters at twelve words for 25¢; weekend letters at twenty-four words for $1.15; and a reduction in the press rate from 10¢ to 5¢. The commercial world, claimed the *New York Times* was "all but stunned at the radical reductions made."[47] The other major cable organizations, including the imperial Pacific Cable and the Eastern Associated Company, also adopted the new classes of more affordable messages almost immediately, so that it did look that one of the great principles of the cable system—the belief that the cables had always to be kept available for high-end users—had finally been broken.

Four concerns soon became apparent to those reformers who hoped for a social revolution in access to long-distance communication. First, the regular business of the cable companies still had priority, since the often absurd regulations surrounding the sending and delivery of deferred messages was admittedly designed to ensure that regular customers would not use it. This was particularly apparent on the Australasian route where weekend cables sent from England on a Friday could not be legally delivered in an Australian city until the following Tuesday morning. Second, the minimum word charge on night and weekend cable letters was soon deemed by the British Dominions Royal Commission to be excessive and socially limiting, and, indeed, even the overall charges were still high (for example, a weekend cable sent from Canada to Australia in 1913 at a minimum charge of $2.90 was equivalent to over $50.00 in 2001 purchasing power).[48] Third, the press did not gain too much, either. In today's terms, it still cost about $1.00 a word to send a press message from, say, Canada to London. And in the view of Eugene Sharp in his 1927 study of international news communication, cable companies did not give lower press rates for altruistic reasons but because the high press volume avoided idle lines "but beyond the saturation point press matter is not always given full consideration."[49] Finally, since weekend and night cables had to be in plain language, there was no chance of cutting costs through use of codes, and, in any case, as the Eastern Associated Company showed when they developed a "Social Code Book" available only in their offices, the average customer not surprisingly had no familiarity with the notion of coding. Over-all, the British government appears to have bent over backward to accommo-date the interests of the cable companies while allowing them to look like reluctant altruists. As for the reformers, they could point to growing use of the new tariff categories and some increase in the social use of cables, but had, nonetheless, been given remarkably little for all their efforts.

At least one leading reformer was not taken in by the new tariffs, as a June 1912 article in the *New York Times* revealed: "Nothing is more promotive of good international relations than cheap communications, and Henniker Heaton is their prophet. Deferred rates and weekend cable letters are an abomination to him. He wants messages between nations to be both instant and cheap, and derides the idea that there is any necessity for there not being both when capital has provided facilities for the carriage annually of 325 mil-lions of words which are only utilised for twenty millions."[50] By this late point in his life, Heaton saw the world as ideally tied together by penny-a-word messages—a true early vision of the Internet—and the hated cable monopolies smitten by the technological sword of wireless as developed by his friend Guglielmo Marconi. The *New York Times* approved, noting that "it is this writing on the wall [wireless] which promises the best hope of penny cable-grams."[51] Both Heaton and the *New York Times* were being wildly optimistic about the new technology, even in the long run. But it is highly relevant, in the context of the next section of this chapter, that Heaton and other global media reformers had so attracted U.S. attention and approbation —"the services which he has rendered in the long years now passed are a contribution to international welfare never to be forgotten."[52] Alas, Heaton's personal tragedy was to come when, as a dedicated pacifist, he was trapped on mainland Europe at the outbreak of hostilities in 1914, so ill that he died before he could return home.[53]

SOCIAL COMMUNICATION AND THE BRITISH EMPIRE: PUBLIC SERVICE AND THE PACIFIC CABLE BOARD

So, if Henniker Heaton's vision was not to be, what did the cable reformers gain from the new tariff categories? We have suggested rather little, but this does not mean that nothing was achieved. Thus, in 1913, John Denison-Pender told the Dominions Royal Commission that 92 percent of the deferred mes-sages from Australia were commercial, and just 8 percent social; for weekend messages, the social amounted to about 17 percent. These latter categories

TABLE 8 Increase and Decrease in Percentage of Messages Carried by the Pacific Cable to and from
Australasia by Rate Category and for Select Periods, 1903–27

YEAR	ORDINARY	GOVERNMENT	PRESS	DEFERRED ORDINARY	DEFERRED PRESS	OTHER REDUCED	ALL WORDS
1903–14	+78.6	+158.6	+461.8	—	—	—	+260.4
1921–27	−24.3	−27.1	−7.2	+72.4	+176.1	+172.9	+38.9

Sources: Britain, Pacific Cable Act, 1927, 10, unnumbered table

were, he noted, more fully used when there was a "temporary" European
settlement as in India or, to take an equally interesting case in point, among
East Indian workers and business people in, for example, Fiji, that is, among
diasporic communities who desired to stay in touch with home and whose
businesses depended on the maintenance of long-distance social networks.[54]
Yet, while the absolute amount of long-distance social communication by
cable remained small, just how relatively large the changes actually were can
be seen by contrasting the admittedly modest figures referred to by Denison-
Pender with the fact that social messages had accounted for an even punier 2–
3 percent of all messages in the past. Yet, rather than relying on the Eastern
Extension Company to further social communication within the British Empire, that task increasingly fell to the Pacific Cable Board.

The imperial Pacific Cable was the first public service–oriented long-
distance communications network, and its operation as such meant that the
impact of the reforms were far greater for it than for the Eastern Extension
Company and other cable companies. Indeed, the Pacific Cable Board, in
marked contrast to its earlier years, showed much enthusiasm for the new
categories. For example, 46 percent of all words carried internationally, in
1913–14, by the Pacific Cable were in the cheap rates categories, including 17
percent for deferred press messages (see table 7). During World War I, all
cable systems were heavily used by overseas troops and government personnel, and, as table 8 shows, in the 1920s, the great bulk of business for the Pacific
Cable came in the form of cheaper rate categories.

Overall, then, one of the most significant outcomes of the 1912 rate reforms is that despite being hedged about by cumbersome restrictions, they
did open previously closed minds to the prospects for greater price flexibility
and helped put the Pacific Cable firmly on the path that its earlier advocates
had so strongly, and tenaciously, pushed for.

CORPORATE CULTURE, THE PUBLIC INTEREST, AND
ANGLO-AMERICAN RELATIONS, 1912–15: COOPERATION
AND COMPETITION OR CONFLICT?

Two other main issues surrounding the Atlantic cable deal and its outcomes
still require attention. The first is the importance of appreciating the extent to
which Theodore Vail embraced not just the notion of an all-inclusive, efficient, and rational international communication system as a engine of progress, but also the idea of a "public service corporation." The latter was shown,
for instance, in the acceptance of the new cheaper message rates, and departed
from the earlier steadfast opposition to anything which could associate Western Union with being a "public utility" and hence open to either government
regulation or even ownership. According to Sklar, there was a method here,
the belief by corporate executives, including Vail, that government regulation
and rule by experts in cooperation with the corporate sector could actually
augment the power of large corporations by providing quasi-legal justification for their policies and by explicitly giving them a sense of public purpose.[55] That is, taken together, the linking of government regulation, corporate power, and the public interest, even if unwittingly, reinforced corporate
power, although Vail overplayed his hand in the failed 1913 bid by AT&T to
unify telegraphs, cables, and telephones all under one corporate umbrella.

Theodore Vail was succeeded as president of Western Union in April 1914
by another astute executive, Newcomb Carlton, who was very much in tune
with the times. Carlton headed the company throughout the rest of the period
covered in this book (i.e., until 1930) and played a decisive role in shaping the
contours not only of Western Union's corporate policies but of the United
States more generally. While contemporaries such as Charles Bright argued
that Britain's loss of control over the Atlantic cables to Western Union posed a
considerable threat to national security, this was not a view that Western
Union had ever adopted, and Newcomb Carlton followed steadily in that path,
perhaps even more so, willingly meeting British national security concerns
where necessary and cooperating rather than competing with the British-
based corporate titans that still remained at the realm of the global communication system. Of course, as in the past, there were squabbles galore involving
competing U.S. and British interests, but if anyone worked hard to ensure that
tensions did not foment outright conflict, it was Carlton. Parenthetically, not
all such corporate leaders were on the same amicable terms with Britain.
Indeed, this was certainly not the case with respect to the nationalistically

inclined James Scrymser and his Central and South America Company (and its successor, the All-America Cable Company), which pushed relentlessly between 1891 and 1930 to wrestle control of South American communication markets from another one of the Pender affiliates, the Western Telegraph Company, as explored in the next section of this chapter.

Meanwhile, the fact that collaboration rather than rivalry carried the day between Western Union and the British government was on full display during the war. Two prime examples stand out. The first involved Britain's desperate need of U.S. dollars for war purposes in 1916 and the Western Union's bid, in the same year, to buy the two leased British cable companies outright. The most important thing about this case is that the British Post Office, the Admiralty, and the Treasury did not oppose, but strongly supported the Western Union's offer. The Treasury's intervention on the side of Western Union was especially strong, as it indicated that it might force a sale by imposing a punishing "penal tax" on the predominantly British-owned Anglo-American Company if its directors proved difficult.[56] The close relationship between Western Union and the British government was also evident in a more sensitive issue: extensive wartime censorship of the cables. In particular, accusations that British authorities were intercepting and making improper use of American diplomatic and commercial messages were legion during the war, most notably by the Central and South America Company and its close circle of allies within certain U.S. foreign policy circles. Immediately after the war, these charges were carried into the heated debates on the future of the American approach to global communication policy. The accusations gained considerable traction in some circles and galvanized the push, most notably among Navy officials, to carve out a greater role for American-owned firms in global markets. And as Headrick notes, "So tight was Britain's grip on world communication, that it could not only block, or read at will, the most secret messages of its enemies, it could even use that information without revealing its sources. Never before or since in history has communications power been so concentrated and so effective."[57]

The accusations of wartime censorship were unpersuasively denied by the British, while Newcomb Carlton remained cagey on the issue, although he did reluctantly admit that such had been the case but then pleaded that it was more a case of "Keystone Cops" than the sinister portrait painted by his adversaries. To illustrate the point, Carlton closed his 1921 testimony to the U.S. Congressional Hearings that had been convened to cover this and a raft of other issues with the bitingly sarcastic comment that wartime British censors were, even if loyal citizens, "sometimes . . . afflicted with arrested mental development" because of the unbelievably "stupid things" they did to Western Union traffic.[58]

The claims and counterclaims are interesting for the rhetorical constructions they create, but the fact of the matter is that censorship was a routine practice of war practiced by all. In fact, censorship over the wires (and the airwaves) was immediately implemented by the United States upon its entry into the war in 1917.[59] The surprising thing about this particular standard-bearing case in global communication history is the one-sided attention that it has received by media historians ever since, with the obligatory multiple mentions of British censorship and indiscriminate references to disgruntled Americans, *tout court*, with hardly an indication that such matters were routine and that even in U.S. circles, opinion was deeply divided as to their morality. Perhaps the decisive difference in this case is the role played by the All-America Company, and its supporters among the tightly knit hawkish wing of American foreign policy circles. Mostly high-ranking Republicans under the administrations of McKinley, Roosevelt, Taft, and, later, Coolidge, these individuals were also arch-rivals to President Wilson's more liberal brand of internationalism and that was a decisive difference that figured prominently in the discourse on the politics of global communication in the United States.[60] In this context, the issues of British censorship and control were dragged into the All-America Company's no-holds barred attempt to compete in South America against the Western Telegraph Company and to cast these efforts in terms of a great patriotic struggle between the United States and Britain for the control of global communication.

SOUTH AMERICAN COMMUNICATION AND MEDIA
MARKETS: COLLABORATION, COMPETITION, AND
CRISIS, 1890–1914

Boom, Bust, and the Precarious State of South American Cable Communications

Our critique of "conventional knowledge" should not be misread to mean that the All-America Cable Company's views were without substance. Indeed, the company's predecessor, the Central and South American Company, under the stewardship of James Scrymser, fought a pitched struggle for control of South American communication markets with the Eastern Associated Company affiliates, most notably after 1891, when the early period of uneasy cooperation between the firms collapsed into a period of bitter rivalry that would last for four decades. That long-term struggle deeply shaped the corporate culture of the Central and South American Company and exacted a heavy toll on both sides, foreclosing access to several key markets for Scrymser's firm and, for Pender and his successors, contributing to the liquidation of the West Coast of America Company in 1896 and wholesale reorganization of the conglomerate's corporate structure with the consolidation of its affiliates in the Western Hemisphere under the umbrella of one new corporate entity (and the biggest in the Pender corporate empire), the Western Telegraph Company, in 1899 (see chapter 2).

The Western Telegraph Company (or its predecessors prior to 1899, to be precise) controlled access to South America almost to the same degree as the Eastern Extension Company, the Far Eastern branch of the Pender family communications empire, did in Asia. The only real differences were that the Western Telegraph Company's concessions in South America allowed at least some competitive inroads on a city-by-city basis, were set to expire nearly two decades earlier (1913 versus 1930) than in China, and were challenged by a far stronger phalanx of would-be American, European, and South American interests. Later American observers and many media historians refer to the long-standing efforts to dislodge the Western Company from its impregnable position in this era as the "cable wars." However, that phrase is too crude, even inaccurate. For one, almost all rivals between 1891 and 1914 decided to cooperate rather than compete with the Western Company, except that is, the Central and South American Company. Highlighting competition and conflict between Scrymser and Pender and his successors (Pender died in 1896), almost completely obscures the extent to which collaboration among the British and European cable companies carried the day. And second, the cable wars frame assumes an overly simplistic portrayal of the Central and South American Company as the erstwhile underdog in a "David versus Goliath" struggle for control of South American communication markets. Yet, such a portrait hardly fits a reality which was evident, despite the company's own rhetoric, in its own long-term monopolistic concessions on the west coast of South America, and in a solid niche which it had carved out for itself in some of the most important markets on the Atlantic coast of South America (for example, the River Plate markets) after refusing to renew its joint purse agreement with Pender which expired in 1891. It also out-maneuvered the Eastern Associated Company to acquire the Transandine Telegraph Company's telegraph line from Santiago to Buenos Aires .[61]

19 "The Stock Exchange at Buenos Ayres, Argentine Republic" in 1893 in *Illustrated London News*,
October 7, 1893, shows the frenetic activity of a stock exchange regularly subject to booms and
busts, the trajectory of which contemporaries often attributed to the instantaneous movement of
messages by telegraph and cable technology. Reproduced by permission of the Illustrated London News
Picture Library.

While at first blush the Transandine acquisition appeared to give Scrymser's company only a minor toehold on the Atlantic coast of South America,
its position was actually quite strong, since it immediately signed deals with
two important independent communications companies in Argentina and
Uruguay: the Oriental Telegraph Company and the River Plate Telegraph and
Telephone Company. The agreement with the latter was essential. The telephone systems it owned in Rosario (Uruguay) and Buenos Aires as well as the
handful of cables it possessed between Buenos Aires and Montevideo gave
Scrymser direct access to the leading banks, trading houses, ports, and other
important customers headquartered in this dynamic hub of the world economy and heart of Britain's informal empire. With access to two of the most
important commercial cities, Buenos Aires and Montevideo, on the east coast
of South America and an unobstructed communications channel between
them, the Central and South American Company was well poised to compete
with the Western Company, at least for service to North America.[62]

20 "The Brazilian Insurrection: Daily Scare in Rio de Janeiro." Periodically shaken by economic and political crisis during the early years of the new Republic of Brazil, the citizens of Rio de Janeiro anxiously turned to telegraphic news for information on the latest events. The case sketched here refers to the Conselheiro insurrection, 1893–97. That the telegrams are posted at a newspaper office reveals the close ties between telegraph and cable services and the daily press in South America, as elsewhere. From *Illustrated London News*, October 21, 1893. Reproduced by permission of the Illustrated London News Picture Library.

Scrymser's early forays into the Atlantic coastal markets of South America coincided with a period of major economic advance and state-led modernization initiatives as well as rapid growth in the number of cable connections linking the River Plate to Europe and North America between 1884 and 1893. The consequence was a dramatic expansion in the capacity of the South American communications systems, and rates to Europe and the United States plummeted by half from their extraordinarily high levels of between approximately $4 and $7 a word.[63] And while some of the communications companies benefited, the new cables and telegraph lines just as often created a glut of "bandwidth" which saddled less-fortunate companies with the burden of the prior financial outlays. After about 1893, however, such developments appear to have come to an almost complete standstill, with real momentum only reoccurring around 1910 as the expiration date for the Western Telegraph Company's exclusive south of Rio concession drew closer. But meanwhile, and a good index of declining interest in South American investment in the 1890s, even Jay Gould's efforts while at the helm of Western Union to enter South America, under the guise of the Dom Pedro Segundo American Cable Company, faded out after having peaked just as the River Plate boom reached its zenith.[64] The fact that Gould, one of the wealthiest men in America, withdrew was also symbolic as an indication of, as yet, limited U.S. interest in funding projects in South America; as such, Scrymser's company was the exception rather than the rule. U.S. investment and commerce did expand rapidly southward between 1898 and 1914, mostly to Central America and the Caribbean, but remaining relatively modest in comparison to British and European stakes in South America until after the war.[65]

As a report by the Argentine government in 1891 put it, the boom had been built on top of a regional banking system filled with corruption, dubious loans to politicians and government officials, rampant stock market speculation, fraudulent mortgages, and myriad other woes, all of which caused the River Plate and Brazilian economies to tumble and venerable international financial institutions such as the Barings Bank in London to almost collapse. The withering of cable company investment was not unique, but reflected broader trends as construction on ports, hospitals, municipal infrastructure, and other public projects, that is, the material infrastructure of modernization, ground to a halt. Fantasies held just a short while ago by the Argentinean elite about Buenos Aires's inevitable ascent to become the "Paris of South America" were abruptly replaced by the rough and tumble of revolutionary politics in the 1890s. Brazil, Argentina, Uruguay, Chile, and Venezuela all underwent revolutions or major political reforms between 1889 and 1892.[66]

COMMUNICATION, CRISIS, AND SUPERPOWER NEWS (1890-1914)

The intermingling dynamics of economic crisis and revolution also had the effect of decisively elevating the importance of South America in the global news system. Indeed, concerns swelled about the role of news organizations and their impact on international business and financial markets, foreign policy, and public opinion. Some indication of these changes and their impact on foreign policy and public opinion were registered by Charles W. Eliot, Harvard University president, on the occasion of the Venezuelan Crisis in 1895: "There has been brought forcibly to our notice a phenomenon new in our country, and perhaps in the world, namely, the formidable inflammability of our multitudinous population, in consequence of the recent development of telegraph, telephone, and bi-daily press . . . our population is more inflammable than it used to be, because of the increased use in comparatively recent years of these great inventions."[67]

While Eliot's comments were typical of elitist disdain for the masses at this time, they did register a salient new force in world affairs: public opinion. Concerns about communication, public opinion, and the news intensified between 1890 and 1914. That was evident as postrevolutionary governments sought to establish their legitimacy within the international community and as the global superpowers, in particular, France (1890s), Germany (1900s), and, later, the United States (1910s) adopted more concerted efforts to elevate their political, business, and cultural influence in such countries. While Argentina, Brazil, and Uruguay had long been well within the zones of Britain's informal empire, French conceptions of enlightenment and republicanism had a major impact on the culture and politics of Latin American elites.[68] This influence resonated even more strongly at the end of the nineteenth century as France's own revolutionary moments of 1789 and 1848 became touchstones for, as Rhoda Desbordes observes, the imagination and language of republican democracy in South America.[69]

The idea that foreign news and culture could influence South America's political development was not in and of itself novel, but such notions became considerably more salient between 1890 and 1920. During the first decade of this period, the French government urged the consolidation of the cable communication industry in the country, offered more generous subsidies (or, more precisely, revenue guarantees), and passed new laws to tie the industry together. The aim was to expand the development of cable communications systems in the transatlantic economies and, incidentally, to the west coast of Africa, insofar as serving France's imperial and colonial interests there coincided with augmenting its commercial, diplomatic, and financial interests in South America. Three fundamental changes were also undertaken by France's premier global news agency, Havas, to enlarge its role in South America. The first step occurred in the 1890s when Havas broadened its private telegram service to launch a range of new *services spéciaux*, which were specially coded (to save money and gain secrecy) and made available to the South American press. The *services spéciaux* were not just supplied to the big metropolitan dailies spread out along the Atlantic coastline and served principally by the European cable companies, but were sent by the mostly state-owned telegraph systems to inland newspapers as well. Second, in 1890 Reuters and Havas changed the cartel agreement so they could swap their respective interests in Egypt and South America—"zones of informal empire" where the two agencies had cooperated in the past. Under the new terms, the news markets of Egypt were handed over to Reuters and those of South America to Havas.[70] The third innovation was unveiled by Havas's new general manager for South

America, Charles Houssaye, in 1904. At only 28 years of age, Houssaye, according to Desbordes, "made a fresh start . . . by signing more than one hundred contracts . . . with the richest and most important South American newspapers, including those of new promising cities in the interior such as Sao Paulo, Porto Alegre, Santiago de Chile."[71]

As Havas consolidated its expansion in South American news markets between 1890 and 1904, Germany embarked on similar but less significant initiatives shortly after the latter date, and so too did the United States, although not substantially for about another decade, as we have noted. The problem for Germany, however, was that South America news markets had been ceded to Havas, while those in the Caribbean and Central America had become the exclusive preserve of Associated Press after changes to the ring circle agreement in 1903. Squeezed between the American empire in Central America and Havas's monopoly in South America, the news markets of the Western Hemisphere were closed to the German-based agency, Wolff. In response, in 1908, the German government organized the Transatlantisches Büro, with an annual subsidy of roughly $25,000, to supply the press and markets of South America with a commercial and diplomatic news service. While the news agencies protested against these incursions, their criticisms were brushed aside by German officials.[72]

American news agencies followed, but not significantly until under the combined influence of President Woodrow Wilson, Dollar Diplomacy, and Pan-Americanism in the second decade of the twentieth century. Indeed, reflecting the vital importance of communication and culture to the U.S. version of Pan-Americanism on offer at the time, Wilson told the Pan-American Financial Conference of 1915 that "we cannot know each other unless we see each other; we cannot deal with each other unless we communicate with each other."[73] Lying behind Wilson's call were perceptions that American interests were ill-served by relying on foreign agencies like Havas for the distribution of U.S. news in Latin America. Frank Noyes, the president of Associated Press, explicitly expressed these concerns to the U.S. Congress in 1913: "The . . . lack of news is strikingly apparent in the relations of the United States with the Central and South American nations. These countries secure their news of the United States by way of Europe, and it consists mainly of murders, lynchings, and embezzlements. The antipathy to the United States by these countries is undoubtedly largely due to the false perspective given by their newspapers. If in truth we were the kind of people they are led to believe we are, they would be fully justified in their attitude."[74]

Following the well-worn path set in the previous two decades by France and Germany, a growing body of opinion in the United States believed that the key to redressing these perceived misrepresentations lay in a much expanded role for U.S. communications companies and the three American international news agencies, Associated Press, United Press, and the smaller International News Service, in foreign news markets.[75] But that, however, was much later, and during the period of our focus in this section of the chapter (1900–14), there were no such concerted actions, just tentative first steps.

Some of those first steps emerged as the Associated Press, the biggest of the three U.S. agencies, began to expand into Europe after signing its first agreement with the other big three foreign news agencies, Reuters, Havas, and Wolff, in 1893. It was not until it renegotiated the terms of its membership in the news cartel in 1903, however, that Associated Press enlarged its operations in the expanding American "zones of informal and formal empire" in the Caribbean, Central America, and the Philippines.[76] Before that, none of the U.S. news organizations had a permanent presence in Latin America. Some steps toward that end were taken in the 1890s, but the main trend was for individual newspapers to send correspondents episodically to cover the outbreak of political and military crises besetting the region, as the *New York Herald* did during the revolt against the Chilean Balmaceda reforms of 1891;[77] as a bevy of general and business-oriented newspapers did during the Venezuelan crises of 1895–96; and, crucially, as the mainstream press covered the insurgency against the last vestiges of the Spanish Empire in the Caribbean and the Philippines during the 1890s.[78] These events are best known today for spawning, especially in the heat of the Spanish-American War of 1898, the awful phase "yellow journalism" and ill-considered quips by callous news barons, such as William Randolph Hearst's instructions to the war artist Frederick Remington: "You furnish the pictures and I'll furnish the war."

However, rather than dwelling on callous media moguls and jingoism, the more vital lesson is that it was probably these events on which the U.S. media first cut its teeth in terms of international news. There are two other key qualifications here. First, that far from being uniformly prone to periodic outbursts of jingoistic diatribes and supportive of imperialist adventures, there were consistent and fairly strong currents in American public opinion that were highly suspicious of both jingoism and imperialism, and which tended to see military interventions, independence movements, and political revolutions in Latin America through the prism of their own political and cultural myths about how similar events had shaped American history and

values. Second, this prism was not only something that Americans super-
imposed on top of their interpretations of Latin American politics but one
which politically savvy Latin Americans used to their advantage as well. Thus,
to take just a few examples, Chilean officials intercepted cable communications between American diplomats in Washington and Iquique during
the revolt against the Balmaceda reforms (1891) and successfully used what
they learned from those messages to help shape the course of events in their
favor. And later in the decade, the Cuban *independistes* and the Venezuelan
government both used modern "publicity" methods to present their case
directly before the court of American public opinion, newspaper editors, and
the U.S. Congress.[79] Indeed, during the Venezuelan Boundary Crisis with
Britain (1895–96), the Venezuelan government hired a twenty-year veteran of
the U.S. foreign civil service and former minister to Colombia and Venezuela,
W. L. Scruggs, who, in John Britton's words

> launched a successful anti-British propaganda campaign . . . which surprised both
> U.S. and British officials because of the intense emotional outpouring from the U.S.
> public. That same public accepted in large part the anti-Spanish propaganda cam-
> paign engineered by the Cuban exile community in the U.S. Employing the U.S.
> wire services and press, the Cuban rebel organizations (or juntas) located mainly in
> Florida and New York, captured the sympathy of U.S. newspaper editors and
> presumably their readers by stirring anti-Spanish feeling in the U.S. to gain support
> for their revolution (although in the long run, the Cubans came to regret the extent
> of the U.S. involvement on the island after the Spanish-American War).

In each of these cases, as Britton wryly observes, "the Chileans, Venezue-
lans, and Cubans exploited the communications system for their own ends, in
a sense turning imperialism on its head by exerting their influence within the
domestic politics of the United States through its print media." Britton's
observations, while made in personal correspondence with the authors, are
supported in their details by David Pletcher's thoroughly documented account of the Venezuelan crisis and by several fascinating recently published
studies by American foreign policy historians such as Emily Rosenberg, Frank
Ninkovich, and Russell Johnson. Together, the analysis offered by Britton and
these sources offer valuable insights into how communications, culture, and
the media decisively influenced Latin American politics and the politics of
(anti-) imperialism in the United States. These sources expand the range of
cases studied, but the general picture that emerges is one in which the participants often found a sympathetic audience in the United States for their cri-

tique of foreign power and imperial expansion, although the firmament in which such sentiment was anchored tended to be organized around the ideals of U.S. political and cultural myths and patriotism rather than well-thought-through commitments to anti-imperialism.[80] Regardless, the main point here is that a basic disinclination toward imperialism was visible in the United States political culture from the initial expansion of international news coverage by the American press and at precisely the moment when the United States was undergoing its own rise as an imperial power. In sum, the connections between the politics of (anti-) imperialism and international news were intertwined from their respective moments of conception.

Investment, Commerce, News, and the Imperial Impulse: Expanding the Southern Transatlantic Communications Infrastructure, 1902–14

One of the crucial lessons from the first two decades of the twentieth century is the power that exclusive rights and ties to national power elites gave to global communication companies. Indeed, the Western Telegraph Company leaned on such measures to exact stern commitments from would-be rivals which required them to uphold rates and to collaborate in the joint interests of all in return for market access. While Scrymser's Central and South American Company had long abandoned any desire to cooperate with the Pender group and relentlessly pounded against what must have seem like a brick wall for nearly a quarter of a century before finally seeing the Western Telegraph Company fortress begin to crumble (around 1915 to 1922, as chapter 9 explains), none of the other European cable companies that entered South America between 1902 and 1914 were quite so obstinate. In the end, these years saw a continuation of long-standing practices and justifications which essentially argued that the structured management of change and competition was vital to the rational organization of communication markets and the economic viability of all the companies involved.

The result of these relatively stable conditions, unlike conventional accounts of the events and times, is the remarkable extent to which the Western Telegraph Company succeeded in forming alliances with several other firms, notably the South American Cable Company, which was gradually transformed from a British-owned and -managed company to a thoroughly French company between 1901 and 1915. Then there was the German South American Telegraph Company, which was formally registered in 1908 but which extended the role that German cable manufacturers Felten and Guilleaume had built up in the global communications business since the 1890s. These developments unfolded through a sequence of events that fit firmly within the recognizable features of the global communications business in the first decade of the twentieth century. While these points are developed further in the following paragraphs, they are introduced in summary form in table 9.

The formation and role of the Western Telegraph Company during the course of these events was absolutely decisive. This could be seen as the Western Telegraph Company, or its corporate affiliates, used its exclusive landing rights both in South America and in the Azores, a series of Portuguese islands in the Atlantic that emerged as a key hub in the transatlantic communications economy in the late-1890s. The location of the islands well off the southwestern European coast made them a natural and critical intersection in the transatlantic communications system, not just between Europe and North America but to Africa and South America as well. While they had never assumed such a position in the past, changes in policy, the global economy, and the revival of interest in South American communication markets changed that. Various affiliates of the Eastern Associated Company leaned heavily on their rights in the Azores to regulate access to Europe, Africa, and South America communication markets, thereby maintaining rates and their lion's share of the market.[81] It was a key new locus of control in the global communication system; and particularly after World War I, it became an international irritant that drew strong rebukes from several governments, especially the United States, and from new corporate rivals.

The strength of this new lever of control was demonstrated by a new contract signed in 1899 between the Western Telegraph Company and many of the French, German, and American companies that had an interest in the cable systems, existing or planned, that connected together the four corners of the transatlantic communications economy (imagined as a quadrant, with Europe in the top right-hand corner, Africa in the bottom right, and North and South America occupying similar positions but on the opposite side of the "box").[82] This contract, in turn, built on the dramatic expansion of the northern transatlantic cable system which had occurred in the last fifteen years of the nineteenth century (see chapter 2) and anticipated a similarly large expansion in the cable systems reaching South America and, albeit incidentally and on a more modest scale, to the west coast of Africa. In short, the contract represented the crystallization of a transatlantic communications economy, not yet complete, but in the process of emerging. With so much at stake and all crammed into one legal document designed to structure rela-

TABLE 9   Continuity and Managed Change in South American Communication Markets, 1899–1914

KEY DATES	KEY ACTORS	MAIN INSTITUTIONAL PROCESSES AND DYNAMICS
1899	Western Telegraph Company created as largest affiliate in Eastern Associated Company.	Culminates trend visible since late 1880s. Amalgamation does for Eastern Associated Co. in Western Hemisphere what Eastern Extension Co. does for it in the Far East, i.e., consolidates and rationalizes control.
1899	Western Telegraph Company South American Cable Co. French Cable Co. German Atlantic Telegraph Co. Commercial Cable Co. Other affiliates of above companies.	Creation of a "supercartel" based on control of landing rights to regulate access to South American markets and Africa by British, European, and U.S. cable companies. Set rates, divide revenue, and manage competition.
1902	South American Cable Co.	Created by Gutta Percha Telegraph Works Co. (1891) (Pender affiliate). Acquired by French Post and Telegraph Authority (1902). Cable to Recife via Africa (1905). Gradual shift of control from London to Paris (1902–15).
1903–11	Felten and Guilleaume, create South American Telegraph Co. (1908), and acquire River Plate Telegraph and Telephone Co. (1911).	Membership of other Felten and Guilleaume firms in southern transatlantic "supercartel" (1899). Acquire River Plate Telegraph and Telephone Co. (1911). Joint purse with Pender (1903–14). New cable to Recife via Africa (1909–11).
1909–14	Central and South American Co.	Operations on west coast of South America from 1882 and in River Plate from 1891. Only real rival to Western Telegraph Co. after latter date. Revival of interest in Atlantic coast of South America circa 1909–13 pending expiry of Western Telegraph Co.'s "south of Rio" landing rights. Thwarted by legal battles until 1919.

Sources: Various, as cited in text

tions among several of the world's biggest cable communication companies and their respective governments, it is not surprising that the results were exceedingly complex. That said, however, the arrangements were described about as clearly as they could be twenty years later by John Goldhammer, the secretary of the Commercial Cable Company, during the course of U.S. Cable Landing Licenses Hearings (1920):

> The Portuguese Government preferred to deal with all the companies through the medium of one company [and] would not grant a concession . . . direct to the German Cable Co. or ourselves. . . . Now, it must be remembered that when we sent our traffic for South America to the Azores . . . we were active competitors of the [Central and South] American Company. . . . All the messages . . . collected in the United States [by] the Western Union. . . . for points served by the [Central and South] American Company were handed to the latter, and vice versa, all the traffic which the [Central and South] American Co. collected in South America for the United States were handed to the Western Union Co. On the other hand, all the traffic which the Western [Telegraph] Co. collected in South America for the United States came to our line at the Azores through our contract with the German company, and in return all the traffic we collected in North America for South America we sent by the Azores.[83]

In short, just as there were two rival cartels controlling the Euro-American communications market, so too were there two alliances controlling the communications markets of South America, one organized around the Western Telegraph Company and the other around the Central and South American Company and Western Union. The Western Telegraph Company, however, towered over the rest of the companies involved in South America because only it had direct transatlantic cable connections to the continent (in fact, it had three cables by 1899), exclusive rights in the Azores, and vital concessions that gave it, and only it, permission to establish cable connections within and between cities in Brazil, Uruguay, and Argentina. Yet, at the same time, there were three cracks in this otherwise seemingly impenetrable wall: first, the concessions in South America did not preclude competing cable connections between the continent and foreign destinations; second, some of the Western Telegraph Company's most valuable concessions were set to expire in 1913, notably the south of Rio concession; and third, the constant threat of competition from Scrymser loomed in the background. In addition, the fact that the South American economies were in a phase of revival by the beginning of the twentieth century magnified the combined impact of these challenges to the Western Telegraph Company's dominant position.

French Cable Companies and the Western Telegraph
Company's Hegemony in South America, 1901–07

It was this combination of factors that compelled the Western Telegraph Company to put together the institutional framework described above and which became absolutely decisive to its attempts to manage competition and change over the next fifteen years or so. That could be seen immediately in two events. The first was the decision by the Eastern Associated Company to sell the South American Cable Company to the French Post and Telegraph Authority in 1901. The company had been unprofitable since its creation in 1891, but now, and with France having spent the last decade consolidating its new approach to global communications and eager for more independent links to South America and Africa, the Eastern Associated Company decided to sell the South American for roughly $1.8 million in 1901.[84] France now had its second direct cable to South America, the first having been established many years earlier when the Compagnie Française des Télégraphes Sous-Marins laid a new cable from the northern Brazilian city of Salinas to Venezuela and Haiti and then to New York in 1889. While both initiatives were indisputably borne of a greater desire to gain more autonomy from British controlled cables, that both cables were hardly independent could be seen in a number of facts: they both relied on the Western Telegraph Company system for connections to the rest of the Atlantic coastal cities of South America; the operatives of both French firms had accepted the terms of the 1899 agreement; and finally a new division of revenues and slightly cheaper rates were mutually agreed upon as part of the sale of the South American Cable Company.[85]

Although the sale of the South American Company in 1901 meant that ownership had undoubtedly passed from British to French hands, the company continued to be an amalgam of French state and British private corporate interests until 1915. Before then, its headquarters continued to be in London and its chairmanship firmly in the hands of the Englishman Robert Gray, as it had been since 1891. In fact, rather than abrupt changes, there was a gradual Francophonization of the company with a formal name change to La Compagnie des Câbles Sud-Américains (South American Cable Company) taking place in 1906 and as British staff were replaced with French workers over time through attrition. Yet, the fact that not all friction had been eliminated could be seen in many instances. Most important, while the Western Company permitted newcomers to have access to South America, its concessions still prohibited them from establishing their own connections between cities. Thus, to take one case in point, the South American Cable Company's plans for a new cable between Dakar (a French colony) and Recife, with crucial extensions to the

River Plate, had to be scaled back in the face of opposition from the Western Telegraph Company. The cable was ultimately laid in 1905, but with no extension to the lucrative River Plate markets.[86] Two years later, a new French company emerged with another proposal, but this time for a cable between Senegal and Buenos Aires, that is, right in the heart of the most lucrative communications market in South America. A delegation representing the firm led by the entrepreneur and Parisian city counselor, Henri Turot, with backing by the French Foreign Ministry, was favorably received in Buenos Aires and Montevideo. A new company, the Société d'Etudes du Câble Franco-Argentin, was registered in Paris in 1907 and an agent appointed in Buenos Aires. However, the company then fell quiet and was subsequently formally wound up in 1912, likely because its plans were blocked by the Western Telegraph Company's (and through it, the Eastern Associated Company's) concessions.[87] Thus, while the Eastern Associated Company was willing to negotiate limited access to South America, it is equally clear that it had no intention of relinquishing its control over what were, in effect, parallel national communication systems running between the most commercially vibrant cities of Brazil, Uruguay, and Argentina. That sowed the seeds of discontent among rival cable interests as well as within business and government circles in South America, but it did not yet constitute a full-blown struggle for control of global communication.

Felten and Guilleaume, German Cables, and the Western Telegraph Company's Hegemony in South America, 1902–10

Of course, more than just French companies were taking the initiative, and Jorma Ahvenainen, in his exhaustively detailed study, *The European Cable Companies in South America*, identifies a whirlwind of activity after the end of the nineteenth century which reaches its zenith between 1910–14, just as the Western Telegraph Company was about, at least so it seemed, to lose its exclusive concessions. Yet, there were at least three additional ventures proposed between 1902 and 1910 that are worthy of mention: one by the British-based Edward M. Coll and Company, owners of the Anglo-Argentine Tramways Company in 1906; a second in 1908 by José C. Paz and his son Ezequiel Paz, owners of one of the wealthiest and most influential dailies in South America, the Buenos Aires–based *La Prensa*; and a third, by Felten and Guilleaume, whose interest in a direct cable to South America emerged as a natural extension of their general transatlantic system as early as 1902.[88]

Only the Felten and Guilleaume initiative gained momentum over the course of the decade. The fact that it was poised to become a significant player in these events could already be seen from the 1899 cartel agreement described earlier. According to the terms of that agreement, all German messages bound for South America were to be sent via the German Cable Company (a Felten and Guilleaume affiliate) to the Azores, from where they would be carried to their final destination by either the South American Cable Company or Western Telegraph Company.[89] An additional joint purse between the Western Telegraph Company and Felten and Guilleaume signed in 1903 further spelled out the share of revenues for messages between Emden and Recife, with just under 60 percent going to the Western Telegraph Company and the rest to Felten and Guilleaume. Momentum gained during the next two years as Felten and Guilleaume and the German Post Office, on one hand, and Denison-Pender, on the other, discussed Germany's need for its own cable connections to South America by way of the Azores. And, as the French interests had done before, Felten and Guilleaume and its German government supporters clearly indicated that they preferred to cooperate rather than compete with the Western Telegraph Company and South American Cable Company. Seen from a broader perspective, the events, far from ushering in the "cable wars," were shaping up to look like a three-way collaborative venture among British, French, and German interests, and American ones as well, if we consider the interests of the Commercial Cable Company (but not those of Scrymser), as we probably should. However, in this there was a glitch, because while Denison-Pender, now managing director of the Eastern Associated Company, personally stated that he was agreeable to much closer operating agreements between all of the parties involved, other directors at the company were not so inclined.[90]

Despite the adverse response, Felten and Guilleaume remained undeterred. Indeed, the firm's prospects seemed to improve again in 1908 as Emil Guilleaume caught wind of the La Prensa bid to create its own cable from Buenos Aires to London. In an attempt to combine their efforts, Guilleaume rushed from Berlin to Paris late in the year to meet with José Paz, La Prensa's publisher, but the much-hoped-for meeting failed to materialize and the potential for a joint German-Argentinean cable evaporated.[91] In another instance, Guilleaume went to Spain in 1906 to seek permission to use the Canary Islands as an alternative to the Azores. He followed that up by meeting with an old business associate, J. Vieweg, managing director of the Dutch-based East Africa Company, to discuss combining his interest in cables to South America with meeting Germany's imperial needs in Africa. Those efforts brought the German Post Office and its Foreign Office deeper into the enterprise than they had been in the past. Within a year, and despite formidable opposition from the Eastern Associated Company's affiliate, the African Direct Company, which had served France and Germany's imperial communication needs since the mid-1880s, Felten and Guilleaume emerged with permission in hand in 1907 to land cables at the Spanish island of Tenerife and also at Monrovia, on the west coast of Africa, thereby bypassing the Azores altogether. Nothing now stood in the way of the German company laying a cable directly from the west coast of Africa to Recife or some other city along the east coast of South America. The Western Telegraph Company appealed to the British Foreign Office to intervene but was informed that there was no basis upon which the latter could or should oppose the German cable.[92]

Perhaps with the force of inevitability apparently behind the German drive to have its own cables, by 1908 a revival of a more cooperative spirit among those involved could be detected. This was evident as discussions on the cable situation in South America took place in London in October 1908 between Emil Guilleaume and Robert Gray, the chair of the French South American Cable Company. That meeting was followed by another with the secretary general of the French Post and Telegraph Authority, J. Bordelonque, who was working closely on this occasion with Denison-Pender, largely because of the ties the two had developed through the close alliance between their respective companies. The prospects for a satisfactory solution were improved by the election of a new slate of French ministers more inclined to cooperate with Germany on a range of issues in 1909. The settlement of their dispute over Morocco later that year was one index of the new spirit of cooperation in the air, and two deals in January and March 1910 between their respective post, telephone, and telegraph authorities pointed in a similar direction. The first deal was mainly technical in nature while the second paved the way for German cables to be brought to the west coast of Africa, where they were to connect to France and Germany's west African colonies. The deal also contained a highly unrealistic plan for extensions across the Belgian Congo to German East Africa. And all this was incidental to the main thrust of the deal: a German cable from Senegal to Brazil and the setting out of terms upon which that development could be harmonized with the existing operations of the South American Cable Company. In sum, much as had been the case from the mid-1880s onward, so too in 1910 were plans to extend the African communications infrastructure contingent upon the project of collaborative empire and derivative of economic trends in South America. And to assuage any concern that Britain might have, French and German officials made statements to the effect that it had no need to worry since no hostile changes to the current balance of power were implied by the deals, although the thinly concealed mutterings of a few French officials to the contrary might have suggested otherwise.[93]

While the formal zones of empire in Africa continued to be dealt with as the imperial superpowers saw fit, in the zones of informal empire located in South America things were changing. While the wheels of (in)justice had been greased by corruption, bribery, and ties between foreign and local elites in the past, this was no longer so obviously the case during the first decade of the twentieth century, at least insofar as Brazil, Argentina, Uruguay, and Chile were concerned. In these nations, established political and legal processes were seen as more likely to deliver fair judgments than not. As a result, diplomatic intervention on behalf of private corporate affairs was something to be used more hesitantly.[94] Between 1908 and 1913, this was evident as matters in South America, especially in Brazil, were brought to a head by a relatively rapid sequence of events which seemed set to tumble the old order. Or were they?

As mentioned just above, the fortunes of the Felten and Guilleaume–backed bid to enter South America gained a great deal of traction between 1908 and 1910 as old alliances with the South American Cable Company were resuscitated. The company's prospects improved further when its representative, W. J. Spoerer, with the support of the German minister in Rio de Janeiro, Franz von Reichenau, gained Brazilian landing rights in 1908 in return for accepting four conditions: reduce rates to about 75¢ per word from the current level of roughly 86¢; solely use the Brazilian state-owned telegraph system for connections between the end of its cable network in Recife to other points in Brazil and beyond; cooperate with the existing cable companies; and ensure that its rights contain nothing in the way of a monopoly.[95] Felten and Guilleaume readily accepted and in August 1908 formally registered a new company: the German South American Telegraph Company (Deutsch-Südamerikanische Telegraphengesellschaft). The company was initially capitalized at about $2.5 million (in relative terms, about one-quarter the capitalization of the much larger Western Telegraph Company but larger than the South American Cable Company).[96] Felten and Guilleaume put up a third of the company's capital, five of Germany's leading banks provided another 43 percent, and the rest was supplied by a handful of German business interests. The German Foreign Ministry also provided it with a loan of $650,000 and an annual revenue guarantee of $750,000, with the company required to pay 75 percent of its annual revenues up to that amount to the German Post Office. These latter details were, in essence, an elaborate and roundabout way of using cable revenues to subsidize the German postal system while simultaneously allowing the privately owned German South American Telegraph Company to repay the annual revenue guarantees it received from the Foreign Ministry.[97]

With everything seemingly in order, between the end of 1909 and May of the following year the German South American Company began to lay its new cable from Emden to the Spanish island of Tenerife and then to Monrovia on the west coast of Africa.[98] However, just prior to laying the final transatlantic section to Brazil, H. Pfitzner, a director at the German South American Company, visited South America and reported back to other directors that the Brazilian telegraph system was in such poor shape that it could not be depended upon for reliable service. Pfitzner also argued that the commercial viability of the company required more than just access to one Brazilian city. Instead, the company needed its own cables to other important cities in Brazil and preferably to Rio de Janeiro, Montevideo, and also Buenos Aires. The fly in the ointment was, of course, that the Western Telegraph Company was no more amenable in 1909–10 than it had been in the recent past with respect to permitting direct access to more than one city along the Atlantic coastline of South America, especially insofar as the River Plate was involved. The fact that Felten and Guilleaume's Brazilian concessions required that the company negotiate any desired additional rights directly with the Western Telegraph Company seemed to erect an insurmountable obstacle. The idea that the company could deal directly with the Western Telegraph Company, as we have seen, was hugely problematic. While Felten and Guilleaume's interest had been accommodated to a considerable degree in the past, its relations to the Western Telegraph Company since about 1905 had been rather prickly. In fact, as the expiration date for some of its most important concessions drew nearer, the Western Telegraph Company assumed a more hostile and defensive attitude toward any challengers to its hegemonic position within South American communication markets. The upshot was that the German South American Telegraph Company now appeared to be at an impasse.[99]

Defending Monopoly: The Role of Law, Bribery, Diplomacy, and Industrial Espionage in the Struggle to Liberalize South American Communication Markets, 1909–14

The Brazilian concession granted to Felten and Guilleaume opened the floodgates for a slew of new initiatives as would-be new rivals redoubled their efforts to erode the Western Telegraph Company's concessions and to ensure that, at the very least, its south of Rio concession would not be renewed after it expired in 1913. In 1910, the Brazilian government again demonstrated its resolve to liberalize access to its international cable services market by announcing that it would not renew this concession. Appeals to Brazil's Federal Court by the affected company the next year failed to reverse that decision, and other moves soon followed to open up the Brazilian market in international cable services, while the tide of opposition against the Western Telegraph Company continued to swell.[100]

The Western Telegraph Company responded vigorously, as an Eastern Associate company would. In the space of just a few years, 1909–13, it was involved in legal action against rival concessions; it was bribing employees of the Brazilian Foreign Ministry and the Ministry of Public Works to act on its behalf; it was using the good services of the British ambassador, though hesitantly, for fear of the negative response which such an action might provoke with Brazil; and, not least, it was engaged in industrial espionage.[101] Taken together, all of these actions revealed two important elements: first, the extent to which the company still sought to use its old methods of influence peddling, bribery, and ties to local elites and to British consular officials for interventions in its support, but second, the waning utility of such measures in this context. That the Brazilian state and some of its officials still fell for the old methods was nicely illustrated in an example related by the general superintendent in Rio, William Robertson, to the chairman of the Western Telegraph Board in London. In this case, the friendship between the family of one of its managers, Mr. Dunlop, and Dr. Bulhões, the minister of public finance, led the latter and some of his friends in high places to squash an attempt in 1908 by the Chamber of Deputies to suppress the Western Telegraph Company's local delivery service because of its competition with the public telegraph service.

A few years later, however, Robertson was actually commenting on the honesty of the Brazilian judicial system. Indeed, as he suggested, it was far better to rely on the courts than face strong rebuke within Brazil if it became public knowledge that "we had invoked the support of the British Foreign Office."[102] In Robertson's estimation, such intervention should be held in reserve and only resorted to if an appeal to the Brazilian Supreme Court proved necessary. This balance between relying on the local Brazilian courts and British diplomacy was especially evident with respect to a case involving a concession that had been granted to a renegade ex-company employee, James

Reidy. The Reidy case was so important because immediately after obtaining the concession, Reidy had turned around and tried to sell it to anyone who had a modicum of interest in South America. That was like drawing moths to a flame, and it caused the company no end of grief for the next three years as it tried to sift through the deluge of rumors and intrigue that came to surround the concession. Unable to distinguish between real and imaginary threats, in late 1911, Robertson wrote to the secretary of the Western Telegraph Company Board in London of an unknown entity called the Farquhar-Legru group which, he claimed, was based on European money and American management. Representatives of this group, he claimed "say they intend to have cable service of their own either in agreement with Western or in opposition to it. They are taking up the idea in exactly the spirit in which the American-owned Atlantic cables were laid. Our position is now that occupied by the Anglo-American Company. We shall, I trust, be able to profit from the Anglo Company's experiences."[103]

With the benefit of hindsight, there is no evidence that any such group of this name posed a real threat. Yet, not all those attracted by the Reidy concessions were of such a wispy nature. Indeed, Robertson and the corporate management in the Western Telegraph Company were so worried precisely because Reidy was attracting serious interest from the French Cable Company, from Emil Guilleaume, and, returning to the stage after a long period of relative quiescence, James Scrymser. In short, there was no mistake that the Western Telegraph Company did have cause for worry. Not that it need have in this particular instance, since when its real potential rivals looked carefully at the Reidy concession, they found it worthless.[104] Nonetheless, and leaving nothing to chance, the company turned to industrial espionage. Thus, in March 1910, as Scrymser arrived in Rio to check out the Reidy concession he was tracked down by Western Company agents through a passenger list, and apparently, as "our friends in the [Brazilian] Foreign Office" informed Robertson, found the concession useless.[105] Through the same methods, Robertson mistakenly reported a year later that Scrymser had succeeded in getting an agreement from the Argentinean government for a cable to Brazil.[106]

The concern over Scrymser's supposed concession in Argentina was also important because that country was, after Brazil, the next most important theater of contestation in the struggle to open up South American cable markets. Indeed, between 1907 and 1910, there were no less than four serious bids to lay new cables either directly (with connections to Africa, of course) from Europe to Buenos Aires, or from there to Montevideo and points in Brazil: the South American Cable Company (1905), Société d'Études du Câble Franco-Argentin (1907), German South American Telegraph Company (1909–11), and the Central and South American Company (1910–19). However, in stark contrast to the liberalizing actions taken in Brazil, the Argentinean government was more inclined to maintain and even extend the Western Telegraph Company's control over the country's international communication services. The matter was not only a solid indicator of the strong grip of Britain's informal empire in Argentina at this time but also a reflection of the menacing saber rattling which was taking place between Argentina and its neighbors in 1909. The country's government saw in the Western Telegraph Company a useful ally of long-standing which it could trust to meet its needs for network security and military communications. Thus, in 1908–9, negotiations between the Argentine government and the company began with the aim of establishing a more direct connection from Buenos Aires to Europe without passing through Brazil. In January 1909, the Western Telegraph Company proposed to lay a new cable, without subsidy and at reduced rates, between Argentina and Europe by way of Ascension Island. The offer was accepted six months later and while the Argentinean government got its more direct route to Europe and cheaper rates without having to pay subsidies, the Western Telegraph Company also obtained valuable benefits in return: a twenty-five year monopoly on Argentina's international communication connections, sole rights to carry official messages bound for foreign destinations, and a powerful position from which to stem the mounting pressure of new would-be competitors.[107] The new cable was completed and opened for service in April 1910. While the cable had cost the company dearly, at around $5 million, it accelerated the flow and cheapened the cost of communicating with Europe and North America; it also preempted any other prospective effort along these lines. For good reasons, the deal was popular among Argentineans, a factor that helped anchor support for the company's monopoly not just in a stale set of legal documents but in the minds of those who used it; a clever move, the U.S. State Department acknowledged, which "would make the Government loath to change the contract."[108]

But if certain classes in Argentina were thrilled, foreign cable and diplomatic interests were livid. The American ambassador to Argentina, Charles Sherrill, for example, lodged an official protest with the Argentinean government, arguing that the claim violated commercial treaties governing trade relations between the United States and Argentina, to which the Argentine

Foreign Ministry replied with the statement that international cable services were public services and, as such, were not covered by commercial treaties. That was a neat response which cast international communication services in the language of public services with the aim of solidifying a monopoly, a move which did not look all that dissimilar to trends in the United States during the progressive era.

The Western Telegraph Company had also offered a similar deal to the Brazilian government, but not surprisingly was tersely rebuffed. In the end, the important but limited success in Argentina revealed that the company had maintained influential friends but also made some powerful enemies, and this was going to come back to haunt it. Indeed, of the three major eastern South American countries, Argentina stood alone while Uruguay and Brazil continued to push for the expansion of state-ownership over their national communication systems[109] and to liberalize international communication services. The Director General of Post and Telegraphs in Uruguay began discussions on the achievement of these aims in 1913 with the Progressive Era American Postmaster Albert Burleson.[110] A year earlier in Brazil, Deputy Augusto do Amaral introduced a bill in the Chamber of Deputies designed to lay a state-owned cable along the Brazilian coast, a measure which he specifically linked to the Western Telegraph Company's continued obstruction of the French and German companies' efforts to gain direct access to customers in Brazil "this side of Recife" and the River Plate Republics. Amaral complained bitterly that "the Western imposes on German and French companies rates more onerous than terminal and international rates ratified by our government and . . . dominates the international traffic as it sees best for its own receipts, ignoring the enormous prejudice caused to our country and the Republics to the south of us, a prejudice which cannot be remedied by calling into existence other companies."[111]

Amaral's comment on block rates also struck home for interested Americans, since many of the same obstructions placed in the way of the European cable companies by the Western Telegraph Company also applied to the Central and South American Company. Indeed, by comparison, the European cable companies' decisions to assume a more cooperative stance toward the Western Telegraph Company left them in a more favorable position than the U.S.-based Company, and that gap, as we will see momentarily, was soon to become even more pronounced as a new round of deals in 1910 sweetened the European companies' access to the River Plate and to a more tightly co-ordinated and harmonized set of relationships between each of the three main companies operating on the Atlantic coast of South America: the Western Telegraph Company (British), South American Cable Company (Anglo-French but becoming decidedly more French by this time), and the German South American Telegraph Company. By contrast, Scrymser's Central and South American Company in Argentina continued to be treated poorly by the Western Telegraph Company, not only dependent on it for access to customers outside the River Plate Republics but its customers subject to a punishing surtax of 25¢ per word over regular rates. The blatant discrimination of these block rates, the U.S. ambassador to Brazil in 1917 later complained, meant that senders often sent their messages to the United States via the considerably cheaper albeit more circuitous route of the Western Telegraph Company, that is, via England, where they were subjected to wartime censorship "with its diverse inconveniences for American interests."[112] But such was the resolve of the Western Telegraph Company that a combination of stalling tactics and protracted legal challenges delayed a deal to give the Central and South American Company access to Brazil until late 1919. And this deal occurred in the context of maneuvers that pitted the Western Telegraph Company and Western Union against Scrymser's corporate interests and, in turn, all the corporate interests against the views of the Wilson administration. The details of these events are examined in chapter 9. Meanwhile, the theme of the cooperation among the European cable companies in South America before World War I completes this section of the current chapter.

Schizophrenia: Communication, Cartels, and Collaboration on the Eve of Destruction

Ironically, it is now evident that the four years immediately preceding the outbreak of war witnessed the European cable companies drawing closer together. The first indication of this was the fact that Denison-Pender, about the same time as his company's new cable to Argentina began service, re-opened negotiations with the German South American Telegraph Company in 1910. These discussions bore fruit, since the German company reignited initiatives that had remained stalled for the past year and completed the final link in its substantial transatlantic communications system between the west coast of Africa and Recife. When the system opened for business on March 29, 1911, the European rate was immediately reduced to 75¢ per word, and the Western Telegraph Company and South American Cable Company, by prior agreement, quickly followed suit. Whereas rates had been roughly $1.50 in 1890, they had now fallen to half that amount and, furthermore, rates to

TABLE 10   The Explosion of Network Capacity in South American Cable Communication
            Markets, 1874–1911 (by number of cables)

COMPANY	1874	1885	1891	1905	1911
Western Telegraph Co.	1	2	2	2	3
Central and South American Co.	—	1	2	2	2
Compagnie Française des Télégraphes Sous-Marins	—	—	1	1	1
South American Cable Co.	—	—	1	2	2
German South American Telegraph Co.	—	—	—	—	1
Total Number of Cables	1	3	6	7	9

Europe were standardized, creating a kind of uniform "transatlantic communicative space" where the cost of sending a message from London was now the
same as it was in Barcelona, Paris, or Berlin.[113] More than that, the sheer
capacity of the cable networks linking Europe to South America had taken a
quantum leap forward, while those linking the Central and South American
Company to the United States remained much as they had been since 1893, as
shown in table 10.

Table 10 shows the large increase in the number of cables reaching South
America; yet, it under-represents the magnitude of change because it does not
account for the doubling of capacity that would have occurred by 1880 on
account of the introduction of duplex technology or the doubling again of the
capacity of the new fat cables laid after the beginning of the twentieth century.
Accounting for the use of these technological innovations, by the first decade
of the twentieth century we can reasonably estimate that the capacity to send
information to and from South America had probably increased in the range
of twenty- to twenty-five-fold between 1874, when the first cable was laid
by the Brazilian Submarine Telegraph Company, and 1911, when the German
South American Telegraph Company established the ninth intercontinental
link between South America and the rest of the world.

While there was a quantum leap in available cable capacity, the issue of
direct access continued to be a contentious one, but even here there were
strong signs between 1911 and 1914 that things were changing dramatically.
Continually blocked from laying a cable from Recife to Montevideo and
Buenos Aires, Felten and Guilleaume bought the largest urban and regional
network in the area in 1911, that of the River Plate Telegraph and Telephone
Company. At this time, the River Plate Company had five of its own cables

between these two cities as well as urban telephone networks in Montevideo and Rosario. It was, in sum, a critical hub in one of the most important communication markets in the world, and now Felten and Guilleaume owned it. It was a small stroke of genius and one that cemented the rise of the company over a relatively short period of two decades into at least the upper echelons of the tier two global communication players. Once the Western Telegraph Company's south of Rio concessions expired, the River Plate Company would become even more important to the German South American Telegraph Company, in essence serving as a feeder for that system and offering the company end-to-end control over its network in a key area of the world and one in which German stakes were growing rapidly at this time.[114]

The Western Telegraph Company seemed to have resigned itself to the coming expiration of its south of Rio concessions and that it would either have to meet its potential French and German rivals in competition to maintain its dominant position in the River Plate or somehow come to terms with them. The only significant remnants of its old hegemony over these markets was the Brazilian intercity monopoly, and it was the Central and South American Company, outside of the cartel, as it was, that seemed to bear the costs of that. Signs that Western Telegraph would choose cooperation over competition emerged soon enough. At the 1912 World Radio Telegraph Conference in Berlin, John Denison-Pender and the Ministerial Director of the German Post Office met to discuss the German South American Telegraph Company's desire to lay its own cable directly to the River Plate. During a secret, closed-door meeting between all of the European cable companies as well as representatives from the French and German Postal Telegraph and Telephone Authorities, in March 1913 agreement was reached that brought all of the parties together around the idea for a new cable between Rio and the River Plate. The cable would be jointly owned by the French and German companies, with an application jointly submitted to Brazilian authorities in the name of all three parties: the Western Telegraph Company, South American Cable Company, and German South American Telegraph Company, subject only to approval by the French and German governments. That, in the end, never happened because before the new *verständigung*, as Jorma Ahvenainen refers to this set of events, received government ratification, war had erupted and the stunned participants were left wondering just how they had fallen so far out of step with the dangerous pulse of the times. Rather than harmonizing the entire "transatlantic communication system" under the guidance of "private structures of coordination," the entire system was thrown into disarray by the war.[115]

THE ASIAN COMMUNICATIONS SYSTEM, EUROPEAN CABLE
CARTELS, AND JAPANESE IMPERIALISM, 1912-16

If the strongest challenge to circumvent the Western Telegraph Company's
control over South American markets came from Scrymser, in the Far East
similar challenges were mounted by the Imperial Japanese Telegraph Administration. Japan's growing commercial and imperial interests in Asia were the
driving force behind its early-twentieth-century efforts to extend its own role
in the Asian cable system and to disentangle itself from the concessions that it
had signed with the Great Northern Company in the 1870s, as previously
discussed. By the early 1910s, these efforts assumed a far different character
altogether befitting Japan's own national aspirations and its bid to consolidate
its rapidly expanding imperium in Korea, Manchuria, and Taiwan. And by the
middle of that decade, the Japanese regional cable network had grown from a
single line between Yokohama to Taiwan to become a system consisting of that
cable and a half-dozen others: two to the east coast of Russia at Vladivostock,
another to Pusan, one more between Japan and Taiwan, another connecting
Taiwan directly to Shanghai, and finally, after 1915, a direct cable from Nagasaki to Shanghai.[116]

The scope of Japan's cable interests and expansionist agenda gave it far
more clout in renegotiating the terms of the Asian cable cartel. Formerly
required to compensate the Great Northern and Eastern Extension Company-
led cartel for "lost business" due to its Shanghai-Taiwan-Nagasaki cable, that
requirement appears to have fallen in 1913. Indeed, two new agreements in
that year involving Japan, the just-mentioned British and Danish companies,
and the China Telegraph Administration revealed clearly that Japan had interests of its own that were now treated as being on a par with those of its
European counterparts.[117] The clearest "loser" in the deals was China and the
China Telegraph Administration, as its system was bound ever more tightly to
that of Japan as well as to the systems of the Great Northern Company and the
Eastern Extension Company. This was an outcome of these agreements and
also a series of highly questionable loans.

Although U.S. communication interests and investment in telegraph, cable, and telephone systems in Asia paled alongside British, Danish, Japanese,
and German interests at this time, the United States continued, much as it had
in the past, to play a large diplomatic role in the region trying simultaneously
to hem in Japan's growing imperial ambitions while asserting its own, yet all
the while oscillating wildly between the two models of foreign policy and

empire that had characterized this region and global politics for decades: the spheres of influence and straightforward imperialist model of global relations versus the collaborative model of empire and the internationalization of control. That, too, was decisive, because the latter model had largely gone into abeyance after the Boxer Rebellion of 1900 in China and the resulting scramble among the Europeans and Japanese to carve out their own mutually exclusive spheres of domination in the beleaguered country. However, after a decade in which the problem of imperialism only became more severe, the model of collaborative empire was once again trotted out by President William Taft in 1909 and made a cornerstone of international banking and industrial development projects in China.[118] The prospects for that approach, however, were severely jeopardized by the subsequent Wilson administration's erratic swings between both models and the fact that each of the other imperial powers looked upon America's growing interest in Asia with much suspicion. The fact that Wilson's approach to China looked more like a Rorschach inkblot test than a clear-cut statement of policy also allowed the Japanese to define deals which they signed with the United States as giving them carte blanche in their own "sphere of influence," that is, Korea, China, and Taiwan. This was most observable with the Lansing-Ishii agreement of 1916, which Japan read as a straightforward endorsement of its interests in the region in return for recognition of the United States' growing imperium in the Philippines, Hawaii, and Guam.[119] And given that the much promised independence of the Philippines after eighteen years of tutelage was rejected by the U.S. Congress in 1916, Japan appeared to be not far from the mark. Wilson's unilateral withdrawal from the International Banking Consortium in 1913—one of his first major policy initiatives upon taking office—only a few years after it had been endorsed by his predecessor, William Taft, also suggested a lack of substance to Wilson's sketchy policies on international collaboration at this time.

While the relations between China and Western nations such as the United States were important, by far the largest force in the region—aside from Britain's considerable imperial interests—was Japan. And on this score, as Daqing Yang notes, after 1912 Japan was committed to a comprehensive international communications policy in which cable connections to Shanghai— "the international hub of news, finance and commerce in the Far East"—was the centerpiece.[120] Agreements signed by all of the abovementioned corporate parties in 1913 paved the way for a new Japanese cable to Shanghai in 1915, which handled Japanese language telegrams and European-language government messages between the two countries. Yang claims that the Great Northern Company only permitted the Japan to Shanghai cable after securing commitments that it would be compensated for any lost business that resulted. This is possible, but seems at odds with the assertive status of Japanese interests in the 1913 agreements and indeed, the relevant contracts make no reference to compensation for lost business.

The "lost business" requirement probably applied between 1904 and 1912 when Japan was still struggling to gain a footing among world superpowers, but by the latter date, as Eric Hobsbawm acerbically notes, Japan was a wolf among wolves.[121] Indeed, if more evidence is needed of Japan's growing influence, the contracts provide it, noting that the entire Asian market north of Shanghai (in accordance with the territorial divisions set between the Eastern Extension and the Danish-based Great Northern Company back in the 1880s) was to be shared two-thirds to the Great Northern Company and the rest to the Imperial Japanese Telegraph Administration. While not equal, it certainly left little room for doubt as to the rapidly tilting balance of power between the two agencies. To weigh down the balance further, the agreements made Japan's growing power even clearer: no new cables would be laid by either company without the consent of the other, and while the Great Northern Company would have preference in linking Japan to points south of Shanghai, notably the Philippines, it had no monopoly over any such developments. Finally, both parties set the rates for ordinary, government, and press messages. In a nod to the global superpowers' foreign policy agenda and the modernizing backdrop against which these events were taking shape, the door was kept open for the China Telegraph Administration to join the agreement sometime in the future. That never happened.

CONCLUDING REFLECTIONS ON GLOBAL
COMMUNICATION IN THE AGE OF MANAGED MARKETS,
EMPIRE(S), AND SOCIAL REFORM

As Yang concludes, Japan was hardly "a helpless victim of rapacious Western imperialism, or a blind imitator of Western expansionism."[122] In the remarkably short span between, roughly, 1903 and 1916, it had staked out a formidable role in the system of global communication, not at the top, to be sure, but solidly among the tier two regional cable system operators and with far more power than most to assert its own interests.

Another kind of empire was also flourishing as we have seen in these pages, namely that of the now huge tier one and tier two global communications companies. Decisive shifts took place as British corporate control over the North Atlantic cables was largely ceded, by 1912, to two groups headed by Western Union and the Commercial Cable Company. Elsewhere, however, the Eastern Associated Company still towered over all others, with its two biggest affiliates, the Western Telegraph Company and the Eastern Extension Company, dominating communications markets in South America and Asia and using their formidable power to fortify the cartels they had led in each of these regions for decades. And that was vital in itself because it demonstrated that while there was a stronger desire for greater autonomy by national and corporate interests, those interests were still managed through the long-standing private structures of cooperation that had existed in global communications since the outset. Thus, unlike the northern transatlantic economy, there was no transfer of control in South America but instead, between 1902 and 1914, a process whereby the Western Company managed the forces of change by forging alliances with two new companies—the South American Cable Company, owned by the French Post and Telegraph Authority, and the German South American Telegraph Company, owned by the German firm of Felten and Guilleaume—while staving off persistent threats by James Scrymser's Central and South American Company. In Asia, the Eastern Extension Company had overseen a similar process, albeit several years earlier (1902–6), and while the cartel persisted for the next several decades it was clear that these private structures of cooperation were no mere means for squelching rivalry outright, but a mechanism for managing change. And from 1912 to 1916 there was much change, instigated mostly by the rapidly expanding new imperial Asian power: Japan and the Imperial Japanese Telegraph Administration. So, whereas in the nineteenth century it was probably appropriate to speak of a clear-cut British hegemony over global communications, by the middle of the second decade of the twentieth century, that was no longer the case. Now, market shares and spheres of power and influence were parceled out in a more conscious way and with results that chiseled away at the Eastern Associated Company's hegemony over global communications.

And while the Eastern Associated Company and Western Union sought to retain or expand their hold over communication markets by adopting unified structures of corporate control, the governments of Britain, France, Germany, Japan, and, somewhat later, the United States took more concerted efforts of their own to promote the expansion of cable communications systems in ways that were, in varying degrees, more susceptible to the influence of national interests. In Britain, those efforts began with the revival of subsidies but grew to include the state-owned Pacific Cable in 1902 and the acceptance of a larger role for state regulation thereafter. In France, the state promoted the consolidation of private companies in the mid-1890s into the French Cable Company and continued its efforts by adopting new legislation in 1902, acquiring the South American Cable Company in the same year and steadily expanding in South America thereafter. In Germany, much the same trend could be seen as two affiliates of Felten and Guilleaume, the German South American Telegraph Company and the South American Telegraph Company, carved out key roles for themselves in the transatlantic communications economy and as another of its affiliates, the German-Dutch Telegraph Company, expanded into the Far East to meet German and Dutch commercial and imperial communication needs in the region.

Several important observations can be made about these trends by way of concluding this chapter. First, rather than competing with existing companies or charting a completely autonomous course of action, newcomers joined the existing cartels. The point is especially noteworthy given the fact that many of those newcomers were either state owned or heavily backed by their respective national governments. Second, there is not much reference to the United States in the above summary. That, however, reflects the reality that before the mid-1910s the first steps by the American state were more tentative, less coherent, partially preempted by the Commercial Pacific Company, and more restricted to its own "zones of empire" in Central America, the Caribbean, and the Philippines. The limited role of the American state during this time becomes especially apparent if one looks at the most direct form of state intervention: cable ownership. Not surprisingly, given the trends explored in this chapter, the balance between state-owned to privately owned cables nearly doubled from roughly 10 percent in 1910 to nearly 20 percent a decade later, but almost none of that increase was accounted for by the United States, much to the chagrin of American imperial and military bureaucrats who argued that this was a huge oversight.[123] These observations do not constitute a naïve view of American power, but an effort to represent accurately the relative scale of American state and corporate power in global communications at this time. They also establish a benchmark by which the substantial expansion that later did take place, can be assessed.

Finally, this chapter has shown that there was a decisive elevation in the role of news and the global news agencies between 1890 and 1914. While restricted mainly to developments in South America (later chapters look further afield), this chapter showed that changes in state policies affecting the organization of the cable communication industries were connected to changes in the global news business and that those changes largely followed the temporal sequence summarized above. That is, in 1890 Reuters and Havas brokered a restructuring of the global news cartel that left South America solely to Havas in return for gains elsewhere. This step allowed Havas to vastly expand its position in South American news markets, a position left largely untouched until after the period covered by this chapter, except for some significant concessions made in 1903 to permit a large role for Associated Press in the American zones of formal and informal empire, and, after 1908, by the unilateral but quite minor incursions of the Transatlantische Büro, a German news organization created by the German government. In short, this was a crucial period in which networks and news were connected in vital ways that had not been so observable in the past, and as governments assumed a more integrated approach to the network and content industries in their policies. It was also visible in the growing perceptions of the impact of global news on foreign policy, public opinion, diplomacy, and markets—hence those Americans who worried about the influence of fickle public opinion on foreign policy, such as Charles Eliot, the president of Harvard, and those who concerned themselves with the quality of news about American society distributed in foreign countries, such as Frank Noyes, Associated Press president.

And as a final word on this matter, the press was not just becoming, as noted in the quotation by Walter Lippmann at the beginning of this chapter, the decisive intermediary between the world outside our experience and the images in our head, but had assumed, especially in the northern transatlantic economies and the British Empire, a vital role in rekindling the politics of global media reform around 1906. The role of the news organizations in such events helps shed light on how broader trends in this age of social reform intersected with the politics of communication in ways that managed to eke out a small measure of benefits "for the millions" in the pithy but pungent statement by the *New York Times*. This was not a change to be taken lightly, and it was indeed the first inclination that people's interests and not just those of large corporations, stock market speculators, governments, and the press would be given at least some consideration. All in all, the years 1906–16 represented a very significant turning point in global media history.

Wireless, War, and Communication Networks,

1914–22

> Whether wireless telegraphy will or will not displace the cables is a
> question which only time can decide. The view that it will soon be one
> of the principal means of communication over long distances is unpop-
> ular in England where over $300 million is already invested in cables.
> —MARCONI, "The Progress of Wireless Telegraphy"

> What we are suffering from in this country [Britain] are the atmospherics
> which exist between the Post Office and the Marconi Company.
> —ROBERT DONALD, Chair of Empire Press Union, 1923
> (Britain, *Imperial Economic Conference*, 1924)

> Navy Responsible for Formation of 100 Percent American
> Commercial Companies.—*Navy in Peace*, 1931

A small cartoon guide produced in Britain in 1986 is entitled *Media and Power: From Marconi to Murdoch* and refers to Guglielmo Marconi's genius and "unerring instinct for publicity" which made him a household name in many Western countries by the late 1890s.[1] Marconi's rise involved both a challenge to existing interests and a concerted effort from the beginning to stake out a dominant position over the new wireless electromagnetic waves, primarily through the registering of patents galore and an unending stream of legal actions against any and all who appeared to tread even slightly on the company's patents.

The uneasy interplay between new and old media of communication was already on display in 1901 when Marconi transmitted his famous transatlantic message from Newfoundland to England. From then on, aided by the Marconi publicity machine, the wireless craze went into high gear.[2] Indeed, as a 1909 U.S. financial review noted, in twelve years, the global wireless industry had grown from 1 system, 2 stations, and 16 patents to 36 systems, 1,250 stations, and 2,489 patents. The distance covered by wireless messages expanded from 14 to 3,496 miles and—the major point of the review—capital investment had grown from $50,000 to a "phenomenal" $132 million.[3]

This chapter focuses centrally upon the new medium of wireless communication, first during its early years prior to World War I, followed by attention to its role as an agency for the distribution of international news, information, and propaganda during that war. It concludes with some observations on the creation and early corporate actions of RCA between 1919 and the early 1920s and on elements of British wireless policy. The chapter relies heavily on U.S. and British governmental investigations of developments in wireless technology and the industry during this time as well as reviews of wireless policy conducted in both of those countries. The secondary literature on wireless technology is truly voluminous, but here, for such literature, we rely mainly on the existing studies of wireless development and policy in such countries as Britain, the United States, and France by Hugh Aitken, Susan Douglas, Pascal Griset, Daniel Headrick, and Jill Hills.[4]

TECHNOLOGY, ORGANIZATION, AND POLITICS OF WIRELESS, 1900–14

The pace of innovation underpinning the rise of wireless communication in the first decade of the twentieth century was brisk indeed. Many individuals and organizations made key scientific contributions to the development of wireless during this formative decade besides Marconi, the most notable being the Danish Poulsen system, the American de Forest, and the German Slaby-Arco (Telefunken) system. But even by the middle of the first decade of the twentieth century the big two were already rising to the top of the global wireless business: the Marconi Wireless Telegraph Company, inaugurated in 1897 with its headquarters in England and subsidiaries established swiftly in the United States in 1899, and Canada in 1903; and the German company Telefunken, created in 1903 through an arrangement between the major communications and electrical equipment companies of Siemens and Allgemeine Elecktrizitäts Gesellschaft (AEG), with considerable support from the German government. Likewise, Marconi had a measure of initial help from the British General Post Office, as well as in Canada, and benefited particularly from agreements with the Italian and British navies and Lloyd's Marine Insurance. For example, Lloyd's, with 1,300 agents and subagents in 1901, made a fourteen-year agreement with Marconi International Marine Communications Company, an offshoot of Marconi Wireless Telegraph, which allowed Lloyd's to use Marconi apparatus at all their stations, *and engage to use no other*, and also not to communicate with ships using any other system, the only exception being in American waters. The Lloyd's agreement was a substantial boost to the Marconi organization, but a bad deal for Lloyd's as it prevented them from establishing stations at their offices in various British colonies, including Ceylon, Barbados, and Jamaica, and was soon modified, following litigation.[5] The agreement was, nonetheless, typical of the Marconi corporate strategy of exerting far-reaching control over the use of its patents and equipment.

Given the international ramifications of wireless interconnections, a World Radio Congress was held in Berlin in 1903. Although no agreement was achieved, the principle of "compulsory intercommunication" between different wireless systems was accepted three years later, with provisos, by twenty-seven countries, including Britain, at another Radiotelegraphic Convention. Significantly, the United States did not sign the 1906 agreement out of deference to opposition among industry and amateur wireless groups, despite the fact that the American delegation to the Berlin conference was largely in favor of the agreement.[6] Meanwhile, however, via the British Wireless Telegraph Act of 1904, that nation's ships were required to have a wireless license from the postmaster general, the aim of the act being "to provide against the growth of a monopoly in the hands of any one company." That the Marconi organization was aiming at a monopoly, there can be no doubt, although the company argued that it was just trying to ensure order through unity of control (an argument that Theodore Vail would have supported). A British select committee recommended in favor of the act's ratification and made a statement which, in one guise or another, was going to come back to haunt Marconi and future British wireless policy: "The Marconi Company . . . cannot be regarded as having any claim to a monopoly of wireless-telegraphy."[7]

Some observers, notably in Britain, viewed the Berlin conference as being intentionally designed more to advance the interests of the flagship of German wireless, Telefunken, than to secure a global public good. During parliamentary debates over the ratification of the convention, it was clear that several British Members of Parliament had taken their cues from commentaries which had emanated from the Marconi public relations machine about the horrors which would emerge from German wireless competition on British shores. Indeed, Edward Sassoon and John Henniker Heaton also joined the fray, contributing their own articles in support of Marconi to popular journals and staunchly backing the company in British parliamentary debates. Heaton, as a personal friend of Marconi, made it clear that he saw wireless as the technological antidote to the cable barons and not to be hamstrung by foreign inspired regulations.[8] However, perhaps the best comment came in debate from Postmaster General Sydney Buxton, when he noted that the Marconi Company had done its best to obtain a monopoly, "but the fact that they have not succeeded made it largely necessary to have this Convention. The only alternative to world-wide monopoly was organization under international convention and regulation."[9]

The Marconi organization's standing in British political circles took a further beating in 1912 after a contract awarded to Marconi by the British government for an "Imperial Wireless Chain," subject only to parliamentary approval, ended in scandal. The Marconi scandal erupted when then Chancellor of the Exchequer David Lloyd George, Attorney General Rufus Isaacs (brother of Godfrey Isaacs, Marconi's director), and Postmaster General Herbert Samuel were accused of insider trading by buying shares in American Marconi from Godfrey Isaacs before trading actually began. The Marconi scandal reached into the upper echelons of the Liberal government but also roused a disgraceful bout of anti-Semitism in Britain. The scandal did lasting harm to the image of the Marconi organization—albeit not to the founder himself.[10] In the end, the sordid affair was cynically covered up, but not before the Imperial Wireless Chain deal was heavily criticized for being unduly considerate to Marconi interests.

Ultimately the Marconi contract was not awarded until 1913, and only a couple of stations were under construction when the General Post Office cancelled the contract in December 1914, without explanation. This was the beginning of a stage of the history of wireless in British politics, based on endless litigation between Marconi interests and the Post Office; a history which *Wireless World* remarked later was so bifurcated—scientific advance on one hand and "spies, plots, intrigue, secret documents, startling revelations, and all the other elements of melodrama [on the other] . . . that it is difficult to say at present whether the history . . . ought to be written by, say, Sir Erskine Murray or William Le Queux" (that is, by a well-known wireless engineer or a contemporary pulp-fiction writer of spy novels).[11]

MARCONI'S CORPORATE EXPANSION AND THE ADVENT OF WIRELESS IN NORTH AMERICA, PRE-1914

Wireless technology underwent a radical transformation between 1900 and 1914, notably in a growing shift from spark to continuous high-frequency radio waves.[12] However, not all companies seized on these new innovations with alacrity, and Marconi was among them. Indeed, its reluctance to embrace new technologies from early on, coupled with the political problems that it was encountering, brought the organization into low financial straits by 1908. However, a newly appointed director, Godfrey Isaacs (1908) made a remarkable turnaround through a vigorous worldwide campaign of patent litigation and other commercially adroit measures which transformed the Marconi organization into an international company with numerous subsidiaries and intercompany agreements. The overall corporate strategy behind the Marconi rise to preeminence is finely described by Hugh Aitken, as follows:

> In its original form this strategy had called for the establishment of corporate subsidiaries in all countries where there seemed prospects for profitable radio traffic or for the sale of radio equipment. In these subsidiaries Marconi personally, the other directors of the parent firm, and that firm itself in its corporate capacity held a substantial though not a majority interest. Each subsidiary held an exclusive license to use Marconi patents on its territory, and in normal circumstances it obtained its equipment exclusively from its British parent. In countries such as France and Germany, where well-entrenched private corporations controlled radio, the British Marconi Company . . . sought alliances through corporate treaties and sometimes the purchase of stock interest. In France, for example, the Marconi Company held a substantial equity interest in the Compagnie Générale de Télégraphie sans Fil (TSF); and in Germany, after 1910 Marconi and . . . Telefunken were joint owners of the company that handled all German merchant marine radio, and in addition had agreed to pool their patents. In countries such as Norway and Sweden where government controlled radio, the Marconi Company sought to become the preferred supplier of equipment, and bid aggressively on contracts for the construction of new stations.[13]

Aitken's description also shows that it did not take long for the large wireless companies to make deals which deeply integrated some of their corporate interests, including patent sharing, cross-stock ownership, and joint ventures. Indeed, we may note that the corporate giant of Siemens Brothers was, in the years before 1914, a major supplier of telegraph equipment to domestic telegraph systems in Europe, and with AEG, had acquired a 50 percent stake in Telefunken upon the latter's creation. Since Siemens was clearly a major multinational corporation, including substantial long-term English interests, it is hardly surprising that Telefunken found little difficulty in reaching an agreement with Marconi on German marine radio, nor indeed that in the early 1900s, Siemens and AEG would reach an agreement with the U.S.-based General Electric (GE) and Westinghouse to exchange technical know-how

and divide markets which allowed these companies to create a global oligop-
oly which lasted until well after World War II.

This was a time of relatively open borders as well as international corporate
consolidation and cooperation in the communications and electrical technology business.[14] And as another index of that openness stood the American
decision to initially allow foreign-owned wireless firms to build stations on its
territory. By 1914 Telefunken, through a subsidiary, built a station at Sayville on
Long Island which was in constant contact with another at Nauen near Berlin,
and another German firm had built a high-power transmitter at Tuckerton in
New Jersey for the French Compagnie Universelle de Télégraphie et Télé-
phonie sans Fil. The fact that General Electric and Westinghouse subsidiaries,
along with the English branch of Siemens, controlled two-thirds of British
communications and electrical manufacturing output in 1912 also points to
the highly globalized nature of the communication business at this date.[15] The
forced divestiture, in 1914, of the latter from the corporate organization in
Germany of which it was a part came as an abrupt signal that things were
about to change dramatically.

But to return to Marconi, the company policy was to establish subsidiaries
wherever this was possible and commercially viable. By 1910, in addition to
the subsidiary companies in Canada and the United States, and a virtual
monopoly over wireless in Britain, Italy, and Canada, the Marconi subsidiaries had entered Russia, Spain, and possessed, alongside other companies,
concessions in Latin America and China; indeed, wireless communications
between Canada and Argentina were reported to have taken place as early as
1910. Yet, by 1914, the Marconi Wireless Telegraph Company of America was
probably the jewel in the Marconi corporate crown. The company started off
as a small enterprise and might have remained that way but for two factors:
Marconi's remarkable ability as an entrepreneur and what Susan Douglas
describes as " the truly maladroit handling of business affairs by Marconi's
American competitors."[16]

Thomas Streeter identifies a third decisive factor behind Marconi's success
in the United States: an early decision to treat wireless "as a service rather than
a manufactured product, and by controlling that service through the control
of critical technology, and through policies restricting how that technology
should be used."[17] By 1912, this combination of strategies and the misfortunes
of others allowed the company to become dominant in the field of American
wireless. It operated a large number of stations for marine traffic; provided
most of the equipment and operators for wireless installation in ships of

British and American registry; had begun construction of two high-power stations to exchange traffic with sister stations in Britain; had just completed stations in California and Hawaii; was looking towards an eventual link with Japan; and finally was testing two stations in Massachusetts for communication with Norway when war intervened.[18]

In the United States, Marconi's corporate strategy was much assisted by the terms of the 1912 Radio Act. The act was initiated belatedly to ensure intercommunication and public safety at sea—the recency of the Titanic disaster was not coincidental. The act also gave the government extensive regulatory authority, a power that was used immediately to make spectrum assignments and in the process relegated the large body of amateur wireless enthusiasts to short wave, which was then thought valueless. These powers also allowed regulators to become arbiters of the national and public interest and were thereby connected to the consolidation of corporate power and its legitimation. In Streeter's view, the act was not a simple triumph of big organizations over individuals, but rather "reflected the triumph of a particular configuration of business organization, technology and state action, a configuration characteristic of corporate liberalism: corporate private-sector cooperation with the public sector, small businesses relegated to a secondary role, and grassroots non-profit activities pushed to the fringes."[19]

While American Marconi, as the dominant firm in U.S. wireless, clearly benefited, so too did two other commercial wireless networks in the United States at the time. One of these, the Federal Telegraph Company of California, following its development of a network of wireless stations in the central and western United States to compete with the telegraph companies, had opened a high-power station which, by 1912, was transmitting news and commercial information between San Francisco and Hawaii a couple of years before American Marconi got in on the act. The other, the United Fruit Company, through its Tropical Radio subsidiary operated a number of land-based wireless stations in the Caribbean and Central America and used wireless to control the movements of its extensive fleet.[20] And then there was the U.S. Navy, not really a commercial rival, but an early user of wireless facilities, and an agency which later became an ambitious promoter of nationalist visions for wireless technology. One should add to this list such companies as GE, AT&T, and Westinghouse, which were also busily engaged in research, buying up patents and using them to break a plethora of small wireless and long-distance phone companies—just as American Marconi itself was doing.[21]

American Marconi's powerful position in American wireless allowed its managers to enjoy a good deal of autonomy from British head office control. It was also clear at an early stage, however, that not everybody was inclined to welcome Marconi interests. Indeed, among the American delegates to the 1906 World Radio Congress was the retired U.S. Navy Commander Francis Barber who was strongly anti-Marconi. Over the course of the decade, he continued to hope that the U.S. Navy "would be able to drive the American Marconi Company out of business" and his strong opinions had an impact on some of his colleagues.[22] Although not all agreed with him, the 1912 Radio Act was seen by the Navy as a chance to "assert certain, if not absolute control, over the spectrum"[23] . . . a view which clashed inevitably with American Marconi's ambitions.

THE U.S. NAVY AND ITS SUPPLEMENTAL ROLE IN COMMERCIAL WIRELESS

In his book on the history of communications technology, Brian Winston writes of the "law of suppression of radical potential": efforts made to ensure that the advent of a new medium does not disrupt a social or corporate status quo.[24] Britain's desire to support its huge cable system, as well as a political ambivalence toward wireless that lasted until well into the 1920s, is a good example of this law. Similarly, in France, the advent of wireless was peculiarly retarded by a complex and inert state telegraph and telephone bureaucracy, the regulations of which were seemingly designed to discourage bright entrepreneurs from creating large-scale, indigenous wireless plans for France and its colonies.[25] Germany, in contrast, had what Winston calls an important "accelerator," that is, a limited commitment to cables and the recognition that wireless could be an alternative in the eventuality that its cables would all be cut by the enemy, should war break out. So too, the United States was neither encumbered by a large, competing cable system nor by a state bureaucracy intent on maintaining its own authority, and indeed by 1914 the United States was certainly well ahead of both Britain and France in wireless installations and technological sophistication. Although the U.S. Navy certainly exhibited a traditionalism which limited its use of wireless for a long period, it pushed strongly and consistently for greater state involvement with the new technology. Thus, in 1904, President Theodore Roosevelt appointed the Inter-Departmental Board to consider the whole question of wireless telegraphy in the service of national government. The board recommended that the Navy should equip, install, and complete a coastal wireless system of stations covering the entire U.S. coastline, its insular possessions, and the Panama Canal Zone. These recommendations accepted, by 1915 the Navy had 250 ship stations and 50 shore stations, including a high-power station at Arlington in Virginia and a station at Beijing for contact between the American legation and U.S. ships in Asiatic waters. Also authorized in 1912–13 was a long-distance chain of powerful Navy continuous wave stations to bind together the United States imperium, with stations in Panama, Honolulu, Puerto Rico, the Philippines, Guam, and American Samoa. This American imperial wireless communications network surpassed anything the British or any other nation could claim, because of its continuous operation and technical efficiency.[26]

Equally important, William Bullard, a naval officer and later an admiral, noted that the early station transmissions were devoted to shipping interests, such as time signal transmissions and weather forecasts. The Navy also began receiving "radiograms" of a commercial nature from ships at sea for further transmission by telegraph, telephone, or cable lines and vice versa, from inland to ships at sea.[27] Bullard also noted that there were several agreements among governments permitting the exchange of messages between, for example, Florida and Nassau in the Bahamas, where the radio system paralleled the cable, and from the naval station at Colon in Panama, where messages were sent directly to Columbia and to Costa Rica, thus avoiding complex long-line telegraph and cable transmission. The tariff for all this, noted Bullard, was the same as for commercial stations: 6¢ a word for regular traffic and 2¢ for press messages.[28] The commercial expansion of Navy wireless services became a crucial element, after 1912 and especially after the United States entered the war, as evidenced in the Navy's much more concerted endeavors to control the wireless spectrum, which is the time when President Woodrow Wilson's Secretary of the Navy, Josephus Daniels (1913–21), aided by his subordinates— most notably, Lieutenant Commander Stanford C. Hooper, head of the Naval Bureau of Steam Engineering, and Bullard—began a campaign to achieve a naval monopoly on wireless in America.

WIRELESS, THE CABLE SYSTEMS, AND THE CUSTOMERS, 1900–14

When early international wireless service sought to gain a market niche in competition with the cable systems, one potential attraction, as we will see below, was the prospect of cheap rates, notably for the press. This was made all the more effective by the fact that, even though a substantial element of cable company business was the flow of international news, news organizations increasingly felt ill-treated by the cable companies. Press rates were lower than for ordinary messages, but not so low as to prevent frequent complaints about the costs of news transmission. But there was also a catch in the lower press rates, namely, that the cable companies made most of their income from expensive full-rate messages, and, therefore, tended to place news material lower down their list of priorities. Indeed, as late as 1920, the American Publishers' Committee on Cable and Radio Communications was complaining that "the great [cable] corporations controlling the world's communications collect tens of millions of dollars annually in tolls without any firm avenue of recourse by those who find themselves the victims of overcharge or vital delays. These corporations set up their own laws from which there appears to be no appeal or which lead those who challenge them into the intricacies of conflicting national courts."[29]

From this perspective, one can appreciate the importance of the aforementioned Federal Telegraph Company's decision to offer long-distance service between San Francisco and Hawaii, commencing in 1912. The Commercial Pacific Cable Company's existing service was both poor and expensive: 35¢ per "ordinary" word from San Francisco to Honolulu, 16¢ per word for the press, and $1.08 per word from San Francisco to Manila. The Federal Telegraph Company was able to compete with a regular rate of 25¢ and a press rate at an extraordinarily cheap 2¢, with a guarantee of a daily press bulletin of 1,500 words.[30]

The achievement of Federal's founder, Cyril Elwell, was technically remarkable, but he was competing with a particularly weak and expensive cable line in a market which was desperate for cheaper alternatives. Thus, on the other side of North America, a different picture is given by the only significant commercial wireless service across the Atlantic at the time, Marconi's service between Glace Bay in eastern Canada to Clifden in Ireland. This service opened for commercial work in 1907, and there is little doubt that Marconi believed that his main clients would be the press. As Dunlap notes of Marconi's policy at this period, "Why not a press service across the Atlantic? Newspapers clamored for it. . . . Wireless speeded news while it was fresh."[31] In 1907, the first wireless press message from east to west across the Atlantic came to the *New York Times* via Marconi and Western Union telegraph wires. It began a commercial relationship between the *New York Times* and the Marconi service that gave the latter valuable publicity and the former an image as a technically savvy newspaper.[32] Thus, between 1907 and 1914, the *New York*

Times boasted repeatedly of its growing use of international wireless services, and, as a case in point, in January 1912 outlined in detail how the newspaper obtained its daily news dispatches from Europe. Thus, in addition to news by wireless from London, news from Berlin and Paris, and soon from Spain, was sent by *New York Times* correspondents via long-distance telephone to London and thence by wireless via Clifden to Glace Bay.[33] Later in 1912, the *New York Times* claimed to get 30,000 words a week by wireless with an average time from London of 30 minutes.[34] Then, with the advent of "deferred rates" which included wireless, messages could be sent across the Atlantic at 4¢ a word, or half the cable rate.[35]

Marconi was so grateful for the support of the *New York Times* that, in 1911, his first message from a new high-power station at Coltano in Italy was sent to the newspaper. Two years later, high-power stations established in Wales and New Jersey allowed direct contact between America day and night and were linked to a deal between American Marconi and Western Union which allowed the speedy movement of wireless messages across the landlines. Yet, for all the positive publicity of the *New York Times*, wireless companies with international services such as the Canadian Marconi Company and the Federal Telegraph Company do not appear to have been profitable during peace time. In 1919, the former was noted as never having paid a dividend and was seen as something of a burden to the parent company in England.[36] Between 1911 and 1914, even the Federal Company benefited more from capital gains on its share values than from dividends paid to shareholders.[37] Commenting mainly on the Canadian connection with Britain, Griset notes that the problem was that "the regularity of connections proved to be disappointing, and the clients few in number."[38] The former point was certainly correct with respect to press service up to 1912—aside from the publicity, the *New York Times* used the service partly because it was cheap, but relied on cable for urgent dispatches—but, in 1912, the paper recognized that much of the problem lay with the landline connections and companies. Hence, it established a direct line to Glace Bay and thereafter, up to 1914, obtained such good service that the main Atlantic cable companies each thought that their rivals were surreptitiously working with the newspaper. In particular, the German Cable Company peevishly claimed that the whole wireless deal was a fraud. In essence then, wireless was poised to become a significant, if still small-scale, alternative to the cables by 1914, with not only several New York papers but also the *Chicago Tribune* obtaining much of their European news from wireless service.

FROM PREVENTING TO CAUTIOUSLY PERMITTING
WIRELESS DEVELOPMENT IN LATIN AMERICA, 1904–20

As Hills notes, in Latin America and even Japan, such had been the experience with the exploitation of the cable companies that wireless looked like a fine alternative that could be controlled without diminution of sovereignty. However, the early response to the threat of wireless by these cable companies was to renegotiate their concessions in such a way as to shut out wireless. Thus, in 1904, when faced by competition from Telefunken, the two companies that controlled Mexico's access to the international cable system—James Scrymser's Mexican Telegraph Company and Western Union—essentially obliged the Mexican government to send all international traffic by cable. Elsewhere, and in the same year, Scrymser cut the Central and South American Company's cable connection with Columbia in order to force the Columbian government to reject a bid for a competing concession from Marconi, and in Venezuela, the French Cable Company emerged after its long-running dispute with the government with rights over wireless.[39]

Taking a slightly less obstructionist tact, the Central and South American Company also acquired the right in Ecuador and Peru to install wireless stations to supplement its existing cable systems, although that would only happen in the distant future. The only really significant U.S. presence in wireless systems of Latin America and the Caribbean in the first two decades of the twentieth century was through the U.S. Navy and the United Fruit Company, whose subsidiary Tropical Radio and Telegraph Company dominated wireless services in Central America and to Puerto Rico, Haiti, and the Dominican Republic.[40] Other examples abound, and they tell us two things: first, that there was an early and widespread movement in Latin America to adopt wireless, and second, that the foreign cable and telegraph companies would use strong-arm tactics to blunt its competition.

Caribbean Colonies and the Cautious Approval
of Wireless, 1910–14

Over time it became clear that such rear-guard defenses of monopoly against wireless would become ever harder to sustain. Indeed, by the mid-1910s it was clear that a niche was being carefully whittled out for wireless in Latin America, but with the overriding proviso that the existing cable systems should not be injured by any new commitments. Meanwhile, British and colonial officials recognized, at a relatively early point, that the two media could have an important complementary relationship in providing speedy long distance communication within and between regions of the empire. For example, in 1910, a British Royal Commission on trade relationships between Canada and the West Indies devoted substantial space to the poor state of the Caribbean cable system and noted that, over the previous eight years, proposals had been made to establish wireless installations in one or more of Britain's Caribbean colonies. The commission's report forthrightly noted that these proposals had to be dealt with carefully "in view of existing interests and the fact that radio-telegraphy was still in an experimental stage."[41] The report also highlighted the West India and Panama Telegraph Company's use of wireless as a supplement to its cable system as a model of how "new" and "old" communications media could be developed in a complementary way. Other examples of where wireless had been beneficially introduced were also offered. Indeed, the General Post Office observed that the Caribbean cable companies' request for increased subsidies had been rejected because of the possibility that wireless communication might ultimately displace the existing cables. However, it did see the potential for gradually introducing wireless in a manner which might actually enhance cable receipts, notably, between islands not served by cable systems.[42]

In the end, the commission's report recognized the value of wireless but without going so far as to advocate full-scale competition between it and cables. Yet even this was all too tepid for the most avid of advocates, notably Canadian Reginald Fessenden, who saw his ambitions frustrated by the excessive caution of the commission.[43] In contrast, Fessenden stated emphatically that he wanted to immediately launch a high-powered wireless system between Canada and the Caribbean, without subsidy, and in competition with the cable systems. The commission mentioned the proposal in their report, but without further comment, an insult at best.[44] Yet, pace Fessenden, the report did recommend that the cable from Halifax via Bermuda to Jamaica should be supplemented or duplicated by wireless and that British Honduras should have a new wireless connection with Jamaica, so, at a minimum, wireless was seen as an important complement to the cables in the Caribbean area. A couple of years later, in 1912, the British colonial secretary went further and told the governors of the Caribbean islands that he thought they should erect wireless telegraph stations wherever possible; they could thereby communicate interisland and with ships and at a cost much cheaper than cables. Unfortunately, the expectation that the colonial governments should pay the tab tended to squelch any local official enthusiasm for the new medium.[45]

National Modernization and the Promotion
of Wireless in Latin America, 1914–20

Latin American governments assumed a more encouraging approach to the development of domestic and international wireless services. As a result, the largest wireless companies began to make substantial moves beginning around 1914. While they originally had visions of running the national and global wireless services of Latin American countries, their activities were ultimately limited to the provision of international services. This was because most of the commercially significant and stronger Latin American nations established state-owned wireless monopolies for domestic services. In some instances, this more assertive stance toward wireless services within countries came rather late in the day, with the result that British Marconi, for example, had to dismantle stations during the war after they had already been erected at Buenos Aires and at Montevideo. There was also another station built by Telefunken in Uruguay that had to be subsequently dismantled when national wireless services became a state monopoly.[46]

While countries such as Brazil, Uruguay, and Argentina did not assert a state-owned monopoly over international wireless services, it was clear that they were bent on ensuring that such services did not develop along the same lines followed by global cable connections, that is, as a monopoly. To such ends, the more powerful countries of South America granted competing and nonexclusive concessions for international wireless services during the second decade of the twentieth century. And revealing that such actions were not to be curbed by any deference to existing cable companies, even in weaker countries, in 1919 Venezuela and Columbia granted British Marconi a concession to build powerful wireless stations in the teeth of opposition from the French Cable Company, which claimed that its earlier licenses gave it a monopoly over international cable and wireless connections. In contrast, however, in other weak states such as Peru, whose flirtations with financial insolvency were regular and notorious, British Marconi signed an extraordinary contract that gave it a monopoly over all of the country's wireless and postal services. Germany also established a very important presence in Latin America before and during the war, including a high-powered international station built by Telefunken in Uruguay.

One could go on, but these examples are sufficient to show that the region was deemed fertile ground for international investment and economic development by foreign wireless and telegraph companies. Yet, prior to 1914, telegraph and cable companies had little to fear from wireless competition—it was still a pinprick in the vast flow of cable communications. Nonetheless, in their consistent bid to thwart the rise of new rivals, the cable and telegraph companies had, at least in Britain and France, the support of their respective governments. As Charles Bright observed, the British Post Office was biased against transatlantic wireless as an alternative to cables even while it delayed establishing a state transatlantic cable on the excuse that wireless might make it unnecessary.[47] The Post and Telegraph Authority of France mirrored the actions of the British Colonial Office in arguing that wireless should not be allowed to compete with existing cables "for reasons of state," and even if developed where cables did not exist should be taxed to ensure that it did not "annoyingly" compete with any cables which might yet be laid! [48] This was an extreme attitude; yet, even sympathetic cable executives and communication experts in Europe and the United States held that wireless was nothing more than a useful adjunct to the cable business. Even Charles Bright stuck to the view of wireless as "an auxiliary" to the cable systems, useful in remote areas where the cable could not reach but unable to compete in efficiency, even if it did in cost.

WIRELESS AND THE INTERNATIONAL DISTRIBUTION OF PROPAGANDA, INFORMATION, AND NEWS IN WORLD WAR I, 1914-18

Chapter 6 broached the subject of the impact of the war on electronic communication, but did not address the use of wireless for the international distribution of information—news, government communiqués, and propaganda—during and just after the war nor the decisive impact of the war on technological advances in the medium. That is the task of the next section.

German News and Propaganda by Wireless to South America during the War

Germany, in particular, seemed to be well aware of the vulnerability of the existing German cables and the likelihood that they would be cut in time of war. Consequently, it made a strong effort to establish long-range wireless services, especially to nonbelligerent countries, such as the United States prior to 1917 and to the large German immigrant colonies in South America. Germany turned quickly to wireless systems during the war first and foremost because its entire transatlantic and Asian cable system was cut and taken over by the allies. Britain's control over many of the world's cables, either through

ownership or landing licenses that afforded the government the right to censor messages, also gave added push to Germany's extensive wartime use of
wireless. Prior to 1917, notes Aitken, Germany was subject to a "cable block-
ade," which included loss of access to the New York money market and
thereby loss of the ability to sell securities to raise foreign currency. Communication with German diplomats and agents also came to depend either on
wireless or circuitous cable routes, such as through Sweden, once all of the
German cables were cut.[49]

The pillars in the German use of wireless were the powerful Telefunken
wireless stations at Nauen and Eilvese which, by the fall of 1915, were regularly
sending commercial messages and military communiqués through to the
stations at Tuckerton and Sayville in the United States, with local chambers
of commerce often acting as intermediaries in the commercial sphere.[50] A
year later, in October 1916, 300,000 words passed between Germany and the
United States, and the Germans prided themselves that their military communiqués were being published almost simultaneously in the German and American press.[51] The Transatlantische Büro, created in 1908, now became crucial
for passing on German messages to the Latin American press. Indeed, although attempts to censor the stream of German wireless messages to Latin
America were made, following U.S. intervention on the allied side in April
1917, stations manufactured and operated mainly by Telefunken in Mexico,
Peru, Brazil, and the Danish West Indies were still passing on military news to
the local immigrant populations and the press.[52] In a personalized historical
account, the long-time chairman of Associated Press, Kent Cooper, claimed
that most South American newspapers were unaware that a German-funded
news agency calling itself Prensa Asociada (Associated Press) was not the bona
fide product, despite its pro-German slant on the American war effort.[53]
Although Cooper's historical analysis is questionable, it is likely that by the
end of the war, Telefunken and other German wireless interests held their own
as instruments of propaganda.[54]

Wireless and the Distribution of News and Propaganda
by France during the War
In contrast to the relative preparedness of Germany, the previously negative
experiences of French wireless organizations with government regulation
came home to roost when the country found itself cut off from its principal
economic and military partners, and not least from its ally Russia, since all of
the European belligerents were busy cutting or blocking cable and telegraph

connections. Only a wireless station situated on the Eiffel Tower was capable of covering the substantial distance to Russia, but frantic measures, in light of the initial German military advance, led to the establishment of other wireless stations which relieved the immediate crisis.[55] Then there was the issue of transatlantic communication, which became particularly pressing once the United States joined the war and its Expeditionary Force left for French soil. In April 1917, immediately upon entering the war, President Wilson placed all wireless stations situated on U.S. territory under the jurisdiction of the Navy, including those of the Federal Telegraph Company, the Marconi Company, and the German and French Stations in Long Island and New Jersey, respectively: 53 stations in total.

While up to this point the cable system had apparently been adequate for the dissemination of official material between France and the United States, the ongoing fear that the cables could be cut led to the demand, in October 1917, by the American Radio-Telegraphy Commission that a powerful wireless station should be established on French soil to ensure regular communication between the two countries for official correspondence. The outcome was a very high-power station at Bourdeaux-Lafayette that was capable of handling 100,000 words a day by early 1919.[56] This station was sold to France by the United States for a very modest price at the end of the war. Its range was global, and hence it became the basis of French wireless communication for private and official messages and news with the furthest reaches of the French colonial empire. However, most of the transmission was unilateral from France to her territories, since only a powerful transmitter in Saigon built in the early 1920s was capable of reply. Nonetheless, added to another long-range station at Lyons also built during the war, it is evident that the conflict did shake the French bureaucracy out of its lethargy.

The aforenamed high-power French stations (except at Saigon) were state owned and, according to a contemporary British report, had limited traffic once the war was over and ran at a considerable financial loss.[57] Thus, perhaps the most significant outcome of the war for France in the sphere of wireless rested in the creation of the Compagnie Générale de Télégraphie Sans Fil (TSF), which secured a thirty-year concession for the erection and exploitation of commercial wireless services between France and foreign countries not served by the state-owned system. It swiftly became a major international organization and joined the other leading commercial wireless companies, such as Marconi, RCA, and even, by 1921, a rehabilitated Telefunken, to offer its services on a global scale.[58] We shall return to this theme later.

American News and Propaganda to
South America during the War

While there was a heavy volume of German wireless information and propaganda to Latin America during the war, the French news agency Havas provided a flood of communiqués to neutral countries, especially in South America (and to the French colonies and southern Europe) which fell within its
"exclusive zone of influence" under the terms of the "ring circle" agreement
among the big four global news agencies. More attention will be given in
chapter 9 to the long-term impact of changes in the global media business
after the war, but here we focus on events during wartime, especially as they
relate to South America and to bids emanating from the United States to
dramatically expand the role of American news agencies and communication
facilities in the region. A key factor in this bid was the fact that some interested
parties in the United States seized the opportunities presented by the war to
cast American reliance on the French-based news agency, Havas, in a negative
light with the aim of carving out a bigger role of their own in South American
news markets.[59] Such views were conspicuous in the personalized account of
events offered years later by Kent Cooper in his book, *Barriers Down*. Here
Cooper claims that there was widespread dissatisfaction with Havas for having done little to expose the Prensa Asociada "fraud" and its role in disseminating German propaganda.[60] While there is no doubt about the U.S.
agencies' desire for greater access to South America, Rhoda Desbordes's recent
study of Havas contradicts Cooper's view of the agency as a dupe of German
propaganda initiatives:

> Havas stopped serving its South American clients through London because of the
> British censorship. It urgently established its South American service in New York
> and started a new Latin American one with the support of the French government.
> By then, the AP's reputation and worldwide influence had already expanded. *When
> Havas refused to give official German communiqués to its South American clients*,
> because the French government had asked it to transmit only the official [Allied]
> ones, one of them, *La Nacion*, the most important daily of Buenos Aires and maybe
> of South America, simply sent a telegram to the AP asking for the service. The
> United States was officially neutral but the AP's engagements with the cartel forbade
> it to give that kind of service to *La Nacion*.[61]

A crucial element in this story is not that U.S. news agencies displaced
Havas for reasons suggested by Cooper, but rather that new alignments were
forged between Havas and, most notably, Associated Press as part of the U.S.

attempts to achieve greater influence in South America. One reason for the changes that followed, as Desbordes indicates, was Havas's decision, with the approval of the French government, to relocate the headquarters of its South American service from Reuters' offices in London to a new office in New York in 1916. U.S. sources on the subject are oblique, but subsequent statements by one of the main participants in these events, Walter Rogers—a wartime propagandist at the Committee of Public Information and subsequently a leading advisor to the State Department—shed light on the issues.[62]

Rogers led State Department–brokered negotiations between Havas and Associated Press, with the aim of loosening the constraints of the global news cartel in order to allow a larger role for Associated Press in South America. A crucial element of these events was another set of deals that Rogers and the State Department helped to broker with the Central and South American Company that gave the U.S. press and news agencies more abundant cable facilities and cheaper rates for their operations in South America. While the source material grows foggy on this point, part of this deal seems to have extended the cheaper rates to Havas, most likely as an inducement to accept the changes then taking place and ultimately to persuade the agency to accept revisions to the "news ring" agreement in 1919 which gave Associated Press a greater role in South American news markets than ever before. In sum, the incentives during the war leading up to Havas's acceptance of the 1919 changes seem to have been cheaper cable rates and encouragement to relocate the center of its South American news operation in New York.[63]

Even though these efforts aimed to gain a greater role for Associated Press in South America, the United Press also benefited. In fact, because the United Press never joined the news cartel it was never constrained by restrictions regarding territorial exclusivity. Being thus unbound, and with copious backing from both Woodrow Wilson and the State Department, United Press entered South America in 1916 and immediately struck deals with several leading news organizations in Argentina, Brazil, and Uruguay.[64] Taken together, the key thing to note about South America at this time is that its comparatively well-developed news markets had become crucial to the propaganda battle for "hearts and minds" that was being waged by all of the belligerents. It was also a continent that was flooded by more news services and a larger volume of news than prior to the war, although just where the lines between propaganda and news lay was one of the great mysteries of wartime and seriously polluted the "global news system" both during and after the war. It was no mystery, however, that Havas's monopoly, maintained since 1890, had ended.

Wireless and the Distribution of News and
Propaganda by Britain during the War

That the lines between propaganda and news had become precariously thin could also be seen in Reuter's growing role as an agency for British war communiqués and propaganda. By 1915, the Marconi Company had built thirteen long-range wireless stations for the British Navy and was in much demand by the British government, military, and press as a channel for information flow. In the same year, Charles Bright was sufficiently impressed with wireless to observe that "this war has already, times without number, proved on the one hand the great value of wireless for disseminating instructions and information widely, directly and speedily."[65] Bright also praised Marconi for intercepting German messages destined for the United States with "false versions of war occurrences."[66]

In fact, such interception benefited the Marconi organization because it was indirectly linked to a deal with the British Admiralty; namely, the Admiralty allowed Marconi to use anything they could pluck from the airwaves to fill the columns of a subsidiary, "Wireless Press," which used its ready access to enemy communiqués and other broadcast materials to compete with Reuters.[67] Helped by this deal, Marconi stations were busy supplying their North American high-power stations with both official and commercial information until the British government cut off all commercial activities in 1917; from then on, they formed part of the Allied military communiqué and propaganda effort, "disseminating British bulletins composed by the Foreign Office."[68] Shortly after the war, however, Reuters, backed by other members of the British press, campaigned successfully to prevent Marconi from being allowed to continue to combine the function of a news agency with those of a wireless carrier.[69]

Reuters, however, stood in the midst of the flow of British propaganda. By 1915, the company was on the verge of bankruptcy because of its failed attempts to diversify into advertising and banking.[70] Indeed, Reuters' financial predicament on account of these misguided adventures was so great that the Marconi Company tried to acquire it in 1916, a bid which the British government forcefully repelled because Marconi still had a smell of corruption around it due to the earlier scandal and because powerful figures in the Liberal government were ardently opposed to any such take-over. Thus, Marconi's bid—which had been shepherded by Godfrey Isaacs—was aborted. That, however, did not resolve Reuters's dire economic circumstances, but the clandestine role played by the government in providing Reuters with finance, its involvement in the management of the news agency, and a series of new contracts for propaganda and imperial news services did. The secret deals that led to such arrangements lasted for the duration of the war and were cancelled in 1919, but they underscored the importance which the British government attached to Reuters as a wartime propaganda tool and Reuters's willingness to be used in such a capacity.

Despite its prior high degree of formal independence from the state, Reuters had already been a key player in government efforts to distribute specially designated news. What distinguished the deals between 1915 and 1919, however, was the role assumed by the British government as a financier of the troubled organization and in its internal management in return for a formal pledge that the agency would further British wartime ends. The 1916 agreement between Reuters and the government provided copious amounts of financial support (roughly $2.75 million) to its managing director, Sir Roderick Jones and another Reuters's official, Mark Napier. Those funds, in turn, allowed Jones and Napier to take a controlling ownership stake in the company, subject only to a mysterious "share 999" worth £1 and held by the Foreign Office.[71] This share gave the government the right to nominate a director with special voting rights amounting to 26 percent of the voting power of the company, the right of veto over the appointment of other directors, and of any resolutions passed by shareholders. Coupled with Jones's and Napier's ownership and voting power in the news organization, this meant that control rested with three parties: Jones, Napier, and the British government. The understanding between the three is clearly spelled out in a letter from Sir Robert Cecil at the Foreign Office, which notes, "It is expected that you will on all matters of public policy or the national interest always vote with the Government Director" and that

> His Majesty's Government will not interfere in the business management of the company as a commercial undertaking but it is necessary that the Foreign Office should be able both to prevent the Company from taking any action which might be contrary to public policy (such as the dissemination of reports prejudicial to the national interest, the employment of undesirable correspondents or other employees, the undertaking or continuation of undesirable contracts with other news agencies, or the admission of undesirable persons as shareholders or directors) and also to secure that the Company's operations and actions are in conformity with the public policy or the national interest, and that information of national importance is properly collected and circulated.[72]

While Roderick Jones would later defend the agency's actions, the fact of the matter is that Reuters's service was seriously compromised in the eyes of

British and foreign journalists by these activities. Indeed, Jones seems to have gained something of a penchant for his role, serving as the director of the British Ministry of Public Information in 1918 and arguing strenuously at war's end that the ties between the government and his agency should continue in order to help Reuters meet the mounting challenges from Havas, Associated Press, and United Press. Indeed, it was the government, rather than Reuters, that insisted on terminating the agreement in 1919. Of course, Jones was not the only "press baron" to be involved in propaganda efforts and indeed, Max Aitken, Lord Beaverbrook, the Canadian-born newspaper magnate, headed the British Ministry of Public Information during this period.[73] After the war, many British journalists, however, were deeply concerned about these relationships and suspicious that the "unholy alliance" between the government and Reuters would carry on.[74] While willing to accept the necessity of propaganda during the war, journalists were concerned that such practices might continue long after it was over. By 1920 concern had grown so high that opposition by the press in Montreal and Toronto to a British and Canadian proposal for a new and subsidized imperial news service which would be funneled through Reuters led to a modification of the plan prior to its introduction. These "media professionals" were rightfully concerned that their craft was being tainted, but also that government subsidies were anathema.[75] In the end, journalists' fear for press freedom under such conditions was neither new nor exceptional and indeed it would emerge time and again in the years ahead.

Wireless and War in the United States: "Navalism" and the End of American Marconi

The growing force of navalism during the progressive era, especially during the two Wilson presidencies, incorporated not just wireless but the whole issue of sea power and the use of that power to project American commercial power abroad. It was also a development that produced considerable tensions at home and abroad, with some in the United States fearing that cooperation between the Navy and business interests was a sure path to imperialism, while other nations worried about an excess accumulation of American geopolitical power as exemplified by heated disputes during the Paris Conference in 1919.[76] Rising navalism also had a huge impact on the development of U.S. wireless during the period between 1913 and 1919, and this could be observed in several ways: first, in the vital supplemental role of the Navy in the wireless transmission of news and commercial messages; second, in the repeated, but unsuccessful, attempts by Josephus Daniels to make wireless a state monopoly; and, third, in the intensification of the Navy's campaign against American Marconi which was to conclude with its absorption into the newly formed RCA.

Nothing illustrates better the interconnections between all three of the above elements than Daniels's statement in 1919 that "the Navy occupies a strong position in the commercial radio field on account of efficient service rendered, and I think presages the way for making this service entirely governmental."[77] By that, Daniels was referring to the Navy's network of high-powered stations in Arlington, Darien in the Canal Zone, San Diego, Pearl Harbor, the Philippines, Guam, Samoa, and Puerto Rico and which was expanded through its take-over of the German high-power stations at Tuckerton and Sayville. The Navy also had wireless monopolies in the dollar dependencies of the Dominican Republic, Haiti, and Puerto Rico. And if the service was to be made "entirely governmental," then, by definition, wireless companies such as American Marconi and the Federal Wireless Telegraph Company would need to be bought out or absorbed. Indeed, in response to American Marconi's bid to buy the Federal Wireless Company of California in May 1918, the Navy Department preempted the company by buying the Federal Company's patents itself. Six months later it also bought American Marconi's low-power ship and shore stations.[78]

Although state ownership was, then as today, a weak straw in the United States, two factors seemed to strengthen the Navy's case. The first was Postmaster General Albert Burleson's long-standing desire to nationalize telegraph, telephone, and cable services, which he had been able to carry out, for a short time, at the end of the war. Second, there was a surprising amount of support among the U.S. press and the cable companies for a rather large, even if supplemental role, for a Navy-controlled wireless system. This reflected the reality that the cable networks had long been unwilling, or unable, to give priority to press material. Hence, when chief executives of the news agencies enthusiastically endorsed renewing the Navy's wireless services during the Cable Landing Licenses Hearings in 1920–21, the statement of the president of United Press suggesting that the cable companies would not object was met with a hearty "Hear, hear!" from Newcomb Carlton, the president of Western Union.[79] For his part, Walter Rogers told the hearings that "we sent a lot of news by American [naval] stations as far as India" and again, "The Naval radio, through its stations across the Pacific in the Philippines can cover Australia and New Zealand, because we sent press during the war that way and could reach both China and Japan."[80] Another witness, Frederick Martin, acting general manager of Associated Press, also told the hearings, somewhat ruefully, that the Navy had compelled Associated Press to provide it gratis with a condensed press service which was supplied to Hawaii, the Philippines, Japan, and China. In sum, it is evident that both press and cable were happy with the supplemental service which the Navy had provided during the war, saw the extension of such services as vitally important, and also, as Carlton noted, as a way to have low-rate business handled by others so that the cable companies could concentrate on "full rate commercial traffic."[81]

The one glaring exception in this military–corporate love fest, however, was American Marconi, which was treated by the Navy and sometimes acted as the corporate bad boy of U.S. wireless. While other private wireless companies obviously resented the Navy's incursions during the war, only American Marconi had loudly protested against military intervention into its activities. As a case in point, in December 1918, Edward Nally, the vice president of American Marconi, argued before a House of Representatives committee that the high-power stations taken over by the Navy had yielded the latter a rich return, and he was particularly annoyed by Josephus Daniels's claim that it was "good business" for the Navy to take over all high-power stations. Retorted Nally, it had also been "good business" for American Marconi.[82]

Why was American Marconi, the strongest wireless system operator in the United States in 1914, not one of those companies with which the Navy willingly cooperated? We have seen that, as early as 1906, important naval personnel were eager to get the company out of the United States, and this nationalist element linked to American Marconi's British roots became a dominating line of argument, although the company had substantial autonomy within the wider Marconi organization and was hardly a favored instrument in the British international and imperial communication system. So, too, the Marconi organization added fuel to the fire by its early focus on creating a global wireless monopoly. Also, the company's reluctance to embrace universal intercommunication principles gave the Navy additional grounds for complaint. Then there was the adamant opposition of Marconi to government regulation, not least censorship, which as early as 1915 had led to the short-term closure of one of its high-power stations by the Navy. And again, the Marconi organization had a reputation for placing profits before technical progress, a priority which was later to get it into grave difficulties, and not one that was likely to be palatable to the authorities in time of war.[83]

Yet, there was another side, not so favorable to the Navy. Notably, American Marconi had been negotiating with GE since 1915 to have an Alexanderson alternator—a technological innovation that superseded either spark or arc transmission—fitted into its New Brunswick (N.J.) station, and the work was just completed in time for the Navy's take-over of the station. In late 1918, this alternator was replaced by a larger and more efficient one, with the result that the station became, in Susan Douglas's words, the "government flagship station," being designed to handle most of America's radio communication with Europe.[84] In this context, one can understand the frustration of Nally and other American Marconi executives when their hard work to update the New Brunswick station was whisked away from them by the Navy and seemingly turned to its advantage. To add to which the continued Navy talk about the largest wireless company in the United States being "a foreign company" needs to be seen in the context not just of American Marconi's stubborn opposition to government regulation, but also in light of its explicit opposition, especially in 1918, to Josephus Daniels's attempts to gain public ownership of wireless. Tit for tat perhaps, but one can reasonably conclude that American Marconi was being given a raw deal.[85]

And a raw deal, it certainly got. Many years later, in 1931, in a self-serving report outlining the Navy's contributions to American society, the Navy offered retroactive justification for its successful efforts to drive down American Marconi. Thus, notes the report, "The Navy never could feel free to give full encouragement to the American Marconi Company because of its affiliation to the British company, as it was the *established policy* of the Government to *encourage* only radio companies controlled by American citizens."[86] The italicized words in this quotation tell a great deal about Navy attitudes as well as about their retrospective justification; the argument that it was established government policy in 1918–19 to encourage (note, not require) wireless companies to be U.S. controlled masks the fact that Navy antagonism to American Marconi had begun early in the century. And as for being British influenced, in 1918 only two of American Marconi's thirteen member board of directors were resident in London, ten were Americans and one was a Canadian. The firm's president, John Griggs, was an American, and "Marconi was a very silent vice president."[87] But, the innuendo and inferences fomented by the Navy were enough to damn the firm's prospects.

As a classic case in point, Edward Nally told the aforementioned House committee in 1918 that the company had planned to begin work on a wireless station in Argentina through its subsidiary company Pan-American Wireless, Telegraph and Telephone as part of its proposed venture into Latin America. To this end, he made a tour of Argentina and several other South American countries during the war, but when he got back Josephus Daniels informed him that since the Navy favored government ownership, the Argentinian scheme should be ended—which it was. Nally tried to reassure the committee that the 25 percent of shares held in American Marconi by British Marconi were nonvoting shares, but as a company statement to stockholders noted somewhat plaintively in 1919: "It has been made clear that Congress objects to foreign assets. This is making work very difficult . . . we have found a very strong and irremovable objection to your Company because of the stock interest held therein by the British Company."[88] The outcome, in the context of some patently underhanded actions and threats by senior Navy personnel,[89] was the incorporation of American Marconi into GE and subsequently into the newly formed RCA in October 1919. A presumably appeased Nally became RCA's president.

With remarkable tenacity, the Navy endeavored, time after time, to get Congress to pass a bill for the government ownership of wireless stations. The third successive time was in 1918, but it failed to gain congressional approval. Daniels tried again in the fall of 1919, though he admitted that no longer did he have much hope that the bill would get through, and he was right.[90] The bill combined a bid for Navy control over international official and commercial wireless communications with some suggestion that private American wireless interests could build stations and operate in other countries; that is, it offered a vision of a U.S. wireless system divided between a dominating state-owned monopoly and peripheral private interests. As Griset suggests, this was not a vision likely to appeal to the latter, and indeed the National Wireless Association representing such interests launched a massive lobby effort against the bill.[91] In the last resort, however, it was Congress that defeated the powerful forces pushing for the consolidation of Navy control over U.S. wireless.[92] Although the State Department initially supported Daniels, the dominant congressional view was expressed by Congressman William Greene of Massachusetts, who observed in November 1918, "Having just won a fight against autocracy [that is, against the Kaiser] we would start an autocratic movement with this bill."[93] In light of some aspects of the Navy's overall expansive agenda and its treatment of American Marconi, he was right to be suspicious.

RCA: "The 100 Percent American Commercial Company," 1919–22

Thomas Streeter notes that RCA's creation provides a breathtaking example of how confident in their power and importance the United States corporate liberal elite had become. In 1919, a coalition of corporate managers and some representatives of the military and of Wilson's administration quietly forged an institutional framework for radio, without consulting Congress or the courts, and, Streeter notes, with "clear indifference to the idea of boundary between public and private sectors."[94] The initial idea was that GE would, in combination with the assets of American Marconi and the Navy's recently acquired Federal Telegraph Company patents, build an effective U.S. international wireless system in which it would hold shares, namely, RCA. However, it proved necessary, in 1921, to include AT&T, the United Fruit Company, and ultimately Westinghouse, in a shareholding and patent-pooling agreement. In essence, RCA became an odd corporate body complete with four main corporate owners and regulations which ensured majority control by American citizens. It was, as the Navy was to boast later, the "100 Percent American Commercial Company."

The United States, having come out of the war as the clear leader in wireless technology, now had a "national champion"—a company which could, through these intercorporate arrangements, become dominant in the sphere of both national and international wireless. For a "uniquely American" company, RCA had a remarkable resemblance to its Marconi predecessor, with the new company taking much from the old, including patents, agreements, and several members from the upper ranks of American Marconi, including Edward Nally and David Sarnoff. Consequently, "The Radio Corporation obtained control of practically every privately owned high power station in the United States together with a number of important . . . patents."[95] Notably, in acquiring American Marconi, the company obtained all of its operations in the United States as well as three-eighths share of the stock in American Marconi's Pan-American Wireless Telegraph and Telephone Company, which was to become a major element of RCA's approach to South America. To all this was swiftly added a series of traffic agreements between RCA and foreign governments and the other big three foreign wireless companies (British Marconi, TSF, and Telefunken), so that, by 1922: "It is doubtful if any company seeking to conduct a similar business in the United States could exist, by virtue of the provisions of the majority of these contracts obligating the foreign companies to transmit all messages intended for the United States only through the facilities owned by the Radio Corporation of America."[96]

Where, one might ask, did the U.S. Navy fit into this swift expansion of private monopoly interests? The answer is "not very comfortably." Certainly, as Streeter suggests, the Navy had acted as a kind of "nursemaid" to RCA since, having failed in their bid to establish a state monopoly over U.S. wireless, they clearly saw a private monopoly as the next best thing. Both Daniels and Lieutenant Commander Hooper indeed juxtaposed "a monopoly for the government or a monopoly for a company."[97] That the fit between the Navy and RCA was far from perfect from the beginning, however, was evident in the fact that even Daniels did not sign the agreement establishing RCA. Hence the statement of Streeter, noted earlier, that RCA was established "without consulting Congress," that is, its genesis was not directly blessed by the U.S. government.[98] Certainly, as nursemaid, the Navy had Admiral Bullard as a representative on the RCA board. However, he left quietly in 1921, never to be replaced. Apparently, his presence soon became an embarrassment to the directors of RCA, and not least to Owen Young, vice president of GE, because Bullard favored aggressive competition with, and even buying out, foreign wireless companies, an idea at odds with the preference of Young and most other RCA directors to cooperate rather than compete with companies such as Telefunken and British Marconi. As another strong naval exponent of RCA, Hooper too quickly became caught between his wish for a wireless monopoly and his concern over the monopoly which RCA had established over equipment sales to the Navy. Ironically, he would have preferred competition for government and naval procurement.[99] In sum, in relatively short order, the Navy was no longer four-square behind RCA, nor particularly influential in its corporate plans. This was particularly so once RCA began to move away from wireless toward radio broadcasting in the early 1920s and dampened still further the hopes of those such as Bullard for a U.S.-dominated worldwide wireless system.

CONCLUDING REMARKS: ACCESS TO COMMUNICATIONS AND COMPETITIVE RIVALRY BETWEEN "NEW" AND "OLD" MEDIA

Early wireless only offered very limited competition with the cable companies before the 1920s. The cable companies, as well as state agencies in countries such as Britain and France, still functioned, using the terminology of Brian Winston, more as a brake than an accelerator on the development and expansion of the new technology. Notably, in Britain, official decisions favored cable over wireless. On the other side, wireless found a ready customer in the press, and in this sense the needs of the press served as an accelerator on the commercial development and viability of wireless systems. And, of course, although major developments in high-power wireless took place prior to World

War I, not least in Germany, the war provided a major stimulus that begot both fundamentally new technological innovations and the absorption of the new technology into the propaganda machinery of all of the protagonists in that bout of destruction. Finally, with the Marconi organization as the archetypical example, and RCA later learning much from that firm's experience, it is evident that the wireless companies quickly learned the benefits of intercorporate cooperation and cartel arrangements as well as the crucial importance of maintaining good, even if uneasy, relations to the nation-state. It is just as clear, especially in Britain and, to a lesser extent, the United States, that the interests of the major corporate wireless companies (Marconi and, after, 1919 in the United States, RCA) were in no way synonymous with those of the state, with each forging relations with the other to advance their own interests.

One other issue deserves brief attention here, namely, the continuation of efforts to promote easier access to long-distance communications for the average citizen. Quite evidently, the costs of establishing long-wave wireless stations made the prospect of substantial competition with the cable systems limited, while during the war, government control of all communication systems resulted in the suspension, for the duration of hostilities, of those cheaper types of service which had begun to offer some promise of wider social access to long-distance electronic communication. However, with Britain in the lead and many other countries following, important concessions in rates were made to the public at the end of 1919 with the intention of bringing communications within the reach of an increasing proportion of the population. Alas, none of the important early reformers—Fleming, Henniker Heaton, or Sassoon—lived long enough to see such rewards for their efforts.

Thick and Thin Globalism: Wilson,

the Communication Experts, and the American

Approach to Global Communication, 1918–22

> Internationalism has come, and we must choose what form the
> internationalism is to take.—GILBERT HITCHCOCK,
> U.S. Democratic Senate leader, 1919 (quoted in Knock,
> "Wilsonian Concepts and International
> Realities at the End of the War")

> The public mind of Great Britain has seriously endeavoured for a number
> of years to exercise sufficient pressure to bear down the iniquities of the
> financially highly selfish cable combinations. It has not succeeded in
> bringing about the desired result . . . but it has paved the way for a uni-
> versal review of the situation. . . . The time has arrived when this freedom
> has become a fundamental requisite in support of a democratic world
> empire, and no league of nations can be established which does not
> regard an unlimited and uncostly flow of international intelligence as a
> basis of its very existence.—ERNEST POWER (NARA, Power,
> Memorandum to Lansing, 1919)

POSTWAR AMERICA AND THE GLOBAL
MEDIA SYSTEM: A BIRD'S-EYE VIEW

Woodrow Wilson has usually been seen either as a dreamy idealist or a cunning leader who knew how to wrap America's postwar expansionist agenda in
universalist garb. And for those who, say in 1919, looked at the international
scene, it was the latter approach that was likely more convincing. The American empire was at its apogee, the Philippines and Cuba recently denied their
independence despite long-standing promises to the contrary, and an unprecedented concerted approach was afoot to expand the U.S.-based communication and media businesses. In its entirety the global media system was
being swept by a wave of changes in technology, markets, politics, and culture,
and at the forefront of this wave was the United States. For most contempo-

rary observers, Wilson's ideas of a new liberal internationalism certainly must have seemed like a thinly veiled mask behind which lurked an ascendant communications and media colossus.

At the beginning of the 1920s, the three major U.S. global cable companies alone—Western Union, the Commercial Cable Company, and the recently renamed All-America Cable Company—were planning to invest more than $100 million in the world's key communications markets—Europe, South America, and Asia—and while there were parallel plans by European companies to do likewise, they were not on the same scale. A major boom in capital investment, much of it American, did occur, but mostly after 1924. Almost all of this expansion took place only after the commercial crisis of the early 1920s had abated and, crucially, governments finished sorting through a raft of global communication issues that had been opened up under the umbrella of the Versailles peace negotiations.

In each sector of the global communication and media business there were the big four tier one global players. In cable communications this included the Eastern Associated Company, Western Union, the Commercial Cable Company, and the All-America Cable Company. The big four global wireless players—RCA, Marconi, Compagnie Générale de Télégraphie sans Fils (TSF), and Telefunken—had also risen to extraordinary heights in a remarkably short period of time. Although never achieving the same scale as the cable companies, either in terms of capital invested, revenues earned, or messages transmitted, these four wireless companies learned much from the cable industry's experience and incorporated such lessons into their own practices. Last, there were the big four news agencies—Reuters, Havas, Associated Press, and Wolff. However, postwar changes to the "ring circle" agreement confined Wolff to Germany, while its former markets in central Europe, the Balkans, and the Ottoman Empire were redistributed to Reuters and Havas according to the geopolitical settlements reached at Versailles and pressures from the French and British governments. In each of these new areas, Reuters and Havas created new bureaus and press agencies of their own, many of which had questionable autonomy from the propaganda machinery of their respective states. And there was also a new rival to the old cartel: the United Press, which opened its own bureaus in Europe, established close ties to new national agencies such as Dentsu in Japan, and had exclusive agreements with some of the leading newspapers of the world (for example, *La Prensa* in Buenos Aires). By establishing such links to national and local media, United Press cobbled together a news-gathering and distribution system which, from 1915 on, constituted one of the most decisive changes in the global news business since its inception in the late nineteenth century. That the "global-local" link was crucial could be seen, for example, in the fact that Reuters and Havas had always fostered close relationships with national news agencies. In Australia and New Zealand, for example, Reuters had signed deals with one of the two main news groups and those deals, in turn, proved decisive in local press competition as well.[1] In Europe, to take just one example, Havas was aligned with the Fabia news agency in Spain and with Stephani in Italy.

Previously peripheral to world affairs, by 1920, the United States was central to them. American financial interests became some of the world's preeminent bankers, financing the expansion of U.S. business abroad and supplying loans for economic development in South America and for reconstruction in Europe. In China and Japan, the United States also became a significant force, although still not on the scale obtained by British commercial and financial interests over the preceding half century. By the 1920s, U.S. financial institutions had become the leading source of new lending, especially to South America. The links between communication and finance tightened and nowhere was this so evident than in the phalanx of communications and media companies—Western Union, RCA, the All-America Company, Commercial Cable Company, and later, the International Telephone and Telegraph Company (ITT)—arrayed along Broad Street in New York, some literally attached to Wall Street. The links propelled the rise of an American-based global communications and media powerhouse; they also paved the way for a speculative bubble that helped decimate the world economy by decade's end. In that event, the media wonder child of the age—RCA—was to see its stock value swing from untold heights to incredible lows—with high levels set prior to the 1929 crash not recovered until 1964.[2]

In the previous decade, the two Wilson administrations (1913 to 1920) consolidated progressive era reforms. Changes to U.S. banking laws (Federal Reserve Act of 1913) allowed for tighter integration of the banking system and permitted such institutions to forge alliances with one another in global markets. American banking syndicates and the American International Corporation, formed in 1915, scoured South America for investment opportunities, while in China, U.S. financial interests rejoined the older financial consortium that had been around in various guises since 1909. These trends were reinforced as new laws sanctioned corporate mergers in return for the acceptance of regulation (Federal Trade Commission Act of 1914) and exempted alliances among U.S. firms operating in international markets from domestic anti-trust legislation (Webb-Pomerene Act of 1918). The impact of these changes on global communications was fundamental.

The first sustained U.S. approach to global communication policy was also forged under the Wilson administration. This approach built on past U.S. practices and the experience of other nations, notably Britain. These efforts culminated in May 1919, when the United States and nine other governments called for a World Communications Conference to be held later that year. The proposed conference inspired a thorough review of the global media system and long-standing negotiations that developed policy solutions to issues that would be dealt with over the next decade. It also revealed how the two foreign policy views used in this study—the realist stress on national interests, power politics, and territorial imperialism versus the structuralist focus on collaboration, systemic power, and capitalist imperialism—still defined global politics. In fact, these two alternatives became key factors that distinguished Wilson's administrations from the subsequent Republican governments of Harding (1921–23), Coolidge (1923–28), and Hoover (1928–33). The politics of global communication from 1918 to 1922 constituted a subset of global politics that registered countries' willingness to seriously entertain the idea of cooperatively organizing a world system through the League of Nations and international law. Such efforts also revealed just how elusive such goals were, as countries continued to pursue realist policies and their own national interests. This was especially apparent as the Wilson administrations oscillated between promoting multilateralism yet acted unilaterally whenever they saw fit. However, more than being just a gap between ideology and an expansionist U.S. agenda, these contradictions were a constitutive element of U.S. foreign policy and durable trends that were, once again, brought to the fore as countries strove to restore the nineteenth-century model of liberal internationalism in a new and improved version fit for the twentieth century.

The idea of empire continued to loom large, but it was not the same as the pre-1914 version. Indeed, the "old-style imperialism" became an object of disdain, a model from which contemporaries had to take lessons on what not to do. But this did not mean that imperialism was eschewed, of course. In some circles it certainly was, since anti-imperialism and decolonization movements flourished in the early decades of the twentieth century.[3] However, in the version of liberalism governing the negotiations at Versailles (1919) and the creation of the League of Nations, globalization did not entail rejecting imperialism but remodeling it. Among progressive internationalists there was a need for a new kind of empire, a world empire of democracy, as Ernest Power of the State Department referred to it (see the epigraph at the outset of this chapter).

On one hand, such a world empire meant that some societies had to be modernized, or Americanized, which was much the same thing in the minds of U.S. contributors to these ideas.[4] Of course, some nations were closer to these goals than others. Some indeed needed to be thoroughly modernized, but few were seen as congenitally incapable of development along such lines. One of the era's more prominent critics of imperialism, the American philosopher John Dewey, got things right in his 1927 *New Republic* article, "Imperialism Is easy."[5] The gist of Dewey's article was that imperialism was so easy because it resonated with myths of modernization and progress, the importance of stable markets, and the sanctity of contracts. The problem was that such ideas easily provided the warrant for the usurpation of sovereignty where such "taken-for-granted" norms were absent, on the grounds that they would be provided in short order by a political and cultural makeover, that is, a program of modernization. In short, modernization was as much a prescription for development as it was a justification for imperialism.

The wave of liberal and socialist-oriented revolutions across Europe between 1917 and the early 1920s also provided a context in which the vision of a world empire of democracy was pitched as a liberal alternative to the status quo and to Marxian-inspired theories of revolution. Close attention to such matters in the reports and correspondence that constituted, in particular, the American position reveals an acute awareness that democracy and global stability had absolutely no chance to succeed on idealism alone, but could only be achieved after thoroughgoing reforms to the world economy, political institutions, communications media, and culture were accomplished. By the early 1930s, things would be very different, as the language of democracy was eclipsed by a myopic focus on markets and technology and as democracy was thrown into reverse by the rise of authoritarianism, especially in Fascist Italy and Nazi Germany. But cavalierly dismissing the discourse of the years between approximately 1918 and 1922 as a "cloud of high-minded rhetoric"[6] misses the rise and fall of democracy completely during the decade.

The United States, most notably, pursued the liberal internationalist option, although the demise of the Wilson administration in 1920 elevated the realist view within that country. The idea of a global system that the United States had a vital stake in upholding and expanding is probably the most important legacy of Woodrow Wilson, rather than the dubious notion that the United States alone had the moral standing from which to proselytize and fill the ethical void endemic to international relations. The main distinction between Wilson and the Republican administrations that followed was that the former pursued a kind of "thick globalism" based on economic, political, and cultural interdependence versus the latter's "thin globalism" based on the supremacy of markets, a minimalist approach to multilateralism, a hard-nosed realist approach to national interests, and a view of culture as a source of misunderstanding rather than the basis of a "new world order."[7] Through this prism, we can gain a vantage point from which to view the changes that took place in the field of global communications. Thus, the next section focuses on the approach of the United States to global communication policy.[8]

THE U.S. APPROACH TO GLOBAL COMMUNICATION POLICY: THE FORMATIVE YEARS, 1918–22

The emphasis of U.S. policy during the period from 1918 until 1922 was, first and foremost, on facilitating the expansion of U.S. trade and investment, but the broader concern was with the viability of the global political and economic system. In fact, the manner in which American expansionism was reconciled with the need to reestablish a stable global system is pivotal to understanding the decade. As the United States sought to expand its role it looked to Britain's experience as a guide. This influence was especially noticeable among a tight-knit group of individuals working either in the U.S. State Department or closely associated with it. The State Department was the lead agency on these matters, crafting policy and overseeing new legislation covering international telecommunications services, the Cable-Landing Licenses Act (otherwise known as the Kellogg Act) of 1921.

The most important person contracted to the State Department's communication policy group was Walter Rogers, past editor of the *Washington Herald*, former official of the Committee of Public Information, which was largely a propaganda tool, and later a repentant propagandist and advisor to Wilson at Versailles. In addition to Rogers, the nucleus of the group consisted of Ernest Power, a third assistant secretary in the department and Breckinbridge Long, who held a similar position. Between them, the group sketched a coherent approach to global communication policy and the first outlines of the "free flow of information" doctrine that has been the centerpiece of U.S. policy ever since. Two decades after leading the campaign to change the global media system, Carl Ackerman, the dean of the Graduate School of Journalism at Columbia University, lauded Walter Rogers as the "leading authority in the United States on international electrical communication . . . [and] the originator of international press freedom both ideologically and practically."[9]

21 Walter Rogers in his office, circa 1920s. His important role in pressing for the internationalization of communication during and after World War I as well as an original source of ideas on the "free flow of information" doctrine has been neglected by media historians. Later, he worked hard to improve journalism education. Reproduced by permission of Rev. Columba Gilliss, Frederick, Md.

These "communication experts" did not see themselves as mavericks, however, but explicitly referred to the paths cut by the earlier generation of reformers: Henniker Heaton, Edward Sassoon, Sandford Fleming, Charles Bright, and Rodolphe Lemieux. The earlier reformers' ideas punctuated the group's work and as such were consistently identified as key instigating elements in the global media debates conducted in the shadows of Versailles. The broader British influence on the group was also reflected in the numerous reports written by Rogers. As he noted:

> The British keenly realize the value of having British quotations published widely and sterling made the basis of trade.... News is a commodity and widely dealt in as such. Competition in the sale of news is keen and frequently, either directly or indirectly, its export is subsidized.... The American press is interested solely in the import of news. However, with low press rates the American reader will receive a wider range of news and the newspapers and news agencies will become less dependent on London, Paris and Berlin. A large export of American . . . news will only be possible when there are low press rates.[10]

The links drawn between news and the pound sterling in the infrastructure of global capitalism sheds much light on the U.S. approach. Rogers and colleagues believed that communication and the free flow of information would be decisive factors in reviving and transforming the global system. The overall image that emerges is one where communication, information, and news play several key roles: as the informational infrastructure of global capitalism; as important markets in their own right; as pivotal to the viability of multilateral institutions (for example, the League of Nations); as vital to expanding the postwar wave of democratization; and as tools for the administration and modernization of empires. The "communication experts" also developed a comprehensive view of the barriers to communication which had built up over the past half century and which were the proper target of any viable critique of the global media—monopolies, cartels, high rates, hidden subsidies, and propaganda.

Like their earlier British counterparts, the U.S. experts became preoccupied with the theme of communication and empire. In fact, the group's papers contain some of the first formal statements on communication and American empire, notably in a 1918 paper by Rogers entitled "An International and Inter-Colonial Communications Program."[11] Written at the height of U.S. imperialism during this era, Rogers highlighted the lack of communication facilities and the paucity of news and information in the U.S. colonies, protectorates, and dollar dependencies. Power expressed similar views, as in one of his major policy papers where he declared that "the . . . possessions . . . can only be Americanized through the development of cable communication with rates so low as to stimulate traffic and encourage a generous flow of intelligence from the mother country."[12] Perhaps surprisingly, the "communication experts" seemed to have very little trouble reconciling imperialism with their expansive views of democracy. This was not, however, because their rhetoric was more ideology than an accurate reflection of real U.S. aims. Instead, as Dewey noted, the reconciliation of democracy and empire was, especially among liberal internationalists, easy for those who saw imperialism in terms of modernization and development rather than conquest and exploitation.

Such views looked remarkably similar to the "Empire Lite" model that contemporary scholars such as Michael Ignatieff advocate to deal with global instability and its roots in failed states. The parallel to Ignatieff in this earlier era was probably Frank Giddings, whose *Democracy and Empire* (1901) inspired Wilson's own views. The same was true of the University of Wisconsin political scientist and later minister to China during Wilson's second term, Paul Reinsch. Reinsch drank deeply from the well of European political theory and international law and translated these ideas into analyses of world politics in terms that were consistent with U.S. political culture. Reconciled in the

realm of theory, such ideas helped legitimate the consolidation of U.S. imperialism between 1915 and 1927, as Congress rejected Philippine independence
(1916) and as Wilson authorized a slew of military occupations in Haiti, the
Dominican Republic, Cuba, and Nicaragua. Thus, in the mainstream of U.S.
foreign policy circles, the "hard power" of U.S. military occupation would
pave the way for reforms which, in turn, would lead to constitutional government, professional financial administration, and social stability (although
U.S. anti-imperialist movements were having nothing of it, seeing imperialism as a repudiation of democratic values and a regression to the model of
international relations set by Europe which, in the discourse of the time, the
United States was supposed to have rejected).

As with French and British colonies, the U.S. imperial system included
some of the least connected, worst served places on the planet. In "International and Inter-Colonial Communications Program," Rogers outlined two
options to improve such conditions: a U.S. government–owned imperial
communication system or the "internationalization of communication"
model. Steps toward the first were well underway. After Congress rejected the
Navy's bid for a monopoly on all wireless services, the Navy was authorized in
1919 to offer what amounted to an imperial wireless system to meet U.S. needs
in the Caribbean and Pacific. Originally passed as an interim measure, the
Navy wireless system was renewed in 1922 and put on a more permanent
footing by the Radio Act of 1927. The idea of a state-owned imperial communications network also anticipated that the Navy wireless system might be
augmented by a state-owned Pacific cable, either a new one or by expropriat-
ing the cable from the Commercial Pacific Cable Company. In the grand
scheme of things, however, the U.S. imperial communication system was
designed to supplement the privately owned cable and wireless networks in
weak markets, much along the lines that had been pursued after the 1890s by
Britain and France.[13]

THE INTERNATIONALIZATION OF COMMUNICATION
AND GERMAN CABLES

The second model, the "internationalization of communication," was more
imaginative than the state ownership model. This model, as Rogers and Power
saw it, would consist of the German-Dutch cables held at the time by Japan,
the cable system operated by the Pacific Cable Board, and a U.S. Navy wireless
system augmented by a U.S. state-owned Pacific cable. Thus, instead of build-

ing rival imperial communication systems that could not be justified according to sound business practice, the "internationalization of control" idea would share the cost of a jointly owned Empire Communications Network among participating governments.[14] The status of all the German cables, previously owned by affiliates of Felten and Guilleaume, was taken up during the Versailles peace negotiations in 1919, but the only decision taken there was that the cables would stay in the possession of Britain, France, and Japan until a final decision on their fate could be reached at the planned World Communication Conference to be held in November 1919.[15]

As a result of these initial decisions, Germany continued to depend entirely on wireless to meet its communication needs or on cable systems controlled by others. The U.S.-based Commercial Cable Company, which had previously formed the core of a loose-knit consortium with the German and French companies, was also adversely affected. The company's subsidiary, the Commercial Pacific Company, had jointly worked the German-Dutch cable in Asia and used it as a backup for its connections between Shanghai and Manila. That arrangement was terminated as Japan took over the German-Dutch cable and made it part of its imperial communication network. The Commercial Company argued that the cables should become its property, not least because of heavy debts owed to it by Felten and Guilleaume, while the U.S. government pushed to "internationalize" them—that is, to put them under the joint ownership of the superpowers, which would operate them as a kind of "global communications utility."[16]

Britain, France, and Japan, however, saw the German cables as "prizes of war." While this did not fit with the U.S. position nor with that of the Institute of International Law, in this case possession was nine-tenths of the law. The French government used the German cables as part of a bid to gain further autonomy from British-owned cable systems and revitalize the operations of two French companies—the French Cable Company and the South American Cable Company.[17] Thus, the government ordered the North Atlantic cable rerouted from offices which had been shared with the Commercial Company to those of the French Cable Company and invested a considerable sum to rejuvenate the 1910–11 cable of the former German South American Telegraph Company. That France was using the opportunity to gain further distance from Britain was also signaled by the fact that the South American Cable Company moved its headquarters from London to Paris in 1915, just one year before Havas relocated the hub of its South American news service from London to New York to avoid British cable censorship (see chapter 6). Yet, despite these changes, the entire

French transatlantic cable system continued to flounder, with the north Atlantic cable continuing to offer poor service, the cable commandeered from the former South American Telegraph Company never restored, and the French Caribbean cables abandoned totally in the mid-1920s.

The absorption of the second German Atlantic cable by Britain into the Pacific Cable Board system was also significant because it displaced the Commercial Company as the final transatlantic link in that organization's operations. This constituted another blow to the company's revenues, and it complained bitterly of its treatment at the hands of the British government.[18] The Commercial Company's woes were compounded in 1916 as the Eastern affiliate, the Eastern and Azores Telegraph Company, terminated its landing rights on the Azores, thereby hobbling its access to South America and its anticipated expansion in Europe. Buffeted by this barrage of factors, the Commercial Company shed its agnostic approach to national identity and began wrapping itself in the American flag as it pitched the German cables issue and access to landing rights in the Azores as part of the broader campaign for the U.S. control of global communication. The company's argument that it had a right to all the German cables clearly clashed with their de facto ownership by Britain, France, and Japan. Its connections to the Azores, so claimed the company, should also be restored. However, these claims received surprisingly little support from the State Department, which preferred to "internationalize" both the German cables and access to the Azores.[19]

Rather than acting on behalf of the Commercial Company, the State Department was more concerned with the exclusive concessions that allowed companies to control access to foreign markets. The most important example was the Eastern and Azores Telegraph Company's control over the Azores, but other Eastern Company affiliates, as we have seen, were in a similar position in South America and Asia. Of the three, the Azores were most important because the islands had developed after the end of the nineteenth century as an alternative junction between Europe and North America. The islands were slated to become more important in the 1920s as all the transatlantic companies—Western Union, the Commercial Company, a reconstituted German Atlantic Cable Company, and the new Italian Submarine Telegraph Company—sought to use them as a essential base for their European and transatlantic operations.[20] Thus, instead of trying to further the specific interests of the Commercial Company, the State Department promoted the more general objective of opening up this critical hub through the "internationalization of control" policy: an approach where access and resources—offices,

traffic management, interconnections, and so forth—would be jointly managed as a kind of "global communications commons."

The fact that the Commercial Company had been chastised and threatened
with having its Pacific cable taken over by the government, or internationalized, indicated its low standing in U.S. political circles. This situation might be
accounted for by the fact that it had squandered much of its political capital
when it opposed the government's take-over of the U.S. communication system at the end of the war. While Newcomb Carlton and Theodore Vail, the
heads of Western Union and AT&T, respectively, had, without hesitation,
turned over their systems to the U.S. Postmaster General in 1919, Clarence
Mackay, president of the Commercial Company, railed against these moves in
the press, Congress, and the courts.[21] One also wonders if the action of the
company in fronting for a French company (the United States and Haiti
Company) in the United States and registering two of its own affiliates—the
Direct West India Cable Company and the Halifax and Bermudas Company—
in Britain was not also a bit too "globalized" for the times? It was one thing to
be free from state control, but maybe the company had taken its self-pro-
claimed cosmopolitan corporate identity a bit too far?

Evidence that this was at least part of the problem was apparent in a slightly
different but related context. Although literally on the other side of the world,
that context had to do with the operations of the company's affiliate, the
Commercial Pacific Cable Company, in Asia. The Commercial Pacific Company's cable service to the Philippines and China had long been abysmal. The
Newspaper and Press Association and National Foreign Trade Council heaped
scorn on the company's service and the U.S. State Department lambasted it
for laying no new cables across the Pacific since its first cable in 1904.[22]
Disclosure in December 1920 that the firm had ceded three-quarters of its
ownership to the Eastern Extension Company and the Great Northern Company in order to enter the Asian communication market further eroded the
firm's standing in U.S. circles.[23] Coupled with complaints from colonial administrators in the Philippines that the company always prevaricated when it
came to meeting their needs further jeopardized confidence. All of these
factors ultimately led the State Department to call for the Commercial Pacific
Company to either be turned into the central pillar of a state-owned American imperial communication network or become a contribution to the "internationalization of communication" model.[24]

With its weakened and bête-noire status, the Commercial Company was in
need of a political rehabilitation program in order to regain its credibility. To

this end, during the five years 1918–23, it appointed the U.S. foreign policy hawk and former counselor to the State Department Frank Polk to its board of directors; hired Charles Evans Hughes, the former Supreme Court Associate Justice (1910–16), Secretary of State in the Harding Administration (1921–25), and later U.S. Supreme Court Chief Justice (1930–41), as its chief legal counsel; assumed a more nationalistic stance; and proposed several new cables, notably to serve American imperial interests in the Philippines. Ties to British and Canadian elites were maintained, but there was a definite tilt toward aligning the firm with the more hawkish wing of the U.S. political establishment.

The importance of the Commercial Company's experience also lay in the fact that it was not the only major U.S. communication company to be at odds with the Wilson administration. Indeed, relations between U.S. communications companies and the administration were generally fragile by 1920, because they were all subjected to the State Department–led review of the global media system. Among other things, the review highlighted the fact that Western Union and the Commercial Company had maintained prices on the transatlantic route for over thirty years, although a few new services—weekend, letter, and deferred rates, for example—were introduced in 1913–14 to increase the social uses of communication. Despite these minor nods to public service, the two companies had steadfastly refused to lower rates and had even stymied attempts by Britain to regulate rates through the introduction of appropriate clauses in its cable landing licenses. Instead of complying, both of the companies simply refused to sign any new licensing agreements containing such measures. While their position had carried the day, Britain, France, and the United States all agreed to try and implement a more standardized approach to rate regulation at the forthcoming World Communication Conference, originally set for late 1919 but constantly delayed, and finally abandoned, for reasons explored below.

Like the cable companies, RCA and the commercial wireless companies were also brought under the microscope. In particular, the State Department was concerned that RCA and the other main global wireless concerns—Marconi, TSF, and Telefunken—were attempting to create a worldwide wireless cartel. Their creation of a consortia to develop international wireless services in South America in 1921, attempts to do the same in China, and plans to form a worldwide "wireless trust" in 1921 all provided solid evidence that these earlier concerns were firmly based.[25] Demonstrating that it was monopolies, not foreign companies, that were the target of reforms, Rogers raked

RCA over the coals for joining with other firms in a bid to "obtain a world wide monopoly," warning that "if buccaneering tactics are tried, a world-wide row may be started. In such a war, the United States is likely to suffer, as [will] other countries."[26] Yet, showing the depth of the emerging animosities, RCA's executive replied with some nasty remarks of its own about the Wilson administration's agenda: "So the underlying thought was the same as that expressed in the Covenant of the League of Nations, a complete subordination of nationalism to what many believe to be utopian views of internationalism which would prove to be to the detriment of the United States. . . . It [a proposed Universal Electrical Communications Union] . . . is based not only upon principles of government ownership, but also upon a government by commissions at the head of which is a super commission created along the lines of the League of Nations."[27]

Despite its rhetoric, RCA was not against globalization, just the kind advocated by Wilson. It did not want open markets, multilateralism, open media, and democracy, but the thinner version of globalization, where private firms organized cartels and managed markets on their own, just as the cable industry had done for the past half century. As such, proposals for a coherent global regulatory framework were anathema to RCA. Never one to withdraw from a tussle with corporate interests, Rogers retorted: "One of the surest ways of bringing about government ownership will be for the American communication companies to go out and try to rape the world, for reprisals will come so thick and fast that American business will cry for relief. . . . The Radio Corporation's Memorandum . . . achieves the distinction of being 99.99/100% inane."[28] The rhetoric was entirely consistent with the progressive era view that stable markets required a viable structure of regulation and state oversight. The only difference was that now solutions previously adopted within countries were being fought over at the global level.

The specter of government ownership raised by RCA was a red herring. It had been floated in two papers written by Rogers and Power in 1918 and mid-1919, at the height of the Navy's push for a wireless monopoly. Events since had led to the thorough repudiation of the idea, partly due to the dire experience with it in the context of the war and partly because Congress was concerned that, as the most ardent advocate of state ownership, the Navy's ambitions threatened to militarize the U.S. economy and bolster delusions of expanding the U.S. imperium. The creation of RCA itself reflected the rejection of state ownership. To the extent that the idea did remain, it was part of the "internationalization of control" model and as a supplement to the private companies' role in nonviable markets. Leaving no ambiguity on this issue, Rogers stated that "the . . . program . . . leaves to the companies what can be reasonably expected of them and places upon the government the burden of meeting other essential needs."[29]

The rift between the State Department and RCA underscored the pervasiveness of the cleavages between the U.S. communication companies and the Wilson administration. The key point, however, is to stress the limits to analyses that grasp the politics of global communication as if they were driven mainly by national economic interests. Moreover, attempts to understand these politics without assessing the Wilson administration's efforts to reconcile U.S. expansionism with an attempt to reconstruct the global system as a whole, or without recognizing the fact that Wilsonian internationalism was a derivative species of liberal internationalism sui generis, are seriously flawed. Indeed, the struggle between internationalist views of the global system versus realist conceptions of national interests set the agenda at Versailles in 1919 and that, in turn, shaped the politics of global communication. The "liberal globalist" agenda did not triumph, but it was a decisive factor, and one that is obscured badly by the reduction of the debates to a crude realist view where the state is seen as little more than the handmaiden of capitalist interests in the sphere of global politics.

THE WORLD COMMUNICATIONS CONFERENCE

As noted, in May 1919, the heads of state of ten countries (the Council of Ten) meeting at Versailles initially called for a World Communication Conference to be held in November of that year. The goal was "to consider all international aspects for communication . . . and to make recommendations . . . with a view to providing the entire world with adequate facilities of this nature on a fair and equitable basis."[30] The conference was to be as inclusive as it was comprehensive, only Costa Rica, Turkey, Germany, Austria, and Hungary were specifically excluded. It would cover several issues: (1) the distribution of German cables; (2) an international agreement on cable landing rights and radio concessions; (3) the unification of telegraph, cable, and wireless conventions under a single global regulatory commission; (4) the need for an international communications network for the distribution of news; (5) the need to expand facilities throughout the world; (6) the harmonization and reduction of rates; and (7) the need to promote "understanding between the peoples of the world."[31]

The conference proposal was highly significant because it hit all the notes in the chorus of Wilsonian internationalism. That view, in turn, saw the restoration and transformation of nineteenth-century liberal internationalism in the 1920s as turning on three things: expanding technological and economic interdependence, establishing a deeper web of multilateral institutions as the basis of greater political interdependence, and, finally, the importance of an open global media system and the free flow of information to cultural interdependence.[32] Such ideas underpinned a broad and imaginative approach to foreign policy in which global communications loomed large, not just as idealism and in terms of world public opinion (which was important), but as markets, mechanisms of multilateral integration, and means of deepening cultural interdependence.

At the outset, U.S. participants saw the conference in the broadest terms possible. As a result, they linked issues dealing with the communications infrastructure to a vast array of issues relating to journalism, propaganda, culture, news agencies, democracy, and so forth. A sense of such views was well reflected in the following statement by Ernest Power:

> There is only one way in which our present civilisation can be advanced along lines . . . which will minimize any such regrettable misunderstandings as may produce friction and ultimately war. . . . Popular good-will depends upon mutual understanding, and mutual understanding upon the interchange of intelligence, so that, unless such intelligence is freely accessible to the people at large, *the mere machinery to be established in connection with any proposed League of Nations will be futile.* The nations of the world must appreciate each other mentally, and the only aid to such appreciation is the free flow of . . . information through the press and through other public and private channels."[33]

The "mere machinery" of global diplomacy, as Power notes, was not enough. Just as vital was the need to nurture a substratum of cultural interdependence through which processes of technological, economic, and political integration could be linked to experience. In Habermasian terms, only by linking the systems world (markets, the state, technology, and instrumental reason) to the life world (communication, culture, and personal experience) could globalization "go deep." That is, the prospects for globalization in the twentieth century turned on new mental constructs, the ability of people to "appreciate each other mentally," as Power put it.

While crystallized in the shorthand of Wilsonian internationalism, these ideas were neither new nor exceptional. Instead, similar ideas had been circulating among European legal scholars associated with the Institute for International Law and numerous other sources for a decade or more. There were also important continuities here given that the Institute of International Law had endorsed similar principles in the United States' first statement on global communication outlined a half century earlier (see chapter 2). While such principles had lain dormant over the years or had been used in a highly selective way, the 1918 and early 1920s version no longer consisted of mere rhetorical statements but a set of policies to be anchored in global institutions.

The response among other countries to these ideas was mixed. Yet, the fact of the matter was that the conference had been called for by the Council of Ten, not just by the United States.[34] However, as the U.S. views took shape, Britain scrambled to rein in the agenda and defer the date. It also proposed that a preliminary conference be held in Washington in late 1920 and that a more technical focus be given to the event.[35] Japan adopted a similar position. Japan had worked out a series of agreements with China, Britain, and the United States between 1915 and 1918 on a range of issues touching on its approach to communications in Asia and was little disposed to have matters which it considered settled dragged into the anticipated World Communication Conference.[36]

France, however, was less concerned about the broad agenda than the failure of the United States to ratify the Treaty of Versailles or join the League of Nations. Although eager to recast the global communication system to further its own objectives, the French government feared that a weak U.S. delegation would allow Britain to dominate the conference.[37] Surprisingly, however, even vociferous domestic opponents of Wilson's foreign policy agenda, notably Senator Henry Cabot Lodge, authorized U.S. participation in the conference.[38] Several important business groups in the country, including the National Foreign Trade Council, the American Manufacturers Export Association, and the Merchants Association of New York also strongly supported the conference.[39] U.S. news organizations enthusiastically supported the conference as well. Several of the leading U.S. news media—the *Chicago Tribune, New York Times, New York Herald, New York Tribune,* Universal Service, United Press, and the International News Service—created the Cable-Using Newspaper and Press Associations to present their concerns in the interdepartmental committee set up to hammer out the U.S. position at the World Conference. Not inclined to criticize solely along national lines, the group wanted to abolish all monopolies and supported the creation of global regulatory standards that would harmonize rates and ensure that those who controlled the communication networks maintained enough capacity to meet the needs of the press.[40] In this, they found much support from the State Department, whose position all along was that "*all* . . . exclusive rights should be abolished . . . and cable landing rights thrown open to the citizens of all nations on an equal basis."[41] The State Department also wanted this anchored in international law. As Power stated, exclusive concessions "make it . . . impossible for . . . companies . . . to access the conceded territory under existing conditions, unless they enter into an agreement, . . . such terms usually including the upholding of the cable rates and the holding up of the public. . . . Exclusive concessions seriously interfere with the benefit to be derived by the public at large from the world's cable system."[42] The U.S. communication companies, however, opposed the creation of a global regulatory regime, although they participated in interdepartmental committee meetings in a bid to define the U.S. position at the conference.

U.S. cable companies and news organizations were closer together on the issue of direct access to foreign markets. Both implored the State Department to prioritize access to leased lines and the ability to open offices in foreign cities so they could deal directly with customers. In other words, the cable companies and news organizations sought to liberalize foreign telecommunications and news markets. For the cable companies, the issue was end-to-end control over their networks and access to users. The U.S. news organizations' position was also bound up with other issues dealing with the news cartel, and its principles of territorial exclusivity, which impeded their own access to foreign news sources and markets. It was on such issues that American views on communication, democracy, and the inadequate representation of the United States in the foreign press blossomed, as the following internal memorandum from Power reveals:

> It is generally realized . . . that . . . trade would certainly be furthered by making American thought and purpose better known and if American market reports were accessible when still fresh. It is even so from a political point of view. Very little American news reaches the Orient. Occasional items that are sent by Reuter from London via Eastern cable, press bulletin items are received in Japan and China through Kokusai, a Japanese agency, and the United Press from New York or San Francisco, but such brief notices concerning the United States . . . present a distorted perspective to their readers and are altogether inadequate.[43]

The quote is an excellent example of how easily U.S. policy makers slipped from a focused discussion of cable and wireless issues into a whole raft of wider concerns. It was an incredibly broad view that aimed at precipitating a discussion of the global media system as a whole, but it was precisely this breadth that Britain and Japan, in particular, saw as potentially swamping the conference with tendentious issues. However, the British government was keen to resolve the issue of monopolistic concessions, although there was a good deal of difference of opinion as to the appropriate actions. The General Post Office followed its well-worn path and recommended that "the policy of the British Government should be . . . to give British companies such support as they can whether or not they are working for monopolistic concessions."[44] The Committee of Imperial Defence largely agreed, arguing that "the companies would hesitate to incur the considerable capital expenditures required to establish these services . . . if exposed to competition and rate-cutting."[45] However, rather than adopting such a protectionist stance, the "British Delegation . . . was instructed to make it plain . . . that Great Britain is . . . opposed to exclusive concessions, although she cannot put pressure on her nationals to abrogate exclusive concessions which they already enjoy."[46] This, ultimately, was the "consensus position" adopted at the preliminary Washington Conference in 1920.

Ultimately, the full conference was never held. By late 1920, the idea of an omnibus review of global communications had been transformed into a series of mini conferences. No longer universal in conception, nor inclusive in participation, only six countries—U.S., Britain, France, Japan, China, and Italy—took part. Even these conferences were neutered as delegates were limited to stating their desires rather than adopting robust international instruments: treaties, laws, resolutions, etc. In many ways, the outcomes reflected a triumph for Britain and Japan, which had sought a less powerful conference from the beginning. The demise of Wilson as a formidable force in American politics magnified France's concern that the conference would likely do more to prolong Britain's domination than to transform global communication. Even China saw the conference as more of a threat than a means to recasting the problems which affected that country within the "international development of China" framework.[47]

The inauguration of the new Republican administration of Warren G. Harding in March 1921 also reinforced the drift toward a confined agenda. As the following statement from Joshua W. Alexander, the last Secretary of Commerce in the Wilson administration, to Breckinridge Long at the State Department indicated, that drift had already gained momentum on the eve of the autumn elections of 1920 that brought about a changing of the guard in U.S.

politics: "Any commitments to create a global communications network under joint ownership of various governments for the distribution of news must be mindful to not encroach on the role of U.S. news agencies and be pursued only after gaining a clear view of the private companies' position on such matters. Also must be mindful of U.S. traditional policy of relying on private enterprise . . . [and] not to imply commitments at the preliminary conference that cannot possibly be met."[48]

The new Republican administration reinforced that drift. A rising star within its ranks, Herbert Hoover, became Secretary of Commerce and played a decisive role in domestic and international communication affairs throughout the 1920s, but had little inclination to follow through with the kind of conference that had been anticipated by the Wilson administration. Never one to let go of a vision, Rogers briefed incoming officials in the State Department and let them know that he was available to participate in the conference should the need arise, but thereafter he moved out of the political limelight.

The demise of the full conference makes it far too easy to dismiss what did happen and ignore the events that followed the preliminary conferences. Headrick takes just such a position when he suggests that America was quite capable of masking its "expansionist aims in the rhetoric of self-pity [but] government policy, in the end, played almost no role in the rise of American power in telecommunications. Instead it was America's wealth aided by changing technology."[49] But whether expansively interpreted or tightly confined, the idea of the conference did sustain an unprecedented discussion of global communications policy and brought about the most thorough review of the global media system ever undertaken up to that point. Moreover, even though the United States moved away from policies that had been laid out under Wilson, sturdier themes and general policy norms were staked out that did shape that country's approach to these issues for the next decade and, arguably ever since, especially with respect to the "free flow of information." And through all the twists and turns that took place from 1918 until 1922, foundations were laid for solutions that shortly afterward actually did transform the world's three most important communications markets: South America, Asia (notably China), and the Euro-American communications market. And as these were worked out on a case-by-case basis and through new domestic legislation, the global media system was changed—as the next two chapters demonstrate.

Communication and Informal Empires:
Consortia and the Evolution of South American
and Asian Communication Markets, 1918–30

> I am perfectly clear in my judgement that if private capital cannot soon en-
> ter upon the adventure of establishing these . . . means of communication,
> the Government must undertake to do so. We cannot indefinitely stand
> apart and need each other for lack of what can easily be supplied, and if one
> instrumentality cannot supply it, then another must be found which will. . . .
> So soon as we communicate and are upon familiar footing of intercourse,
> we shall understand one another, and the bonds between the Americas will
> be such bonds that no influence that the world may produce in the future
> will ever break them.—WOODROW WILSON's opening address to the
> Pan-American Financial Conference, 1915 (quoted in Triana,
> *The Pan-American Financial Conference of 1915*)

COLLABORATION AND CARTELS VERSUS
CONFLICT AND COMPETITION IN SOUTH AMERICAN
COMMUNICATION MARKETS

As of 1914, Britain was still the largest source of long-term capital investment
in South America ($3.8 billion), an amount that was far ahead of French ($1
billion), German, and U.S. investment in the region.[1] U.S. investment in
South America remained marginal between 1914 and 1920 at around $20
million per year, even as European and British investment fell to a trickle.[2]
American exports to Latin America, however, surged and groundwork was
laid for the expansion of U.S. interests through several Pan-American Financial Conferences and other U.S.–Latin American initiatives. The goal of expanding U.S.-controlled communication infrastructure and news ranked at
the top of the Wilson administration's priorities. In fact, so concerned was the
administration with the paucity of communication facilities, as revealed by
the epigraph above, that it bludgeoned U.S. cable and wireless interests into
action by threatening that if they failed to rectify the situation, the government would do so itself.

Wilson's hectoring was directed as much at getting U.S. companies to act as it was to put the South Americans, British, and Europeans on notice that conditions were intolerable. The U.S. Navy ratcheted up the rhetoric further by registering its own desire to provide communication and maritime transportation facilities in the region. Yet, the Navy's rhetoric should not be confused with U.S. policy in toto. Indeed, that this is so could be seen by the fact that the Navy often outlined ambitious visions of its role only to see them beaten back by a circumspect Congress. Moreover, on closer inspection, the main thrust of the Navy's statement was that South American governments, as general policy, should develop national wireless systems under the control of state ownership. Yet, even here, the Navy's statements on state ownership were formally repudiated within a few years, and by 1924 attempts by Latin American governments to adopt a uniform policy of state-owned national wireless systems was sternly rebuked by the American Delegation to the 1924 Inter-American Committee on Electrical Communications.[3]

The United States managed its expanding presence in the region through a range of "soft" and "hard" power. At the soft end of the spectrum, the expansion of U.S. investment, especially after 1920, was combined with U.S. experts to lead administrative and political reforms, notably through the Kemmerer Missions. Edwin Kemmerer was a Princeton economics professor who led teams of experts to study and retool the financial and administrative systems of Columbia, Guatemala, Chile, Ecuador, Bolivia, and Peru. The results were mixed. When the reforms failed, the United States took direct control over the customs and financial machinery of governments in Cuba, Haiti, Dominica, Nicaragua, Peru, Ecuador, Honduras, and Bolivia for either part or the whole of the period from 1916 until late in the 1920s. The practice was modeled along the lines adopted by British and European finance since the late nineteenth century, and American anti-imperialists resented it as such. In fact, numerous U.S. military interventions came on the heels of situations where European governments were already interceding, or preparing to do so, on behalf of threatened investors, as in Venezuela (1913), Guatemala (1913), Haiti (1914), Honduras (1910), and Mexico (1914). In the American view, its own interventions were designed as much to cut off the possibility of European occupation as they were to restore order and protect property in nations beset by chronic instability. As a result, the Monroe Doctrine and the Roosevelt Corollary (first announced in 1904) took on a more muscular approach to a still relatively new U.S. foreign policy objective: debt collection. This "dollar diplomacy" was prominent in smaller Central American and Caribbean countries, while the

stronger states of South America—Argentina, Brazil, Uruguay, and Chile—
were treated more as "normal states" and able to resist such neocolonial
policies.[4] Indeed, South America was recognized as an area where the private
sector should take care of itself, even if it did need an occasional stern prod-
ding from Wilson.

Reflecting these patterns, the Navy's imperial wireless network homed in
on the formal zones of the U.S. empire as it claimed a wireless monopoly in
three Caribbean dollar dependencies in 1917: the Dominican Republic, Haiti,
and Puerto Rico, while the private wireless companies were dominant elsewhere. The French Cable Company protested the Navy's encroachment on its
exclusive Caribbean concessions, but such protests were brushed aside. The
fact of the matter is that by 1920 that company's concessions were worthless.
This was underscored as two U.S. cable firms—the All-America Cable Company and Western Union—laid new cables to Haiti, the Dominican Republic,
and Cuba around that time. A few years later, the All-America Company tried
to acquire the French company's Caribbean system outright, but its cables
were abandoned completely before a deal could be consummated.[5]

Matters would not be so easy in South America, where more was at stake,
and vested interests were far sturdier than the chastened French Cable Company. It was here, and especially in Argentina, Brazil, and Uruguay, that the
Wilson administration pushed hardest. The efforts began to bear fruit after
1916. At that time, the State Department brokered a deal with the All-America
Company for cheaper rates for U.S. news organizations. It also shepherded
negotiations between Associated Press (AP) and Havas, as noted earlier, that
led to a revision of the news cartel agreement in 1919. The revisions were
important not just because they expanded AP's role in South America but
because it was clear that AP and the State Department were seeking not to
dismantle the global news cartel but to expand AP's role in it. Indeed, this was
the policy of AP president, Melville Stone. In return for gaining access to
South American news markets, AP agreed to allow Havas to distribute a condensed service to its subscribers and to compensate Havas for "lost business."

Other intangible support was also used to help United Press (UP) and AP
expand their role in South American news markets, including letters of introduction from U.S. officials and even President Wilson. While sources are
ambiguous on the point, it is probable that the State Department compensated the All-America Company for lost revenues and arranged to have the
cheap rates extended to Havas as an inducement to accept revisions to the
news cartel. Rogers only added to the appearance of mystery when asked

about which governments subsidized the news agencies by responding: "I know a good many cases but it is a dangerous question."[6] Indeed, all of these arrangements were bound up with issues tied to propaganda, and they threatened to undermine the quality of international news altogether. In fact, such concerns emerged as a key issue in U.S. media politics, with left-leaning American journalists like Oswald Garrison Villard of the New York Evening Post and later the Nation, rebuking Wilson for using the U.S. news agencies as an arm of U.S. propaganda efforts.[7] Even Rogers turned from his role at the Committee of Public Information to become a vocal critic of propaganda, writing a series of articles on "tainted news" for the popular press and making the elimination of subsidies and censorship a key plank in his advocacy of the free international exchange of news.[8]

The All-America Company, at the State Department's behest, reduced rates for U.S. news organizations considerably. Whereas the British-based Western Telegraph Company and the French-based South American Cable Company charged Havas 25¢ per word to carry news from Europe to Buenos Aires, the All-America Company's rate was reduced to 10¢ per word, with a 2¢ drop rate for cities passed by its system along the west coast of South America. The cheaper rates and revised news cartel paved the way for a vast expansion of U.S. news in South America. United Press first entered the market by establishing an agreement with La Nacion in Buenos Aires in 1916, before switching to La Prensa, another daily in the same city, and one of wealthiest papers in South America two years later. By 1920, UP served twenty-three dailies in Argentina as well as five in Rio de Janeiro and another in Sao Paulo, Brazil. By 1919, twenty-five of the foremost papers in South America subscribed to AP's services, and the agency opened news bureaus in many of the continent's principal cities.[9]

Despite this impressive expansion, the U.S. news agencies still believed that their efforts were impeded by two factors. First, since the cable company's network did not reach directly into cities, the news agencies depended on interconnections with poorly run state-owned telegraph systems. The second problem stemmed from the fact that the All-America Company was unable to keep up with demand. As UP and Universal News Service claimed, they would immediately double the amount of news they distributed in South America if the cable company could cope. In Rogers's assessment, the All-America Company was originally very generous in its support of the reduced rate, but as soon as its networks began to groan under the strain, it turned to discouraging the press in favor of regular paying customers. William Hawkins, president of UP, similarly claimed that he had no doubt that the cable company originally had good intentions but that it had quickly come to view the press "as more or less as a burden. I am sure that has come to be the case with the All-America Company in South America, because I have it on the authority of the officials of that company, and we are more or less in the attitude of supplicants; the American press is practically in that attitude toward the cable companies of the world."[10]

Most U.S. cable companies agreed with such assessments. There was also another angle to the issues stemming from the fact that much of the collection and production of news by U.S. news media was conducted from the news capitals of Europe: London, Paris, and Berlin. Previously, there were several routes between Europe and the United States, at one end, and South America, at the other. However, by 1916, many had been crippled by the events mentioned previously—the cutting of the German cables, the precarious status of the French Cable Company, and the elimination of the Commercial Cable Company's landing rights at the Azores. In addition, the All-America Company's service was further compromised by the breakdown of its alliance with Western Union in 1916 (for reasons which will be explained below). In the face of the All-America Company's inability to meet their needs, by 1920 AP and UP turned to sending 3,000 words a day each from London to South America by way of the Western Telegraph Company—more than they sent over the All-America Company's system.[11]

The surprising thing about such arrangements, especially when considered in the light of All-America's tendency to strain the story through the prism of a great patriotic struggle, is that AP and UP did not really seem to care who owned the means of communication. The main problem was that the Western Telegraph Company had adopted quotas to deal with the spike of messages during and after the war and, like its American counterparts, preferred higher revenue traffic to that of the press. The Western Telegraph Company quota for British news was 1,500 words a day at 12¢ per word versus 150 words per day at 50¢ for the U.S. news media. Above those amounts, the press paid full rates. While even the liberal media critics in the U.S. State Department cast these arrangements in terms of egregious discrimination, once considered alongside the same practices adopted by the All-America Company, they look much different. The State Department reformers were on firmer ground when they cited these problems to justify eliminating all subsidies in favor of standard reduced press rates. That was also the stance of the Cable-Using Newspaper and Press Associations.[12]

THE LIBERALIZATION OF SOUTH AMERICAN
COMMUNICATION MARKETS

A landmark decision by the Supreme Court of Brazil in 1916 eliminated the Western Telegraph Company's exclusive concession for cable communication from Brazil to Uruguay and Argentina to the south.[13] The decision was a boon for the All-America Company, and over the next several years it pushed hard, with strong support from the U.S. State Department, to obtain access to Uruguay and Argentina. Both countries approved the company's license in 1919, and in the next year the new cables were laid between Buenos Aires and Montevideo in Uruguay, and two other cables went from there, one to Sao Paulo and the other to Rio de Janeiro. The company gained another license in the same year for a cable from northern Brazil to Cuba, with the caveat that it could not directly link to the United States. Overall, the All-America Company gained access to between 75 percent and 85 percent of the east coast of the South America cable market.[14] The Western Telegraph Company continued to dominate these markets and would do so for years to come, but now there was at least real competition.

Amidst this broad liberalization of South American markets, the Western Company still possessed a monopoly over the cable system linking cities along the enormous Brazilian coastline, and that concession was not set to expire until 1933. It was for this reason that the All-America Company was forced to lay two separate cables from Buenos Aires and Montevideo to Sao Paolo and Rio de Janeiro. The Western Company continued to place one obstacle after another in a rear-guard defense of its dominant position, and in this the most important method was that of "block rates." These discriminatory rates meant that there was one rate charged for messages going over the firm's coastal network of cables to the United States and other destinations if handled wholly by the Western Telegraph Company (34¢ per word) and another (58¢ per word) if handed over to the All-America Company. The Western Company also made a deal with the Argentine government so that all governmental messages and those received at state-owned telegraph offices were to be handed over to it.[15]

These defensive measures, however, did little to stem the tide. Indeed, the All-America Company was quickly followed by the French-based South American Cable Company, which also applied for new licenses so that it could expand its operations in the River Plate District.[16] This broad liberalization of South American communication markets even brought forth an initiative by Western Union, which seized the opportunity to acquire conces-sions in Chile, Peru, and Brazil.[17] Western Union's thrust was also motivated by an American regulatory decision (1917) that outlawed its alliance with the All-America Company.[18] However, while such an adverse ruling might lead others to withdraw altogether, Western Union, as it had done previously, sought to surmount U.S. restrictions by taking over its former partner alto-gether. Newcomb Carlton, the president of Western Union, justified his firm's attempt to acquire All-America on the following grounds: "Our thought is that cables are auxiliaries to great land line terminals: that when you have a system like the Western Union, . . . that this great collecting and distributing system should work in close and sympathetic harmony with the operation of cables in order that there may be proper operation of the whole."[19]

The main concern, in Carlton's view, was the need for end-to-end control over communication networks. With their alliance severed by U.S. regulators, the key was to merge, argued Carlton. And even if the two companies did not merge, Carlton advised the All-America Company to join the only other U.S. firm that had a North American telegraph network and global presence: the Commercial Company. The Commercial Company's North American tele-graph network, coupled with its vast cable system across the Atlantic, Pacific, and Caribbean, would provide the conditions, Carleton argued, that were necessary for the All-America Company to survive in the 1920s. The company rejected the Western Union's bid, but from 1918 on, its operations became ever more tightly intertwined with those of the Commercial Company, ultimately leading to their merger into a new U.S. global communications conglomer-ate in the late 1920s: the International Telegraph and Telephone Company (ITT)—a point we return to in the next chapter.

As the All-America Company drew closer to the Commercial Cable Com-pany, it never missed an opportunity to heap invective on Western Union. Like a spurned lover, the firm chastised Western Union's actions in seeking to merge as a great betrayal, while Western Union saw the events as a rational response to U.S. regulatory changes and the realities of global communication markets. Unlike the All-America Company, Western Union also had few com-punctions of joining with the Western Telegraph Company. After all, it had come to satisfactory arrangements with British cable interests and the British government for the transatlantic cables, so why not in South America?

From 1916 to 1922, skirmishes between British and U.S. interests, punctu-ated by interventions from Brazil, Argentina, and Uruguay, persisted. The U.S. State Department continued to thematize the problem of "block rates";

local British embassies, with strong support from the Foreign Office, remained intransigent on the issue, seeing them as one of the last bulwarks in the defense of the Western Telegraph Company's continued dominance of South American cable markets. The Imperial Communications Committee also called for a more protectionist stance.[20] Reflecting the magnitude of the resistance, Denison-Pender, in a bid for more British diplomatic support, threatened "to go out of business and leave cable communication to the Americans" altogether.[21]

Yet, while the Western Telegraph Company was engaged in a no-holds-barred defense of its concessions, it was working discretely with Western Union on other arrangements. The results were disclosed in 1920, as the managing director of the Western Telegraph Company informed British officials "that his company has entered into a most satisfactory agreement with the Western Union Company . . . whereby the latter are not to lay fresh cables direct to Brazil but are to link up their system with that of the Western Telegraph Company. The connection is to be made at Barbados and is expected to be in working order within a few weeks."[22]

CORPORATE GLOBALIZATION I, OR A GREAT PATRIOTIC STRUGGLE? THE CASE OF THE MIAMI INCIDENT

In fact, the Western Company had delayed informing officials of the arrangements for almost a year as it and the Western Union tried to cajole the All-America Company into joining them in a cartel for the organization of the entire South American communications market. The gist of the effort was that Western Union and the Western Telegraph Company would leave the west coast to All-America while reserving the east coast to themselves, except for permitting All-America to maintain offices in Buenos Aires. Should the All-America Company not agree to this proposal, then the two others would encircle the continent and engage their rival in an all-out struggle for control of South America.[23]

The All-America Company spurned the offer midway through 1919 after five months of bitter negotiations. In May 1920, Western Union and the Western Telegraph Company began laying their cable from Rio to Miami. However, their actions were abruptly interdicted off the coast of Miami by the U.S. Navy, under instructions from President Wilson, and from the British vice counsel at Miami, who was on board the U.S. destroyer at the time.[24] Why this dramatic action was taken will emerge shortly. As the cable lay useless off the Miami coast, Western Union scrambled to get an injunction in the Federal Courts challenging Wilson's authority to grant or withhold cable-landing licenses, while Congress considered new legislation—the Cable Landing License Act—and convened extensive hearings on the issues. The hearings brought up far more than just the South American cables, as the titans of the U.S. global media companies lined up to express their views on everything under the sun: cable communications, advances in wireless technology, monopolistic concessions, the global news agencies, representations of the United States in the foreign press, propaganda, the World Communications Conference, and so forth.

The All-America Company tried to set the tone for the hearings by couching the issues entirely in the "struggle for control" mold, as the following quote underscores: " It is a vital step in a struggle which is going on between the United States and Great Britain for the control of cable communications . . . and we ought not to . . . be blinded to the broad general fact by any detailed question of what has been going on for the last few weeks in the State Department. . . . There is an international struggle going on . . . and . . . there is danger that we might forget . . . its importance in considering the details of the transactions of the last few weeks."[25]

Continuing in a similar vein, Elihu Root Jr., All-America's legal counsel, summarized the state of affairs as follows: "The British and American lines . . . get what they can for themselves and, where it is possible, break up the monopolies of the rivals, and so far the British Foreign Office has actively assisted the British lines and the American State Department has actively assisted the American lines."[26] The comment is extraordinary in terms of what it reveals and, just as important, what it glosses over. This is evident in the rhetorical strategies of Root, who invites participants to skip the details in favor of his preferred narrative: the struggle for control. Also notable was the extent to which he aligned the firm and the State Department as if the two were joined at the hip and singularly dedicated to one goal: expanding U.S. economic interests. The fact that Root was speaking at all was intriguing as well, insofar as it illuminated connections between All-America and the hawkish wing of the U.S. foreign policy establishment which was the main alternative to Wilsonian internationalism, that is, conservative internationalists with their strident realist approach to international relations and minimalist commitments to multilateralism and global interdependence. The company had always been tightly tied to this wing of U.S. foreign policy (see chapter 2), and such links were in full bloom at this time. Root was none other than the son of Elihu Root Sr., the Secretary of State in the "walk softly, carry a big stick" presidency of Theodore Roosevelt. Root was not just representing the All-America Company; he and the company were carrying the flag for a particular American approach to foreign policy, noted for its muscularity rather than any sensitivity to wooly liberal ideas.

While subsequent historians of the global media, from Leslie B. Tribolet in the 1920s, to Joseph Tulchin in the 1970s, Daniel Headrick in the 1990s, and Jill Hills in 2002, have generally adopted the Root/All-America Company narrative framework as the definitive word on the subject, the surprising thing is just how little traction this perspective had during the hearings. As we have seen, the U.S. news agencies did not couch their analysis in such terms and, besides pointing to some very real problems that could be attached to discriminatory rates, they were no more likely to criticize the Western Telegraph Company than they were the All-America Company. In fact, they were rather grateful to both companies for having gone a long way to meeting their needs during the most urgent of circumstances. They also rebuked each for not going far enough. Walter Rogers was also clear, when he stated: "I think the fundamental problem involved here [is] that the monopoly owned by both companies [All-America Company and the Western Telegraph Company] . . . do to a certain extent block each other, and independent of any national interest, or interest of a private company, it is quite unfortunate that some of the South American countries have tied themselves up by granting monopolies. It does prevent the fullest possible development. There is a good deal to be said in favour of getting all of the monopolies away."[27]

Further examples could be piled up like leaves in autumn, but the most important statements for understanding the events in South America were those offered by the representatives of Western Union: Newcomb Carlton and William McAdoo. In contrast to the All-America Company, Western Union argued that its cable system was an instrument of commerce, not a weapon of politics. Any notion that things should be otherwise were sternly beaten back by Carlton throughout the hearings, notably in the following passage: "We will not permit ourselves to be used as a coercive measure. If the United States, or the State Department, desires to have the contractual relations between the Western company and Brazil altered, there is an orderly . . . method through which they can approach the British company. Certainly we are not that medium, and we will not be used as that medium."[28] Carlton also presented his company's "theory about the telegraphic business—do not duplicate what some one else had done."[29] Western Union's primary aim, it argued, was to link together the two great telegraph systems at each end of the cable; its own and that of the Western Telegraph Company.

The differences between Western Union and the All-America Company were also evident in the character of their legal advisors. Instead of a conservative internationalist like Root, William G. McAdoo, former Secretary of the Treasury in the Wilson administration, presented Western Union's version of the events. McAdoo, in turn, stressed the consistency between the companies' actions and policies that Wilson had promoted: informal negotiations, the need to expand the means of communication, and substantial rate reductions. On the matter of rate reductions alone, success seemed to be in the offing, as All-America dropped its rates from 85¢ per word to 50¢ in anticipation of the new competitive route to the United States.[30]

Yet, there was another aspect that McAdoo skirted as he glossed over the inconsistencies between the Western Union and the Western Telegraph Company's actions and Wilson's policies, namely, Wilson's insistence that such matters had to be settled by international agreement. The fact that the two companies had tried to lay the cable in May 1920 as preparations for the Washington Communications Conference were being finalized constituted a brazen attempt to foist a fait accompli on the United States and others who still believed that the conference might yet produce tangible results. This was the crux of the issue, as indicated by the Undersecretary of State, Norman Davis:

> One of the principal questions on which we hoped to get a satisfactory agreement in this . . . conference . . . was the elimination of monopolies, and the President considered it very important that we should await the results of that conference before endeavoring to outline the conditions under which we would permit the landing. . . . The President decided that consideration of the application for landing the cable should be postponed pending a discussion of the subject . . . by the . . . international conference.[31]

Given Wilson's unflinching commitment to expanding multilateralism and international law as the basis of the global political economy at this time, the companies' actions had zero chance of success.

The Western Union and Western Telegraph Company's jointly owned cable remained stalled until 1922. In the interim Britain and the United States continued to skirmish over block rates and proposals to divide South America into mutually exclusive "spheres of influence." Newcomb Carlton also sparred with Walter Rogers in the pages of the *New York Times* over the delay and the subordination of markets to global politics.[32] In the end, the problems were resolved once the Western Telegraph Company renounced its preferential agreements on the east coast of South America and All-America did the same on the opposite side of the continent.[33]

The crucial turning point was the passage of the Cable Landing License Act by Congress on May 27, 1921. The new law reaffirmed the president's authority to grant licenses and delegated that authority to the State Department. The sale of U.S. cable companies also became subject to review to ensure that they comported with national security interests. In addition, the law did not prevent mergers, cartels, or other combinations, but required companies contemplating such moves to submit their plans for review. In short, the act authorized the State Department to distinguish between "good" and "bad" monopolies and cartels and thus, like the Webb-Pomerene Act of 1918, did not see consolidation as inherently bad. Overall, the act constituted a "reciprocity regime," whereby access to U.S. markets depended upon similar rights being available to U.S. firms in foreign markets.[34]

Britain adopted similar measures, not through new legislation, but by a more vigorous use of its existing regulatory regime. And in that context, Britain—to the chagrin of Western Union and the Commercial Cable Company which had balked at these terms for years—made license renewals contingent on the companies' acceptance of a "control of rates" clause. The new regime, in turn, was organized so that renewable licenses were granted for seven-year periods and set to expire simultaneously so as to allow for a general policy review. The General Post Office also used the circumstances to negotiate rate reductions of 15–20 percent and pushed, without success, to require Western Union and the Commercial Cable Company to observe the international regulations.[35]

CORPORATE GLOBALIZATION II: SOUTH AMERICA AND THE WIRELESS CARTEL, 1921–24

Somewhat ironically, just as cable markets in South America were being liberalized, the new sphere of international wireless services was being developed as a cartel. During the earliest phase of wireless development beginning around 1914, this looked unlikely given that most Latin American governments had issued multiple licenses to numerous foreign companies for the development and operation of international wireless services. Indeed, for about five years a deluge of competing concessions were granted to foreign firms throughout Latin America. Many governments in the region had created state monopolies for domestic wireless services, but there seemed to be little inclination to do so for global wireless services. Such trends were consistent with the opening of cable markets, and Latin American governments seemed to be keen to avoid repeating their experience with cable monopolies in the new sphere of global wireless services. However, the fact that the big four global wireless firms preferred a cartel to competing indicated that it would not be Latin American national prerogatives that were decisive. Indeed, the big four—RCA, Marconi, TSF, and Telefunken—signed two agreements on October 14, 1921, to form the Commercial Radio International Committee, a cartel that was set to last until 1945. The agreement reflected classic progressive era thinking about the need to avoid "ruinous competition," but without any reference to the corollary of state regulation, either nationally or globally, as the following quote from the founding agreement testifies:

> Whereas it is realized that the construction of a number of stations for long distance communication in substantially the same locality in South America . . . would be . . . contrary to the interests of the commercial and general public as involving a wasteful use of the relatively small number of bands of wave lengths available for the wireless communications of the world, and as involving also a total expenditure out of all proportion to the volume of traffic which can be reasonably expected; and whereas the parties accordingly desire to cooperate in the establishment and operation of such a number of such stations in the South American Republics as may be necessary for the handling of external communication in the best and most efficient manner.[36]

The agreement created a board of nine directors, with two appointed by each company and the ninth by RCA. No role was provided for local interests among the directors, a clear indication that even by the standards set by the cable companies, whose affiliates were often tied into local elite circles through directors, the interests of South American countries were subordinate to those of the cartel. The first managing director was E. J. Nally, former president of RCA, and this placed the RCA at the apex of the cartel. Indeed, as RCA Vice President Owen Young remarked, "No effective action can be taken without American approval, thus carrying the . . . Monroe Doctrine into the field of communications."[37]

While the balance of control did fall to RCA, Young's claims were unpersuasive. Power within the cartel was structured hierarchically and relationally rather than as a simple appendage of national corporate and state interests.

That the consortium's headquarters were in Paris, not New York or Washington, also contained more than just symbolic significance. Indeed, Admiral Stanford Hooper of the U.S. Navy, one of RCA's original sponsors, criticized the cartel on the grounds that it did not resemble at all the "Monroe Doctrine of Radio." As Hooper groused in 1921, the organization "can only result in an international pool, which, directly or indirectly, will result in submerging the interests of each country with the others in the group."[38] Secretary of the Navy Franklin D. Roosevelt expressed the same point two years later, stating that the consortium "cannot be considered a bona fide American interest deserving the assistance and protection of this Government."[39] Thus, RCA was now seen in some quarters more as a threat to national interests than an asset in the struggle for control of global communication.

These issues became intertwined with national communication policies in the United States in the early 1920s, as RCA came under pressure from a raft of antimonopoly groups and a congressionally authorized investigation by the Federal Trade Commission into its alleged "radio monopoly." Yet, it was also on this shifting political terrain that RCA gained standing among those close to Herbert Hoover, the assistant secretary of commerce in the Harding Administration and a rising star in U.S. political administration. Hoover blessed the AEFG (America, England, France, Germany) consortium. It was a realistic response to a tangled mess of concessions in South America. It was also one simple way of reintegrating German commercial interests into the global system—an important aim of U.S. foreign policy in its own right.[40]

The terms of the cartel were familiar: patents held by each of the participants were to be jointly worked, the development of international wireless services in South America would be cooperative, and prohibitions against cooperating with rivals would be put into place. The only exception to these general rules were those that were made for another American company—the Tropical Radio and Telegraph Company, better known as the wireless arm of the United Fruit Corporation. The company was controlled by the Keith Family, with Minor C. Keith originally setting up its telegraph arm in the 1870s on land that he had acquired during one of the periodic financial crises that swept the Caribbean and Central America. These long-standing interests did have weight in the U.S. strategic calculus, and United Fruit's wireless system worked hand in glove with the Navy wireless system to meet U.S. needs in this, the heartland of its "zones of informal and formal empire." But to return to the AEFG consortia, the big four wireless companies agreed to share ownership equally in the Argentine National Wireless Company, the Brazilian

National Wireless Telegraph Company, and any others they created up until 1945.[41] The first of these two operations, one in Buenos Aires and the other in Rio de Janeiro, began in 1923 and 1924 respectively. Deals were made with the cable companies to establish what was euphemistically referred to as "equality of rates," thereby softening the rivalry between the "old" and "new" media.[42]

THE AGE OF DISORDER CONTINUES: COMMUNICATION AND EMPIRE IN CHINA, 1916–30

One of the most important things about the AEFG consortium was that it became the touchstone for efforts to sort out the development of wireless in China in the 1920s. As the U.S. Federal Trade Commission's 1923 report observed:

> The communication situation in the Far East, and especially in China is not dissimilar from the situation which existed in South America. . . . The British Marconi Co. and the Chinese Government own together the Chinese National Wireless Co., which . . . has a concession for international wireless communication; The Federal Telegraph and Telephone Co of California hold a concession from the Chinese Government for wireless communication; and the Mitsui Russan Kaisha hold a concession . . . which will allow them to build and operate stations capable of carrying on an international service. . . . The Great Northern Telegraph Co., a Danish Company, hold a complete monopoly for all of the external communications of China under cable concessions which the Chinese Government gave to the telegraph company in 1896. The . . . concessions [are] a subject of diplomatic negotiations, not only between the representatives of the American, British, Japanese and Chinese Governments, but also between these parties and the Danish Government.[43]

Remarkably, even the U.S. Navy agreed. As Edwin Denby, the Secretary of the Navy under the Harding Administration, noted: "Co-operation will insure not only that the most modern stations in radio practice will be made readily available in China, but also in the principal countries of the world, stations will be made available for the carrying on of communications with China immediately. Such an advantageous position for China could not be brought about by dealing with China alone as a segregated unit. . . . *It would insure an open-door policy in communication, cooperation of all of the principal Governments, and bring to China the . . . resources of all of the principal wireless companies of the world.*"[44]

PRECURSORS TO THE INTERNATIONALIZATION
OF CONTROL

The decision that wireless services in China should be organized cooperatively coincided with the gradual acceptance by the "superpowers" that the "internationalization of control" model was best suited for conditions in China. The consortium approach gained momentum in 1909 as several banking interests—from Britain, France, Germany, the United States, and, after 1912, two Japanese and Russian financial institutions—formed the International Banking Consortium. Frank Ninkovich captures the importance of these trends when he argues that by 1909, the Taft administration believed that the consortium approach would "remove development from the framework of competitive . . . imperialism . . . The cooperative developmental logic . . . was so compelling that it would be picked up in the early 1920s as the centerpiece of the Harding administration's China policy."[45]

It was a long road between 1909 and the wireless negotiations of the 1920s, but a brief review of that period offers a glimpse of the tensions besetting the model from the beginning. Britain and Japan remained suspicious that the model masked U.S. aims to bolster its influence in Asia. Regardless of their respective governments' positions on policy, British, Japanese, and U.S. corporations also struck their own deals with Chinese officials for the construction of railroads, telephone systems, and other business ventures. This also suited Japan because it welcomed the infusion of cash that went into joint ventures between Japanese industrial interests, on one side, and British and American interests, on the other, for the development of business in China. The fragmented nature of the Chinese state compounded the problems, with corrupt officials playing foreign interests off one another in a bid to sediment their own position within Chinese politics and to gain leverage in negotiations with foreigners. Japan was willing to collaborate but also used military means, the "loans for control" model of imperialism, and bilateral agreements to pursue its own interests. The U.S. position was all over the map, simultaneously promoting international cooperation while signing bilateral deals with Japan that seemed inconsistent with "official" U.S. policy. In addition, just as the International Banking Consortium's loans began to flow in 1913 to China, Wilson suspended U.S. participation in the consortium and denounced the loans: "The conditions of the . . . loan might conceivably go to the length in some unhappy contingency of forcible interference in the . . . great Oriental State, just now awakening to a consciousness of its power and its obligations to its people. The conditions include not only the pledging of particular taxes . . . to secure the loan but also the administration of those taxes by foreign agents. The . . . loan . . . is obnoxious to the principles upon which the Government of our people rests."[46]

Wilson, however, later changed his mind and in 1918 permitted U.S. banking interests to rejoin the consortium. Blueprints for the "international cooperative development" scheme were further outlined in that same year in a report by Paul Whitham of the U.S. Department of Commerce, and J. E. Baker, an American advisor to the Chinese Ministry of Communications. The blueprint called for the unification of the country's railway network, new technical advisory boards staffed by foreign experts, and economic development projects financed by a reformed Banking Consortium. Perhaps surprisingly, the outline was supported by the government in Beijing and the shadow government headed by Sun Zhongshan (often referred to as Sun Yatsen). Sun, however, preferred a more direct role for the United States and, crucially, offered an ambitious plan of his own to the American minister to China, Paul Reinsch: the International Development of China model. That plan placed greater stress on the need to cultivate local expertise and a specific timeline for when "control" would revert to China rather than vague pledges that this would occur whenever the foreign powers thought China was ready.[47] It was, however, largely ignored.

COMMUNICATION, MEDIA REFORM, AND THE
POLITICS OF GLOBALISM

In many ways, however, Sun's proposal represented the zenith of liberal republicanism in China. From the beginning of the Qing Dynasty reforms (1898) through to 1919, China's urban elite developed a program of social transformation that contrasted sharply with conservatives who resisted change at every turn and which went well beyond just adopting new technologies and business techniques, as had been the focus of earlier modernizers. By the 1910s, the movement had been summarized in a neat little Chinese aphorism, "Mr. Science and Mr. Democracy," representing the view that capitalism, technology, and science would renew the material foundations of China, while democracy would remodel politics, culture, interpersonal relations, and society as a whole. Representing trends in social theory and philosophy evident elsewhere, democracy was defined as a kind of cultural communications medium where flows of interaction and information would make society more open,

22 Kang Youwei (often spelled K'ang Yu-wei), 1858–1927. He was a towering figure in Chinese intellectual circles and a leading advocate of modernity, democracy, and global cosmopolitanism in his early years. In 1904, he was appointed Minister of Education, a position he used to reform the outmoded system of education that had ossified under the Qing Dynasty. His influence on future generations of open-minded Chinese journalists, educators, and intellectuals was significant, although his stature was later tarnished by his support for replacing the republican order that had been created in 1911 with a restored constitutional monarchy. Courtesy of the University of California at Berkeley Libraries.

dynamic, and egalitarian. No longer just tools of commerce or weapons of politics, "modern communications" were seen as helping to "facilitate the world-wide spread of Enlightenment."[48]

This "new cultural movement" was amplified and diffused through a range of journals, magazines, and newspapers as well as at leading Beijing and Shanghai universities.[49] More students studied abroad and were coming into contact with the philosophical currents of the time, from anarcho-utopianism and socialism in Europe to John Dewey's ideas on communication and philosophy. In fact, Dewey was in Beijing meeting with former students as Wilson was in Versailles negotiating the treaties. As the spoiled fruit of Versailles was announced, the streets of Beijing exploded into one of the most decisive moments in modern Chinese history: the May 1919 protests. The sense of betrayal stemmed from the acceptance of Japan's claims to Shandong province and the fact that Chinese attempts to influence the outcomes of Versailles had been futile. The results of Versailles reversed the reformist tide.

These currents were refracted through the expanding Chinese press of the time. The growth of the press was nothing short of spectacular between 1905 and 1920 as the number of newspapers printed in China quadrupled from 200 to 800.[50] At the forefront of these developments were Kang Youwei and his

23 Liang Qichao (often spelled Liang Ch'i-ch'ao), 1873–1929. Essayist, journalism educator, and radical modernizer, he was one of the most important Chinese intellectuals of the early twentieth century. Exiled to Japan in 1898, and following in the footsteps of his mentor, Kang Youwei, his writings deepened support for the reformers' cause at home and extended it abroad among the Chinese diaspora and foreign governments. After returning from exile, he began to write on press and "media" development in China, often through comparison to Japan, and became a renowned journalism educator. His influence on the May 1919 movement, one of the most important events in modern Chinese history, was considerable. Courtesy of the University of California at Berkeley Libraries.

protégé, Liang Qichao. Kang was a leading force behind the "new China" movement that swept the country during this period, and many of his radically modernist views focused on the role of communication in the transformation of Chinese society and in the constitution of a cosmopolitan world order. By 1904, Kang had been made the Minister of Education, and Paul Reinsch identified him as one of China's leading advocates of modernity and democracy.[51] For his part, Liang pushed the new China agenda and press reform for the first three decades of the twentieth century. He came at the issues not only as a student of the new China movement but also from the angle of media educator and professional journalist who sought to put the Chinese press on a sounder legal, professional, and economic footing. Reflecting his mentor's influence, Liang developed an evolutionary view of press history in China and Japan. In this view, the press had gone from being an extension of individuals, to an extension of political parties and factions, and finally to become a reflection of the nation-state. Yet, there was a fourth stage that he saw as imminent, but not yet realized: a world press. In Liang's view, global communication and international news services now made that possible; it was also desirable as a pillar of the kind of cosmopolitan globalism that both he and Kang held dearly.[52]

GLOBAL COMMUNICATION AND THE FOREIGN-OWNED
MEDIA IN CHINA

As Liang spearheaded the reforms of the Chinese press and dreamed of a world newspaper, the closest thing to a world newspaper in China was the foreign-owned Chinese-language press. This press had a decisive impact on the indigenous press in terms of professional standards, style, technology, and business practices.[53] However, it also had its own concerns that had little to do with the reformers' agenda and much to do with the struggle to shape the evolution of China's domestic and global communication capabilities across the board. By the 1920s, the role of the foreign-owned press was prominent in the coastal cities of Shanghai, Hong Kong, Guangzhou, Tianjin, and Beijing. According to U.S. and British observers, Japanese interests were also actively "buying newspapers and fostering news agencies in Chinese cities with a view to controlling public opinion."[54] But Reuters still dominated the supply of international news to the foreign and Chinese-owned press alike. Shanghai had been the headquarters of Reuters in the Far East since the arrival of the cables in 1872, and by 1911 it served eighteen newspapers in that city alone. By 1920, there were twenty British-owned newspapers in China, and most relied heavily on Reuters for international news: ten in Shanghai, five in Tianjin, four in Hong Kong, and one in Hankow.[55]

Unlike Havas, who had been coaxed into renegotiating the terms of the global news cartel in South America, Reuters refused to budge in Asia. In fact, Reuters became ever more defensive during the 1920s as news from Japanese, U.S., French, and German sources continued to flow into China. The United States and France used their state-run wireless networks to distribute news to the press through their embassies and consular offices in Beijing and Shanghai, as well as to financial markets and commercial interests at cheap rates and often gratis. A few newspapers equipped with wireless receiving stations also received the German-based Wolff news service through Telefunken. Reuters and the Eastern Extension Company complained bitterly that these operations violated the news cartel and the cable companies' exclusive concessions, which the latter now claimed also covered wireless—a point we will return to below.[56]

In late 1921, Britain and Japan tried to block the flow of U.S. news into China by spearheading the passage of a resolution prohibiting it at the League of Nation's Committee on Pacific and Far Eastern Questions.[57] However, the United States ignored the measures, and six months later Congress reauthorized the Navy wireless system's service for another five years. In the run-up to the renewal of the Navy wireless system, Rogers held a flurry of meetings with Frank Noyes, the president of AP, and E. W. Scripps, the owner of United Press and largest newspaper chain in the United States. He also met with B. W. Fleisher, publisher of the *Japan Advertiser*, who was concerned that the possible demise of the Navy wireless service would kill the Shanghai-based Zhong-Mei News Agency that he had jointly developed with the Chinese Press News Agency just twelve months earlier. Fleischer also argued that the Navy wireless service was essential to continuing his existing operations in Japan and his plans to extend service to the press in Guangzhou, Hong Kong, Beijing, and Manila. Scripps reinforced Fleisher's views, as he told Rogers he would expand his operations in Asia and would do so as soon as adequate facilities were available.[58]

They did not have to wait long. The Navy wireless system was renewed in 1922 and again in 1927, and it underpinned a vast expansion of UP's services in Asia. By 1926, the U.S. agency was Reuters' main rival and had created alliances with Dentsu, one of Japan's premier news agencies and main rival to the other Japanese agency that was connected to Reuters: Kokusai. During the last half of the 1920s Reuters also faced a swelling tide from within AP ranks, led by Kent Cooper and V. S. McClatchy, to break with the global news cartel. Cooper and McClatchy were particularly critical of the fact that Asian news distributed in the United States was first filtered through Reuters and Kokusai, an agency reputed to have strong ties to the Japanese government. However, AP's official policy was to revise rather than dismantle the cartel in toto so as to maintain its ties with Reuters while dealing with the competitive threats posed by UP.[59]

INTERNATIONAL CONTROL OR INTERIMPERIAL
RIVALRY? THE CHINESE WIRELESS CONSORTIUM,
1920–22

The question of access to Chinese and Japanese news markets, as well as the issue of wireless in China, also turned on the role of RCA. RCA began service to Japan in 1921 and as it did the Navy wireless service to Japan was terminated. RCA was also eager to enter China and in late 1921 it acquired the only U.S. company with one of the contested Chinese wireless concessions: the Federal Telegraph Company. The Federal Telegraph Company had obtained concessions in early 1921, but the Chinese Government's default on international loans a short time later made American bankers leery about floating any more loans involving China. With its finances in disarray, the company turned to RCA, the only U.S. wireless company with deep enough pockets and experience with the consortium model able to advance the U.S. role in the development of wireless in China. With Mitsui and Marconi pushing it to join the consortium, RCA approached the State Department and Navy to authorize its acquisition of the Federal Telegraph Company and to join in the Chinese wireless consortium.

In a thinly veiled swipe at Rogers and certain members of the U.S. Congress, RCA's Owen Young asked the State and Navy departments to ignore the former's misguided critics of monopoly in order to expedite its deal with the Federal Telegraph Company and the formation of the wireless consortium.[60] The State Department and Navy ultimately did sign off on the RCA's purchase of the concession, but with several conditions. The first was that RCA accept a key role for the Navy's wireless system in expanding the distribution of U.S. news in China. The State Department and Navy's consent also required RCA to allow the U.S. government to use the technology of both firms (Federal Telegraph and RCA) and that RCA refuse to join with the Commercial Pacific Cable Company—the only nominally U.S.-controlled cable route to Asia—in order to fix rates. These informal arrangements were also significant insofar that they were just that—informal—rather than legally binding deals approved by Congress. That was also their weakness given that informal deals could not prevent Congress from opening up such matters to the gaze of antitrust investigations, which is precisely what happened as the Federal Trade Commission was called on to investigate RCA's "radio monopoly." The arrangements, based as they were on a series of tacit quid pro quos between RCA, the Navy, and State Department, rather than formal arrangements, also played rather loosely with established democratic norms within the United States.

Opposition to RCA did not come just from the United States but, most notably, from the Eastern Extension Company. The Eastern Extension Company was consolidating its cable system in Asia and had just invested $26 million in 1921 and 1922 to lay new cables between Britain and the Far East.[61] Not surprisingly, the scale of the venture made the company especially sensitive to pending competition from wireless. The fact that its concessions were due to expire over the following decade made the company especially nervous. Within this context, the company began prodding the Chinese government to extend its concessions. Most of these were set to expire in 1930, but there was one concession covering a cable from Shanghai to Tianjin and then inland, by way of telegraph lines, to Beijing. That system had been built in response to the Boxer Rebellion of 1900 at the behest of the "international community" but had been paid for through a "loan" from the Eastern Extension Company to the China Telegraph Administration. The date when the "loan" would be paid off and the cables handed over to China—1925—was rapidly approaching, and the Eastern Extension Company had no intention to return the cables to China.

China's position was compromised by the fact that it was on the verge of bankruptcy, and this financial stress played right into the hands of the Eastern Extension Company. In order to bring about a more compliant attitude within the Chinese government, the Eastern Extension Company temporarily cut off the government's access to the global cable system and required that all of its messages be paid for in cash. The upshot was that the Chinese government was forced into negotiations for a "new loan" and a key part of those discussions dealt with the company's demands for an extension to its concessions. The new loan of roughly $7.5 million would also be used to offset the cost of the new cables that had been laid in 1921–22.[62] In short, the "loans for control" model of imperialism was being used to get China to extend the Eastern Extension Company's monopoly and to partially fund the largest expansion of the company's cable system undertaken at this time.

This kind of "financial imperialism" was endemic to the struggle for control of China's domestic and global communication capabilities. The loans for control model of development had also been used by Japan to finance the jointly owned cable laid from Japan to Shanghai in 1916 and was evident in each of the wireless concessions possessed by Mitsui, Marconi, and RCA. Indeed, so rife with corruption and dubious financial arrangements had China's communications system become that it is uncertain whether it had any control over the system at all. The entire revenues of the China Telegraph Administration and even its domestic network were either handed over to, or pledged to, the Japanese Telegraph Administration, or one or the other of the British and Danish cable companies. In total, roughly $25 to $30 million in "loans"—just for telegraphs, cables, and wireless systems—had been issued between 1916 and 1922. Foreign governments appear to have done little more than "reserve judgement" on the matters, although charges of bribery and corruption against Marconi were raised in the British Parliament and, in 1917, the United States cancelled a joint venture in China between the International Western Electric Company—the communications equipment manufacturing arm of AT&T—and its Japanese subsidiary, the Nippon Electric Company, on the ground that it was incorrigibly corrupt.[63]

While the British government neither sanctioned nor repudiated the Eastern Extension Company's "loans" with China, it was clear that the company's aims were inconsistent with the government's commitments made at the Washington Conference and its desire, after about mid-1921, to support the cooperative development of wireless in China. Denison-Pender, the chair of the Eastern Telegraph Company, constantly appealed to the Treasury, Foreign Office, and General Post Office to support his efforts to extend the Chinese cable concessions and to help soften the rivalry between the cable and wireless companies. He went even further when he claimed that the cable companies controlled the electronic gateways to China and that they would only retract such claims once several principles were accepted: first, that the "international community" support the renewal of the concessions; second, "that wireless would obtain no preference over cables in regard to rates and other matters"; and third, that the two companies would have the option of joining the wireless consortium.[64] However, the United States made it known that it could not acquiesce to such terms. Even the British government refused to support the company's claims on the grounds that doing so would damage international relations and contradict positions it had accepted at the Washington Conference.[65]

THE CONFERENCE OF EXPERTS AND THE CHINESE
WIRELESS CONSORTIUM, 1922–25

In February 1922, a Conference of Experts was convened to try to sort out once and for all the tangled snarl of issues affecting the development of China's domestic and global communication capabilities. The fact that the cable companies' interests had been given top priority, albeit not precisely in the manner demanded by Denison-Pender, was evident from the outset in several ways. First, the "experts" called on China to renew the cable concessions indefinitely, so long as the cable firms "immediately withdrew any claim to a monopoly in wireless communication in China or between China and other countries." The recommendation also turned on the companies' willingness to yield on the issue of rates. Instead of "equality of rates," as Pender had originally demanded, the conference recommended that "the rates . . . by wireless may be 25 per cent (or in the case of press traffic more than 25 per cent) lower than the corresponding rates by cable, and that the rate for communication by wireless with America may be as low as the . . . rate to Western Europe—these limits to be reconsidered after 1930."[66] Competition with cable was also effectively prohibited, given that the Eastern Extension Company and the Great Northern Company gained first rights to develop any new cables in China, so long as they did so on terms at least as good as those that potential rivals were willing to accept. This, as we have seen in reference to South America, was a feature that had been in place since the 1890s, and one with which the Eastern Telegraph Company was thoroughly familiar.

The acceptance of the cable companies' monopoly reflected trade-offs that had been worked out secretly before the conference. The gist of those deals is that the United States would compromise with respect to its opposition to monopolistic cable concessions in return for Britain lining up behind the wireless consortium and the greater distribution of U.S. news in China. Britain had previously been tepid in its support of the consortium, at times accepting that it was the best option, while at others standing fully behind Marconi's concessions in opposition to those possessed by Mitsui and the Federal Telegraph Company. Even Marconi had come around to accepting the cable arrangements, seemingly in recognition that there was no way to budge the cable companies on the monopoly issue and even less chance that its own exclusive concessions would be upheld against those of its rivals. Besides, having adopted a similar arrangement by its membership in the AEFG consortium in South America, there was little by way of principle to stand on in China. Thus, even before the conference was held, Lord Balfour announced that "there will probably be no difficulty in arranging for co-operation between wireless companies in regards to existing or prospective stations in China for external communication . . . subject to [the] right of China to take over on fair terms in say 15 or 20 years."[67] France and Japan had also already signaled their approval, as well.

The wireless consortium approach advocated by the conference was explicitly modeled along the lines of the AEFG South American consortium. The existing concession holders—Marconi, Mitsui, and RCA—would share technology, messages, and revenue with one another, while international services between China and the rest of the world would be allocated according to the principles of territorial exclusivity that had been set a year earlier by the big four global wireless companies. Provisions were also made for Telefunken and the TSF to join the cartel at a future date, although they had no existing Chinese concessions. That alone embodied a stronger commitment to the "open door" and was designed to counter charges of a wireless monopoly.

While the Chinese wireless consortium largely mirrored arrangements in South America, it also differed in crucial respects. Most notably, in the Chinese version, each of the parties would have equal standing on the board of directors with no extra powers set aside for any of them. But the fact that control would be shared equally by the three wireless companies glossed over the omission of earlier promises that the consortium would be lead by a Chinese director appointed by China (with the approval of the others). So, too, missing were prior commitments to transfer technology or to cultivate Chinese technical skills. And on the most crucial issue, the "experts" stated that "the Chinese government may not for some time to come be in position to . . . develop . . . radio as a national undertaking, [although] provision would nevertheless be made for the possible purchase by the Chinese Government . . . of stations provided . . . by the co-operative undertaking."[68] In short, there were no measures facilitating Chinese control and on the crucial question of just when control would revert to Chinese sovereignty, only vague references to "possible purchases." Recognizing full well that this was a model for the "international control" of China rather than for its development, China denounced the outcomes: no "foreign Powers or their nationals [could] install or operate radio stations in legation grounds, settlements, concessions, leased territories, railway areas or other similar areas, without [China's] expressed consent."[69]

The Conference of Experts represented both the zenith of the "internationalization of control" model in China and its exhaustion. Nowhere was the failure of the model more obvious than in the hardened attitude taken toward the Chinese government by the Eastern Extension Company and British government. By 1925, the Eastern Extension Company leaned relentlessly on its rights, threatening to play the role of spoiler whenever wireless services were seen to pose a threat to its revenues. The company also made it clear that it had no intention to let control over the Shanghai-Tianjin-Beijing cables—in essence, the most commercially viable and strategically important cables in China, outside Hong Kong—be given to Chinese authorities. As 1925 came and went, this was confirmed. The company also had strong support from the Foreign Office and consular offices as it used the issue of the "telegraph loans" as a firm lever to further entrench its control over China's communications system. To underscore the point, the company once again plunged the country into an electronic black hole by cutting it off from the global communications grid and making the government pay cash rather than maintaining the usual practice of keeping an account. As Michael Palairet of the British Legation wryly noted, the actions were tantamount to a "public declaration of the bankruptcy of the Chinese Telegraph Administration."[70]

TWO STEPS FORWARD, ONE STEP BACK:
U.S. CABLE COMPANIES AND ASIAN COMMUNICATION
MARKETS IN THE LATE 1920S

Throughout this time, U.S. interests in Asia remained secondary to those held by British communication companies. Western Union filed applications with the State Department in 1921 to lay cables to China and Japan, and much effort was put into "internationalising control" of the Japanese island of Yap in the meantime, but the proposed cable never saw the light of day. This was largely because Japan insisted on controlling the line from the interconnection at Yap to Japan, which Western Union refused to concede.[71] The Commercial Pacific Company never laid another cable to the Far East during the 1920s, either. However, the State Department did approve the Commercial Pacific Cable Company's continued participation in the Asian cable cartel in 1927. The reauthorization was given as part of the State Department's approval process for the transfer of control of the Commercial Cable Company to the new American global communications conglomerate, ITT. That company, in turn, assumed a significant role in Asian communication markets at this time after acquiring AT&T's manufacturing arm, the International Western Electric Company in 1925.[72] That acquisition made ITT a tier one global communications firm. It played a lead role in the development of communication technology in Japan, which then served as a distribution point for supplying technology to urban and national telephone systems in Asia. The company also took over the Shanghai Telephone Company around 1927. The other major U.S. player, RCA, began offering service to Hong Kong, China, Japan, and the Philippines at the end of the decade (Marconi began to offer commercial wireless services to China in 1929). The most notable advance by U.S. communication and media interests in Asia during this time was by the news agencies and press. All of these themes—the rise of ITT, the impact of innovation on the cable industry, and competition between cable and wireless—are covered in the next chapter in relation to the Euro-American communications market.

The Euro-American Communication Market
and Media Merger Mania: New Technology and the
Political Economy of Communication in the 1920s

> You may rest assured that the sagacity of America, the commercial
> instincts of Germany, the national inspirations of Italy, and the prudence
> of the British and Colonial Governments, would not allow them to risk
> respectively the expenditures of vast sums of capital in cable enterprise
> unless they were confident that cable systems of communication all over
> the civilized world must continue.—JOHN DENISON-PENDER, 1923
> (quoted in Eastern Telegraph Company, *Annual Report*)

Without undue fear of competition from wireless and poised on the cusp of
an unprecedented boom in the growth of the global communication network,
Denison-Pender had good reason for the optimism displayed in the epigraph.
And as a sign that such sentiment was more broadly felt, in the early 1920s
Western Union and the Commercial Company signed agreements with the
New German Cable Company (Neue Deutsche Kabelgesellschaft) to restore
direct links between the United States and Germany. In 1923, the new Italian
Submarine Telegraph Company announced its plans to establish cable links
between European cities and the United States via the Azores and to invest
another $10 million in a cable network to reach Buenos Aires in order to serve
the growing Italian trade and diaspora in South America.[1] Yet standing as a
dike in the flow of these initiatives was the U.S. State Department, determined
to block access to American markets until access to the Azores and European
cities were liberalized considerably. The boom in global communication was
also delayed by the short but deep economic crisis of 1920–22. As Denison-
Pender noted, there was no better barometer of the conditions of world trade
than the receipts of the cable companies, and in the first few years of the
decade, those of the Eastern Company took a beating. Thus, as he noted in the
company's Annual Report for 1922: "The reported falling off in income, due to
trade depression, . . . is reflected pretty generally in all the countries served by
these Companies. . . . In the current year there has been no improvement in

trade . . . and, in fact, there is a consistent decrease, but better trade returns cannot be anticipated until the present chaotic condition of European affairs resumes a more normal aspect."[2] While the big four commercial wireless companies—RCA, Marconi, TSF, and Telefunken—steadily carved out a solid niche for themselves, the cable companies maintained a rather sanguine view of the new medium. And thus, as late as 1926, Western Union expressed a view that could have been drawn from any one of the cable companies: "We . . . think of Radio in terms of a medium of transmission *supplementing* the resources of the cables. It is our belief that the modern high-speed ocean cable will continue to hold its pre-eminent position in the field of international communication."[3]

However, never willing to leave anything to chance, the cable companies took stern measures to backstop their faith. In addition to currying the favor of governments (as in Britain and France) they leaned on their concessions to block the advent to new wireless services. More importantly, they shed their stodgy past approach to embrace technological innovation and organizational reforms. As they did, new high-speed cables were unveiled in 1924, network routing and switching capabilities were automated, the range of services were broadened, and a more aggressive approach to marketing was put into place. Some of the cable companies approached these changes enthusiastically, while others plodded gingerly along, backstopped by pliant governments and with consequences that were easily observable at the end of the decade.

Ironically, the delay caused by the politics of global communication benefited the American companies immensely. Thus, while they waited, the Eastern Company invested $26 million in new networks to the Far East. The problem, however, was that the network technology used was dramatically superseded by high-speed Permalloy cables unveiled and first deployed by Western Union in 1924. Instead of being able to rely on fundamentally new network designs to meet the challenge of wireless, as the U.S. cable firms could, the Eastern Company could only tinker with technological improvements grafted on at the ends of its existing network. That, in turn, meant that it also had to rely more on organizational and marketing reforms to meet the thrust of wireless competition, especially as short-wave beam services were commercialized in 1927. Much of this was masked later in the decade as the Eastern Company sheltered behind three "ideological screens": a technologically determinist view which blamed new wireless technology for the challenges it faced, a second nationalistic discourse that highlighted the threat of U.S. domination of world communication, and a third security-conscious view which stressed the links between communication and security. In short, these three ideological screens obscured the fact that the company's choices in the early 1920s had harnessed it to an outmoded technological infrastructure precisely at the moment when it needed the most advanced communication infrastructure possible. In addition, British policy sheltered the company by allowing it to merge with Marconi, prolonging its concessions in weak markets and making it a linchpin in the country's international and imperial communication system. The result was that very definite differences between the United States and Britain did emerge, and these differences did get framed in the "struggle for control" mold—as the rest of this chapter explores.

EXPANDING GLOBAL MEDIA MARKETS OR COMPETITIVE THREAT? THE MATURATION OF COMMERCIAL WIRELESS COMMUNICATION SERVICES

Markets and Empires
By the 1920s commercially viable wireless services had finally been established. The big four companies did not so much compete for global markets as carve them up into areas of mutual exclusivity and form consortia in South America while trying to do the same, without success, in China. By 1921, RCA had expanded into most European countries and Japan. It had also, as the U.S. Federal Trade Commission Report of 1923 concluded, established a de facto monopoly over global wireless services to and from the United States:

> With the exception of high-power stations owned by the . . . Government, and a station at Miami, Fla, owned by the United Fruit Co., which was affiliated with the Radio Corporation of America, the Radio Corporation is the only one . . . now engaged in transmitting and receiving radio messages between the United States and foreign countries. Under the . . . traffic agreements between it and foreign governments and radio companies . . . it is doubtful if any company seeking to conduct a similar business in the United States could exist, by virtue of the provisions in the majority of these contracts obligating the foreign companies to transmit all messages intended for the United States only through the facilities owned by the Radio Corporation of America.[4]

The company's position as the gatekeeper over the international flow of information through the ether was also buttressed by the U.S. government's insistence that wireless services between countries could only be conducted through designated organizations.[5] For all intents and purposes, in the United

States, that was RCA. Underlying the decision were two concerns. The first, as Walter Rogers testified during the Cable Landing Licenses Hearings, was that the United States should "not let [every] Tom, Dick and Harry to come in and take all the profitable business and then be unwilling to render service to the public."[6] And second, there were concerns that requiring all wireless traffic to flow through RCA would trample on freedom of the press. France and Britain as well as American news organizations protested vehemently against such requirements, but their concerns were brushed aside by the U.S. government. However, instead of complying with these new rules, many American news organizations such as the *Chicago Tribune*, the New York papers (*Tribune, Times, World,* and *Herald*), the *Philadelphia Public Ledger,* the United Press, the International News Service, and Universal Service—which by this time had created the Press Wireless Service—stopped receiving news at their offices and moved their wireless operations to Nova Scotia, where they continued to use the British and French state-owned wireless services.[7]

In Britain, the 1923 announcement by Conservative Prime Minister Andrew Bonar Law that private ownership and British control were the new pillars of British wireless policy suggested that Marconi might play a role similar to the one played by RCA in the United States.[8] Despite a decade of incoherent public policy, Marconi had opened services with France, Spain, Switzerland, and Austria, obtained a controlling interest in Austria's international wireless service (1922), shared ownership with TSF in the Romanian Wireless Company (1922), held a monopoly over wireless (and postal) services in Peru (1921) and Greece (1923), and possessed interests in South America and China. It also had a share in the British Broadcasting Company, in the Dutch broadcasting system, and with the Norsk Marconi Company (Norway) and working agreements in Poland, Belgium, and Sweden. In addition, in Australia, South Africa, and Canada, frustration with the meandering approach to wireless policy in Britain led them to strike their own deals with Marconi.[9]

Nonetheless, wherever Marconi turned, there was the General Post Office (GPO). As the Marconi Company complained, its own "interpretation of the statement of policy made by Mr. Bonar Law did not convey to them the understanding that the Post Office would be taking so important a part in the conduct of wireless commercial services."[10] A year later, the new Labour government conducted yet another review of the situation. The Donald Committee Report of 1924, which set the mold for British wireless policy for the rest of the decade, reversed Bonar Law's policy and advocated "a considerable extension of State activities."[11] At first blush this did not seem to be the case, given that the report recommended that Marconi's lead role in the development of wireless systems in Australia, Canada, and South Africa be preserved. However, this was just deferring to political reality, since any attempt to foist other outcomes on the Dominions would have fanned the growing nationalism that underpinned their moves. The Dominions were also the most lucrative markets in the British Empire and that, Marconi realized, was a benefit not to be taken lightly. The GPO appeared to take up the rear, limited as it was to controlling two superstations—one in Leafield and another completed a few years later in Rugby. However, in actuality, it was a more vital player. By 1922, its wireless press service with U.S. news organizations was still going strong, and it offered general wireless services to Holland, Germany, Poland, Italy, Czechoslovakia, Hungary, Romania, and Yugoslavia. The news service it developed with Reuters was also broadcast forty times a day to Europe and could, in fact, be picked up around the world.[12]

News and propaganda services were usually conducted at a loss and compromised the integrity of British news organizations, notably Reuters, as the Empire Press Union observed. Nonetheless, they were seen as a kind of cultural glue and an arm of foreign policy as the old borders of Europe crumbled and new nation-states, especially in central Europe, were created. They were also vital as the bonds of empire came under strain and in areas recently brought into the British imperium, where nationalism was not accommodated gingerly, as in the Dominions, but had to be combated.[13] Such was the case in the expanded zones of empire, where the state-owned imperial wireless system played the lead role: Iraq, Palestine, Somalia, and Syria (and into Abyssinia and Eritrea, although these were not part of the British Empire). Reflecting the importance attached to news and propaganda, the state had a monopoly over all national and international broadcasting, in addition to wireless telephony. Marconi could compete with the GPO in Europe but had to compensate it for lost business caused to the cable services that the GPO jointly operated with other European administrations. In sum, Marconi had to share the empire system with the GPO and compete with it in Europe and across the Atlantic. The company was also cut off from a range of new services (broadcasting and telephony) and in a position where its operations in the rest of the world were bound by the "territorial exclusivity" rules contained in the big four wireless agreements. It was hardly an enviable position and seriously compromised the company's viability over time.[14] The organization of wireless services in France revealed similar trends. The French government

AUSTRALIA TO ENGLAND
BY WIRELESS 1/5 SECOND

100 DAYS 60 DAYS 30 DAYS

24 "The Marconi Company Conquers the Tyranny of Distance." Sketch contained in the 1922 report of a special meeting of shareholders of Marconi's Australian subsidiary, Amalgamated Wireless (Australasia) Ltd. (published 1924). It is a graphic illustration of the perceived triumph of wireless in overcoming "the tyranny of distance" by instantaneously linking Britain with far-flung parts of the empire. Note the high long-wave wireless towers and multiple aerials. Unintentionally, the rising sun in the background looks ominously Japanese. Reproduced by permission of Tabcorp (Australia).

granted TSF a thirty-year concession for wireless services between France and foreign countries outside the empire wireless system, except for the TSF's superstation in Saigon. By the early 1920s, TSF was offering wireless services between Paris and London, Madrid, Prague, Bucharest, Belgrade, and Christiania as well as to the United States, from Paris to Beirut and Saigon, and from the latter to the French legation in Shanghai. Like Marconi, TSF was also associated with wireless companies in Italy (shared with Telefunken), Poland, Romania (shared with Marconi), Czechoslovakia, Syria, and Switzerland. The French Post and Telegraph Authority also continued its wireless press service in cooperation with the U.S. press, assumed a monopoly over domestic wireless services, and provided services to European countries where TSF was not active: Sweden, Poland, and Bulgaria. The state-owned system also provided the backbone of the imperial wireless network, reaching French colonies in Southeast Asia (Vietnam), the Caribbean and North Atlantic (Martinique, Guadeloupe, French Guiana, and St. Pierre), the Middle East (Lebanon and Syria), and Africa (Reunion, Senegal, Ivory Coast, Dahomey, French Somaliland, and Madagascar).[15]

Through the twists and turns affecting the development of wireless in the

1920s, several patterns emerged. Clearly, a rough and ready division between markets and empires governed the operations of the private companies and their respective governments. Second, governments now played a large role in the distribution of news and propaganda. Third, as Marconi, TSF, and Telefunken expanded throughout Europe they forged agreements with one another and RCA for the exchange of traffic within Europe and between Europe and North America. The participation of Telefunken was crucial in these arrangements because once overlaid on top of the AEFG consortium in South America and the agreements signed between U.S. cable companies and the German Minister of Post and Telegraphs in 1920 and 1922, such patterns revealed that German interests were being quickly assimilated back into the global system. The quick reintegration of Germany was a key plank in U.S. foreign policy and it was on full display in the context of global communication. The trend toward reintegration was deepened as the Commercial Cable Company signed a deal that allowed it to use Telefunken's patents to create a wireless subsidiary for maritime communication in the United States: the Mackay Radio and Telegraph Company. Although initially opposed to the deal, Owen Young's opposition melted as RCA and the Commercial Cable Company worked out new arrangements in 1922 for the collection and distribution of messages in the United States to be sent overseas by RCA's international network. This was crucial, because up to this time, the cable companies had denied RCA use of their domestic distribution system, thereby limiting its access to customers only in cities where it had a direct presence: Boston, New York, Washington, and San Francisco. The agreement revealed that the U.S. cable and wireless organizations were now adopting a more cooperative attitude with one another, likely to soften the impact of competition between "new" and "old" media, although Western Union waited another decade before signing a similar deal with RCA.[16]

The impact of wireless competition on the cable companies was easily visible in terms of the number of messages carried by each and in rates. Thus, by 1923, and despite the fact that investment in wireless networks was negligible in comparison, RCA accounted for roughly 30 percent of all messages in the Euro-American communications market, and prices that had been set well over three decades earlier began to tumble, as indicated in table 11.

Clearly, the development of competition between cables and wireless brought reduced rates, but it is also evident that there was a trend toward the equalization of rates after a period of adjustment. The impact was most obvious in areas where there were no direct cable connections with the United

TABLE 11 The Impact of Wireless Competition on U.S. Cable Company Rates, 1923

DESTINATION	CABLE RATE (PER WORD)[a]	WIRELESS RATE (PER WORD)[b]
Britain	wu[c] and ccc[d] 25¢ (for 35 years) reduced to 20¢ (1923)	RCA[e]—17¢ in 1920, later raised to 20¢ in 1923
France	wu and ccc—25¢ (1920), later reduced to 22¢	RCA—20¢ (1920), later raised to 25¢ in 1923
Germany	wu and ccc—36¢ (1920), later reduced to 30¢	RCA—36¢ (1920), later reduced to 25¢ in 1923
Poland	30¢ (1920)	RCA—25¢
Hawaii	cpcc[f]—35¢ and no letter-rate (1914), later reduced to 25¢	RCA—25¢ and special letter rate
Japan	cpcc—$1.21/wd (1916). Later reduced to 96¢	RCA—80¢ and reduced to 72¢ with half-rate deferred service

Source: Derived from U.S., Report, 1923, 36
[a]Amounts are in U.S. dollars and cents.
[b]Amounts are in U.S. cents.
[c]wu=Western Union
[d]ccc=Commercial Cable Company
[e]RCA=Radio Corporation of America
[f]cpcc=Commercial Pacific Cable Company

States, as was the case for Germany and Norway. Where the position of U.S. cable companies was weak, as in Asia, the impact was also large, with RCA picking up half of all international messages from the United States by 1923.[17] In order to meet this competition, the American cable companies reduced rates in general and reinstated the cheaper deferred, letter, and weekend rates that had prevailed prior to the war.[18] Yet, even when stiff competition occurred, the cable firms' revenues or profits remained buoyant. Indeed, they continued to dominate global communication markets, accounting for around 85 percent of revenues even in 1930.[19] In sum, competition seemed to be expanding the global communications market rather than constituting a zero sum game. That, at least, was the case for the U.S. companies, although matters were not so clear-cut with respect to the British companies, as we will see shortly.

Even after the commercialization of short-wave radio in 1927 the big three U.S. cable companies—Western Union, the Commercial Company, and the

All-America Company—continuously remarked on how a tide of prosperity had come their way on account of competition and because by this time the world economic boom was in full swing.[20] The Commercial Company described the situation this way: "The gross receipts of the companies engaged in the commercial telegraph business in the United States continue to increase, notwithstanding that the competition . . . from other communications media becomes increasingly vigorous and comprehensive. This steady increase in gross receipts should put at rest the fears of some that developments in other forms of communication will . . . eliminate telegraphy."[21] Yet, amid this celebration of the economic good times, clouds were gathering on the horizon. This was evident in the fact that while cable company revenues continued their upward trajectory, growth slowed after 1925. The All-America Company's revenues, for instance, fluctuated between their peak of $8.8 million in 1925 and $9 million in 1929.[22] This occurred even though the companies were carrying a record volume of messages. By 1927, however, even the Commercial Company remarked that it could not afford any more rate cuts to maintain competition.[23] Moreover, while the cable firms were able to eke out respectable economic performance during the decade, RCA's fortunes skyrocketed from $14.8 million in 1922 to $179.6 million at their peak in 1929.[24]

GLOBAL COMMUNICATIONS AND URBAN
MEDIA SPACES, 1921–30

The cable firms faced competition by adopting organizational, technological, and marketing innovations at a pace unprecedented in the past. The actions of the Commercial Company stood out as it tightened its alliance with the All-America Company in 1922 and renewed its long-standing joint purse with the Canadian Pacific Telegraph Company a year later. The alliance between the two companies gave them a network that stretched from Le Havre on the European side of the Atlantic, across the ocean, down the eastern seaboard of the United States, into the Caribbean, and around most of South America. The All-America Company gained access to 3,500 offices in the cities of North America and Europe, while the Commercial Company gained better access to Latin America.[25] The combined revenues of the two firms' international services was greater than that of the Western Union, although the latter's revenues from North America dwarfed those of the other two companies. As the firms synchronized their activities, they jettisoned marginal operations. The All-America Company sold off 60 percent ownership of the Mexican Telegraph Company to Western Union in 1926, which in turn used the acquisition to extend its own continental telegraph network. The Commercial Company sold its two British-based affiliates—the Direct West India Company and the Halifax and Bermudas Company—to the Pacific Cable Board in 1926.

As part of these changes, the Commercial Company consolidated its U.S. and global operations in one building at 20 Broad Street, New York, some doors from those of the All-America Company (64 Broad Street). That change unified the administration of its domestic and global systems in order to create greater efficiencies and improve the speed of its services. The location, literally attached to the New York Stock Exchange, also reflected the impact of the boom in global financial markets that was in full swing by 1925. As the company noted, information flows were pouring into and out of this hub of finance at an unprecedented rate, "bring[ing] the markets of the world within minutes of each other."[26] Western Union pursued a similar course of action, moving the headquarters of the Anglo American Company, which it had leased from British interests since 1911, from London to Western Union's head office in New York (40 Broad Street). Given the headquarters of RCA at 67 Broad Street, ITT a few doors away, and the Newspaper Publishers Association in the nearby World Building, the U.S. global media players' operations were clustered within easy walking distance of one another—which likely fostered social and interpersonal networks of contact and interaction among them.

In the early to mid-1920s, Western Union and the Commercial Company began constructing a regional European network and opening offices in European cities. The Commercial Company acquired a new building "in the heart of the London business district" and leased space in Paris which became "the headquarters from which the activities of . . . the Company throughout Europe will be directed."[27] All of these new offices reflected a radical liberalization of telecommunications policy in Europe and reversed five decades of restrictive policies that had prevented foreign companies from directly serving customers. The Commercial Company also leased a cable from the British and Dutch governments in 1921 to provide service between London and Antwerp, with additional connections to Rotterdam and other commercially significant cities later in the decade. As the Italian Submarine Telegraph Company and New German Cable Company began operations in 1925 and 1926, respectively, they laid cables along the European coastline from Sicily, Barcelona, and Lisbon in the south to Le Havre, France, and Emden farther north. Not to be outdone, Western Union, in addition to its long-standing offices in London, by 1927 opened new offices in Britain, Paris, Italy, Belgium, and Germany.[28] As the U.S. cable firms saw it, access to cities and "end-to-end" network control was crucial to successfully competing with wireless systems, which continued to dominate service to news agencies and the press.

OPEN NETWORKS AND COMPETITION, 1924–30

Control over landing rights in the Azores was, for all intents and purposes, "internationalized" finally after the mid-1920s. The islands were turned into an open hub where cable firms from a variety of countries interconnected with one another and were managed through a cooperatively run office and switching station. In the past, access to the Azores had been grudgingly parceled out by the Eastern-controlled Eastern and Azores Company and in ways designed to uphold rates and manage competition. That system reached its zenith between 1899 and 1916, but after 1924 the last dike in the firm's dominance of these markets was removed. Consequently, Western Union, the Commercial Company, the French Cable Company, the New German Cable Company, and the Italian Submarine Telegraph Company turned the islands into a crucial hub of their operations—a critical junction from which the global communications infrastructure fanned out in all directions toward Europe, Africa, and North and South America.

The impact of these changes was magnified by the fact that the cables laid between Europe, on one side, and North and South America, on the other, between 1924 and 1930 used a fundamentally new technology: Permalloy cables. These new cables were far faster than the old cables and, as such, decisive to the cable companies' ability to compete with wireless. That competition had also been rendered more serious by the commercialization of short-wave beam radio services in late 1926 and early 1927, a breakthrough as decisive to progress in wireless as the new Permalloy cable was in its sphere. The new Permalloy cable technology had been jointly developed by Western Union, the Western Electric Company, and the Telegraph Construction and Maintenance Company. In 1924, Western Union deployed these new cables with a line across the Atlantic to the Azores, where it connected two years later with the New German Cable Company. Over the next two years, the company laid two more high-speed cables, the first to France (1927) and then to Germany, Spain, and Italy (1928). Western Union was the only U.S. firm to lay these new cables during this time. In contrast, and instead of laying its own cable, the Commercial Company leased bandwidth from Western Union. The fact that both firms interconnected on equal terms with each other and with the German Company and the Italian Submarine Telegraph Company likewise revealed the emergence of the "open network" regime, with its idea that communication resources should be shared rather than squeezed as a lever to force would-be rivals into cartels and price-fixing schemes.[29]

Although the cable companies continued to take a relatively sanguine view of wireless, by 1927 they realized that they needed to wring out greater efficiencies through organizational reforms, the introduction of new services, and more aggressive marketing and, crucially, by deploying new technology as quickly as possible. The companies had been doing so for several years, including the Eastern Company, by installing automated routing and amplifier technology in their networks and high-speed automated printers at the ends. While it is difficult to say whether or not the Eastern Company undertook improvements as effectively as it could, the impression that emerges from corporate reports and government documents is that it was not as aggressive in such matters as its U.S. and other counterparts. In contrast, one of the benefits to the two major new companies of the mid-1920s—the New German Cable Company and the Italian Submarine Telegraph Company—is that both used new communication technology.[30]

Yet, even the Commercial Company, despite its dramatic consolidation of organizational control in New York, closer alliances with the All-America Company, vast expansion into Europe, and so forth, seemed to be underrating the extent to which it needed to adapt. Thus, even as beam radio began to be commercialized in 1927, the company remarked that this did not constitute a revolutionary new technology that would supplant the cables. However, now when the firm spoke its words were hedged about by a sense of caution. The company also noted that there were still newer technological developments on the horizon, notably wireless telephony, that it had little chance of competing with. In fact, the first introduction of transatlantic telephone service from New York to London by AT&T and the British GPO occurred in 1927. The cost of service was astronomical at $75 for three minutes. However, while the revenue generated—$178,011 in 1927—was meager, it quadrupled over a four-year period while the revenues of the cable companies and wireless companies plummeted after 1929 as the Great Depression took hold. The fact that the underlying technology which could make global telephone services viable was already being explored seemed to weigh heavily on the Commercial Company's collective mind. Indeed, no longer a repository of faith in the idea that new media would endlessly expand global markets to the benefit of all, by the last years of the decade a more plaintiff and hesitant quality started to pervade Commercial's statements.[31]

The same can be said about Western Union. Western Union introduced yet further price cuts in 1928 and 1929 that reduced regular rates by 16 percent, for the press by 29 percent, and to the deferred and special classes of messages that were designed to encourage social communication. However, the company also noted that while the cable companies continued to dominate the battle with commercial wireless companies, its own revenues for international services were growing slowly relative to its domestic system. However, in the company's assessment, this was nothing to be too worried about because its massive landline system accounted for 92 percent of its revenues; only 6–7 percent of its income came from global markets.[32] The firm's main acquisition of the time—the 60 percent share in the Mexican Telegraph Company—reflected the importance the company assigned to expanding its continental network. This control over a massive continental telegraph system with offices in nearly every North American city constituted a vital difference which distinguished the U.S. companies from their British and European counterparts.

GLOBAL MEDIA MERGER MANIA AND THE LIMITS TO CONSOLIDATION, 1927–30

Case 1: The Rise of ITT

In 1927 and 1928, respectively, the Commercial Company and All-America Company were taken over by a new American global multimedia colossus: ITT. The acquisitions culminated ITT's transformation from a small Caribbean operator of telephone systems into a global multimedia conglomerate. By 1932, the company had more money tied up in its networks than Western Union, although the latter's revenues ($465 million) still vastly outstripped those of ITT ($138.2 million). It dwarfed the only other remaining substantial U.S. global communication player, RCA, whose revenues at this time were a paltry, by comparison, $66.2 million. At the end of the 1920s, roughly 60 percent of ITT's revenue came from telephone systems, 20 percent from international cable and wireless services, and the rest from the manufacture of communication technology. The global reach of the company was also evident in the source of its revenues: 35 percent from Europe, 33 percent from South America, 25 percent from North America, and 3 and 4 percent from Asia and the Caribbean, respectively.[33]

In short, ITT derived far more revenue from foreign markets than from the

United States. With 75 percent of its revenues coming from foreign markets, the company was the quintessential global corporation. The acquisition of the All-America Company and Commercial Company solidified that position as it obtained a global cable system over 50,000 miles in length (second only to the Eastern Company's much larger network of 140,000 miles) that reached from Europe, to North and South America, and finally across the Pacific to Hawaii, the Philippines, and China. As Robert Sobel put it, "The great bull market was going full blast on Wall Street and ITT's stock was considered one of the leading glamour issues."[34] With easy access to finance, the company went on an acquisitions spree. Over the decade it spent $300 million to become a serious rival to other tier one global communications players—the Eastern Telegraph Company and Western Union, most notably—which had taken over a half-century to achieve the same scale.[35]

The firm expanded dramatically in Europe after acquiring the International Western Electric Company—AT&T's international communications technology manufacturing arm—in 1925, including the Standard Telephone and Cable Company (Britain) and Le Matériel Téléphonique (France). These acquisitions allowed it to compete with Siemens (German) and Ericsson (Sweden) for government contracts to modernize European telephone systems, which by this time were in a phase of unprecedented growth. ITT's acquisition of the Commercial Cable Company gave it offices and access to customers in major European cities and certainly heightened perceptions that the firm was poised to dominate European communication *tout court.* Moreover, ITT's expansion in Europe was mirrored by its activities in Asia and, more significantly, in the Caribbean and South America, where it acquired numerous telephone companies between 1918 and 1929.[36] Acquiring the Commercial and All-America Companies connected these disparate local and national communication systems into an integrated global communication system.

In South America, ITT's control of local telephone systems and cable connections allowed it to mount a serious challenge to the Western Telegraph Company and Western Union's domination over cable links between South America, on one side, and the United States and Europe, on the other. This was compounded as the company acquired wireless concessions in Argentina, Brazil, and Chile in 1927 and 1928. The company tied its urban telephone systems into the global cable and wireless networks that it owned to create an unprecedented system of universal intercommunication and where end-to-end control over networks rested firmly in its hands. In this system, every business, government office, or residence with a telephone was, in essence, a collecting and receiving station for the company's global communications network. Altogether, ITT's meteoric rise threatened a major shake-up of the existing cartels. Just as importantly, the company also reinforced the geo-political logic of global communication, as it cozied up to the United States as an instrument of foreign policy and displayed little compunction to meet the needs of other governments, whether democratically inclined or avowedly fascist, as in Italy, where it supplied communication technology and worked with the Italian Submarine Telegraph Company for service between Europe and South America.

Case 2: The Formation of Imperial and International Communications, Ltd., 1926–30

By 1928, Marconi was in near hysterics as the company pointed to ITT's looming domination of world communication. Indeed, the deputy chair of Marconi, Fred G. Kellaway, claimed that ITT had made a bid to take over "the whole internal and external telegraph and telephone services of the United Kingdom."[37] While that never transpired, Kellaway admonished the British government to take this as a serious wake-up call to action, namely, put its domestic and global communication policies on a more secure footing. In fact, Kellaway, along with Andrew Weir, the chair of Marconi, and John Denison-Pender had already begun talks in December 1927 about merging their respective companies. For his part, Denison-Pender echoed Kellaway: unless a thorough response to ITT was forthcoming the Eastern Company would be forced to liquidate and its cables acquired by "foreign interests." As the two firms fanned the nationalist flames, their talks and favorable state-ments by the government caused their stock "prices to soar rapidly," adding $115 million on top of their combined value of $150 million in four months.[38]

A serious consideration of the companies' claims, however, must start with a wider view of the issues at stake. And in that, it was not just ITT that loomed large, but the effect of competition between cables and wireless, especially after the new empire beam wireless service was commercialized in 1927. In a remarkably short period of time the two leading British cable agencies were pleading for relief. Since the Eastern Company's response to new technology has already been given some consideration, and will be returned to below, the next section focuses on that of the only other significant British cable system at this time: the Pacific Cable Board.

Competition, Social Communication, and the Pacific Cable Board The 1920s saw a considerable expansion of the Pacific Cable Board, with the addition of two Atlantic cables and extensions into the Caribbean after it took over three small firms: the West India and Panama Company, the Direct West India Company, and the Halifax and Bermudas Company in 1926. These acquisitions, in turn, reflected a shift in British policy; rather than continuing to subsidize weak private firms in the Caribbean, the Pacific Cable Board would now own and operate the cable system directly. The result of these acquisitions, however, was that the Pacific Cable Board was now tied to even more antiquated technology than the Eastern Company—given that the acquired cables were laid in the late nineteenth century. This tightly circumscribed the board's ability to compete with wireless. However, the organization was not so constrained on its main route between the west coast of Canada, on one side, and Australia and New Zealand, on the other, where it laid a new high-speed Permalloy cable in 1926. The new cable was the crown jewel in the imperial cable system and partially addressed U.S. needs by including connections to Hawaii, a factor that was also absolutely crucial to the long-term viability of the organization.

In addition, the Pacific Cable Board had already laid several shorter cables between Australia, New Zealand, and Fiji in 1922 in response to the threefold increase in the volume of messages that had taken place between 1914 and 1923 and which appeared set to continue into the foreseeable future (see table 6). The vast explosion in the volume of messages also filled the board's coffers, and this was crucial to its ability to finance the new cable. It also underpinned the board's push to expand the range of cheaper international message services—deferred, weekend, and letter-rate services—and to turn the long-distance cables into a means of mass communication. Indeed, the agency had become so successful by this time that it was the leading public service communications network in the British Empire and probably the world.[39] As Laming Worthington-Evans, the British Postmaster General at the time observed, the board was "bringing cable communication within the reach of an increasing proportion of the population [and] was successfully developing . . . the social . . . traffic."[40]

All of the members of the Pacific Cable Board (Australia, Britain, Canada, and New Zealand) initially supported the expansion of the system in the early 1920s. While such views carried the day with respect to the development of new cables in the Australasian region in 1922 as just mentioned, by the time that preparations began a year later for the new cable across the Pacific, the collegial relationships among the board's members had become fraught with tension.[41] Canada opposed the new cable from the outset, arguing that the current wave of innovation in communication technology made committing to any existing standards perilous. In this sense, the Canadian participants to the debate—notably Postmaster General Charles Murphy, were prescient, and their caution deferred a decision until after the new Permalloy cables had already been well proven in Western Electric's laboratories and on the transatlantic route, where, as we have seen, they were first used by Western Union in 1924. The position prevented the board from being locked into a path too early, as had been the case with the Eastern Company, thereby leaving it free to adopt innovations that were necessary to its commercial viability over the long run. By the time the contract was finally let in 1925, Western Union's "real life" experience with the new technology had proven entirely satisfactory. Given that, the Pacific Cable Board's new cable—the longest in the world, then and now—finally opened for service in November 1926. Beyond just the cable, the acrimonious debate it inspired pointed unmistakably to widening fissures in the imperial edifice, as nationalism in the Dominions eroded support for the kind of project that the Pacific Cable Board envisioned.[42]

Within months, the Pacific Cable Board and the Eastern Telegraph Company claimed that cutthroat competition with wireless was pushing them to the verge of bankruptcy. In stark contrast, the *Ottawa Journal* ran an article on February 18, 1927, in which the Commercial Cable Company proclaimed that communication markets were being expanded because of cheaper rates, new services, and competition. Why the difference?

Part of the explanation lay in the fact that "equality of rates" between cable and wireless prevailed across the Atlantic, whereas it did not in most areas served by the British companies. This was a key factor and reflected the deal made between the GPO and Marconi in 1924 in which the latter agreed to build the British end of the imperial wireless system in return for an annual royalty of 6½ percent on gross revenues and so long as the GPO set rates at levels that would maximize revenues and do so without regard to the interests of the cable companies. While service to Canada was exempted from that provision, the GPO appeared to be finally abandoning its historical policy of protecting the cable companies. Unrestrained by restrictive policies, the empire beam wireless system run by the GPO, on one end, and companies in Australia (in which Marconi held a 17 percent stake), India (two-thirds owned by Marconi), and South Africa (wholly owned by Marconi), on the other, set prices at whatever rate they needed to win the struggle for control of global

TABLE 12 The Impact of Competition on British Cable Rates, 1926–27

COUNTRY	PRE-BEAM CABLE RATES	POST-BEAM CABLE RATE	BEAM RATE AT STARTUP IN 1927
South Africa	50¢ (2s)	41¢ (1s 8d)	33¢ (1s 4d)
India	41¢ (1s 8d)	35¢ (1s 5d)	27¢ (1s 1d)
Australia	62¢ (2s 6d)	50¢ (2s)	41¢ (1s 8d)

Source: Calculations are for the cable companies generally and are based on ordinary rates per word. Derived from Britain, *Imperial Wireless and Cable Conferences,* 1928 (*Proceedings I* and *II*).
Note: Amounts are in U.S. cents, with British equivalents in parentheses.

communication. The Eastern Company and Pacific Cable Board introduced rate cuts of their own, but on each occasion these were followed by another round from the wireless companies, leading to the view that a rate war would quickly deplete the cable companies' reserves. The impact on rates just prior to and after beam wireless introduction is shown in table 12, above.

Even with the cut to cable rates, the beam rate still remained 20 percent cheaper. And while the new rates, as could be expected, led to a further jump in the number of people who could use the services, in the cable companies' view, cutthroat rate competition would inevitably lead to financial ruin. Thus, just to take one of many such statements at random, the Pacific Cable Board stated: "The Empire Radio System has now been in operation for seven months and during that period the Board's business has shown a consistent decline. The reduction of receipts directly attributable to the wireless competition is now at the rate of £95,000 [approximately $475,000] per annum, and there is every indication that the worst effect has not yet been realised."[43]

Ironically, the fact that the Pacific Cable Board was the premier public communication service of the time also meant that it was hardest hit by the competition from wireless. This was because competition was keenest for the cheaper "special classes" of messages that had been designed to increase the social uses of communication, precisely the kind where the Pacific Cable Board had carved out an advantage against commercial rivals (see table 6). Buffeted by competition in its most important services and its backers now torn in opposing directions, the Pacific Cable Board found itself on shaky ground. Pressures mounted to protect the national treasuries of the Dominions and to disband the system altogether. The fact that the two media were competing head-on with one another for control of social communication also differed greatly from conditions in the United States, where the cable companies had largely ceded "cheap rates" and the "public service" remit to RCA and the Navy wireless service. This was a decisive point and one which sharply distinguished the "competitive" relationship between "new" and "old" media within the British empire from the more complementary and supplemental relationship in the United States.

The two British cable organizations were also losing ground on the more important measure of income, although the poor quality of the publicly available data, especially for the Eastern Telegraph Company, and the lack of adequate methods of breaking revenue (versus traffic) into meaningful service categories makes it hard to assess their claims. Britain's foremost cable expert of the time, Frank Brown, explicitly commented on the lack of available information and cautioned against generalizing from the conditions of the Pacific Cable Board to the plight of the cable companies as a whole.[44] Nonetheless, and strikingly dissimilar to the U.S. companies, the case presented by the Eastern Company and largely accepted by British policy makers suggested that competition was shrinking rather than expanding the cable market. This could only mean one thing: that rate decreases were so steep that they offset any corresponding rise in revenue. The impact on British-based cable company revenues is shown in table 13.[45]

The evidence on issues of competition led to six main arguments in support of the need for a drastic change in British communication policy. The first three economic arguments, in brief, were (1) that rate wars were shrinking the size of the market, wholly at the expense of the cable companies; (2) that wireless companies could continuously undercut the cable companies on rates; and (3) that shrinking markets and the inherent price advantages of wireless would drive the cable companies out of business. An additional three broader political and technical arguments were also made: (1) that the reliability and secrecy of cables gave them certain advantages over wireless that should be preserved; (2) that the maintenance of cables was essential to the defense of the British empire; and (3) that bailing out the cable companies was necessary to prevent their cable systems from falling into foreign hands.[46] All of these claims will be critically assessed. However, for the time being the point is that they carried the day.

The results offered a bittersweet vindication of Canada's position, and it pushed hard to expedite a merger between the various British cable companies and the Marconi Company in order to relieve itself of the financial burdens of the Pacific Cable Board's rapidly deteriorating situation.[47] The Eastern Company and Pacific Cable Board also used these arguments to jus-

TABLE 13 Impact on Cable Companies' Annual Revenues, 1926–27

COMPANY	PRE-BEAM ($)	LATE 1927 ($)	REVENUE LOSS ($)	REVENUE GAIN ($)
Eastern Telegraph Co.	5,705,000	2,680,000	3,025,000	
Indo-European Co.	455,000	315,000	240,000	
Indo-European Dept.	1,190,000	790,000	400,000	
Pacific Cable Board	1,750,000	1,110,000	640,000	
Empire Beam		2,490,000		2,490,000
Total	9,100,000	7,385,000	4,305,000	2,490,000

Source: Data derived from Britain, PRO. Imperial Wireless and Cable Congress Proceedings (1 and 11), 1928, para. 4.

Note: All amounts are in U.S. dollars.

tify their claim that the only alternative to financial ruin was a regulatory solution that either established "equality of rates," offset their losses through subsidies, guaranteed them a minimum revenue, allowed the cable and wireless companies to pool their revenues, or, as they and also the government preferred, to allow all the British companies to merge.[48] Despite the fact that by this time all of the British cable organizations and Marconi Wireless were now solidly behind the need to merge, there remains one outstanding question, namely, why was Marconi so inclined?

That Marconi was so inclined was evident in the fact that both Andrew Weir and Fred G. Kellaway began merger talks with John Denison-Pender in December 1927, only months after the Imperial Beam Wireless service had begun operations. There were four factors that seemed to be behind the company's initiative: its precarious financial condition, continued restrictions on a range of new services that it could enter, residual antipathy toward the firm in certain quarters, and the fact that the Eastern Company's own financial conditions were still solid. While the new beam wireless service seemed about to improve the company's fortunes, Marconi's financial condition had deteriorated severely between 1921 and 1926. While the company had paid dividends prior to and just after the war in the range of 20–25 percent, these had dropped off to a still rather rich 15 percent in 1920–21 but had thereafter plummeted to 5 percent. This, in turn, reflected the fact that its operations continued to be hedged about by the restrictions set out in the 1924 Donald Report. Those restrictions prevented Marconi from entering new markets that had been set aside for the GPO, but where the commercial prospects were greatest: international wireless telephony and broadcasting services. The ad-

vent of beam wireless technology put those services on a far firmer technological footing, and if Marconi played the merger talks right, it stood a
chance of undoing these policy restrictions and of carving out an expanded
role in the wireless communications of the British Empire and beyond. Yet,
that was not to be, and indeed the fallibility of the notion that Marconi could
use the merger talks to leverage additional privileges from the state was evident in the outcomes and in the antipathy that some officials continued to
display toward the company. An example of the latter can be seen in an
exchange between the assistant secretary of the treasury, Herbert Samuel, and
Fred Kellaway at the merger hearings:

> KELLAWAY: It would be a worse thing for the cable people if they did not come to
> terms . . . we are here to give evidence and that is all.
> SAMUEL: It is not quite so. I think that Mr Kellaway has led us to believe that his
> Company holds the ace of trumps in its hands, and if the Cable Companies do not
> do as they are told it will be the worse for them. My view is that the final control is
> with His Majesty's Government. If the Marconi Company did anything we thought
> was against the interests of the nation you may rest assured that HMG has the final
> control. I do not want to leave you with the impression that you have the whole
> control on the matter . . . the Government has control of the position through your
> Licence: that should be kept in mind.[49]

If Marconi did have any such pretensions, it was a stinging rebuke. But
while its rhetoric suggested it had the upper hand over the Eastern Company,
Marconi was actually very eager to join that company for two other reasons.
First, as a favored son of the British state, so to speak, the cable company
might be able to take Marconi under its protective wing and perhaps accomplish what it was unable to do on its own: gain access to the most important
new markets that had been set aside for the GPO. The company was well aware
by this time that the experience of RCA in the United States already showed
that access to broadcasting and the entertainment industries was one of the
most lucrative sources of revenue for companies of this kind and key to stak-
ing out a position at the heart of the consumer and communications revolution of the 1920s. While motivated by potential opportunities that would, in
the end, remain forever closed to it as the British Broadcasting Corporation
became the flagship of public service broadcasting at home and the flagship of
international services abroad, Marconi was also aware of the threat posed by
the Eastern Company's seemingly unlimited capital reserves. From that flush
position, as at least a few observers noted, far from the cable Goliath being

driven into submission by the wireless David, it was the other way around. The Eastern Company could both draw on its massive capital reserves, and it continued to pay out dividends in the range of 10 percent or more. While the company preferred to shelter those reserves rather than to use them in order to meet the competition from beam wireless services, that was its choice and, as such, meant that it had the discretion to choose to do otherwise.

The Formation of Cable and Wireless, 1929   On April 8, 1929, a new entity in the history of global communication was formed and registered: Imperial and International Communication, Ltd.[50] The new company was a massive and complicated undertaking designed to unify British control over global communications. It consisted of two branches—Cable and Wireless, Ltd. (a holding company), and the Imperial and International Communications, Ltd. (a communications company). Its initial capitalization was $150 million (£30 million), with two-thirds of the capital representing the interests of the former Eastern Associated Company and the remainder the interests of Marconi. The distribution of control in the company, however, was slightly less lopsided, with the cable interests holding 56 percent of the voting shares and 44 percent falling to Marconi interests.[51] All said and done, the company was a massive global communications conglomerate, much along the lines of ITT, although with important differences, as we will see. Altogether, Cable and Wireless consisted of ownership and controlling interests in twenty-six companies in 1929, many of which had been prominent throughout the history of global communications, as the following list shows:

Eastern Telegraph Company
Western Telegraph Company
Eastern Extension Australasia and China Telegraph Company
Société Anonyme Belge de Cables Télégraphiques
Indo-European Telegraph Company
Pacific and Europe Telegraph Company
West Coast of America Telegraph Company
Pacific Cable Board
Halifax and Bermudas Telegraph Company
Direct West India Telegraph Company
Cuba Submarine Telegraph Company
West India and Panama Telegraph Company
River Plate Telegraph Company
London Platino-Brazilian Telegraph Company

Eastern and South African Telegraph Company

Eastern and Azores Telegraph Company

West African Telegraph Company

African Direct Telegraph Company

Marconi Wireless Telegraph Company

Indian Radio Telegraph Company

Marconi Radio Telegraph Company

Wireless Telegraph Company of South Africa

British East African Broadcasting Company

Companhia Portuguesa Radio Marconi

Peruvian Telephone Company

Guayaquil Telephone Company

The company also had noncontrolling stakes in four other firms—Amalgamated Wireless (Australasia), Companhia Radiotelegraphica Brazileira, Radio Suisse, Russian Wireless Telegraphy and Telephony Company—and three others that comprised Marconi's interests in the South American AEFG consortium: Transradio Chile, Transradio Internacional Argentina, and Trans-Oceanic Wireless Telegraph Company.[52]

The company also took over three state-run operations: the Pacific Cable Board, the Imperial Atlantic Cables, and the GPO's Imperial Wireless Service. In order to do so, it had to pay off the debt of the Pacific Cable Board—$6,150,000 (£1.2 million)—and make payments for the next twenty years totaling $1,038,000 (£207,544) per annum.[53] Similar arrangements applied to the GPO's wireless system. Although the GPO had only paid out $1.25 million (£250,000) for imperial wireless stations up to that point, it received an annual lease fee in the same amount for the next twenty-five years.[54] Even here, there was a catch—the reservation of international wireless telephony services for the GPO. The exception proved to be a sore point between Cable and Wireless and the GPO in the years ahead, in fact, so much so that a new round of bickering hobbled the commercialization of that service for years despite its gradual introduction after 1927.[55]

Nonetheless, while the company was privately owned, the government had several powerful levers of control. The government had the right to appoint two directors to the company's board, one of which was to be co-chair of the company and the other to be selected by the cable interests. There were to be two presidents, one representing the cable group and the other the wireless group. One position was slated for John Denison-Pender, but he died on March 6, 1929, shortly before the company was registered, and Andrew Weir, the chair of Marconi with whom Denison-Pender had started the merger talks back in December 1927, was slated for the other. However, with Denison-Pender dead, Weir became the sole president, Sir Basil Blackett was the government-appointed chair, and Denison-Pender's son, John Cuthbert Denison-Pender, was the governor and joint managing director along with another leading light from Marconi ranks, Fred Kellaway.[56] Strict limits were also put on foreign ownership (25 percent) or the sale of the firm and alliances with foreign communication companies were subject to regulatory approval.[57] The Imperial Communications Advisory Committee was created as the new regulator, and its members were appointed by the British and Dominion governments and the Secretary of State for the Colonies. The committee had the power to approve the introduction or withdrawal of services, to give policy direction with respect to the strategic needs of the empire, and, in consultation with the company, to set rates. Rates, in turn, were set with one eye toward securing an adequate return for the company (in this case, 6 percent on a capitalization of $150 million) and the other toward fairness to the public. Any amount above this was to be directed 50–50 toward rate reductions and to the firm's coffers.[58]

The first annual report of Cable and Wireless triumphantly declared that it had already implemented rate cuts in advance of declaring profits and that operations for both cable and wireless had been brought under unified organization by the end of 1929 without a hitch. However, within nine months, the company was hammered by the collapse of financial markets, with 20 percent of its capitalization wiped out by the end of January 1930.[59] A year later, its profits were a slim $340,500 versus the $9.1 million originally anticipated.[60] This profit decline long continued, and hence the expected profits never were translated into rate reductions. In fact, rates were only reduced again after a government ordered reorganization of the company in 1936. The company's weak economic conditions were also indicated by the fact that no new cables were laid until at least 1945 and, in fact, the company petitioned the government to close down unprofitable ones to Africa and another from Ascension Island to South America in 1930. While that request was refused, it was permitted to put militarily significant cables on a "care and maintenance" basis, meaning, as Daniel Headrick notes, "that they could be closed down if they were ready to be brought back on line with a month's notice, or one or two day's notice where there was no radio connection."[61] Above all else, the structure of the firm subordinated wireless to cable communications, thereby institutionalizing the bureaucratic lethargy as well as a questionable view of relations between "new" and "old" media, which had made the merger seem desirable in the first place.

Case 3: The Long-Term Rise of the Consolidationists:
The Proposed Merger Between ITT and RCA

Perhaps the most significant outcome of the formation of Cable and Wireless was the impact it had on the policies of other countries. Indeed, it became a touchstone for similar trends in France, Germany, Italy, and the United States over the next decade. In France, TSF and the French Cable Company formed a joint purse in 1928, with government approval, and coordinated their actions ever more closely. In contrast to British experience, however, the weak condition of the French Cable Company allowed TSF to gain the upper hand in the reorganized French global communication system. Matters were more pronounced in Germany and Italy. In Germany, the operations of Telefunken and the New German Cable Company in 1932 were merged and put under tight supervision by the Ministry of Post and Telegraphs. In Italy, Mussolini's Fascist government unified the operations of the Italian Radio Telegraph Company and the Italian Submarine Telegraph Company and brought them under strict state control.

The American communication companies watched these events carefully. Indeed, Newcomb Carlton of Western Union and David Sarnoff of RCA traveled together to London in February 1928 to discuss the proposed British merger firsthand with Denison-Pender and Kellaway. And while Sarnoff and Carlton bandied about the prospects of merging their own two companies, the option was rejected by Carlton who still preferred competition between cables and wireless to amalgamation.[62] Spurned, but undeterred, top executives at RCA relentlessly plugged the benefits of amalgamation, and just a few days before events in London culminated, RCA and ITT, on March 30, 1929, announced their own intention to form an American communications conglomerate. Despite triumphant announcements in the *New York Times*, several barriers stopped the merger dead in its track: restrictions against the combination of international cable and wireless services in the Radio Act of 1927, congressional opposition, and forceful interventions by Carlton during Senate Hearings convened to consider the matter.[63]

RCA and ITT pushed for the next two years to eliminate the restrictions of the Radio Act, and Senator James Couzens introduced a bill to do just this, but Congress refused to budge. The restrictions in the Radio Act were designed to ensure that neither media of communications could be subordinated to the other and to halt the push to combine all U.S. global communication capabilities into one conglomerate. Unable to move congressional opposition to "total" amalgamation, ITT and RCA were forced to call off the merger in early 1931, but not before taking another whack at the restrictions which had halted their efforts: "The accord made public by the two companies on March 30, 1929, for the consolidation of their respective communication interests when the law permitted has been dissolved. This decision was necessitated by the fact that despite the increasing influence of communication mergers in foreign countries and the obvious advantage to American communications interests from consolidation of their services, no legislative action has been taken to eliminate these handicaps. . . . The . . . two companies have, however, in no way altered their sincere conviction . . . that the unification of American communication services would be to the interest of our country and people."[64]

The companies held to their convictions, and the push for consolidation regularly resurfaced during the next two decades. Both argued that rather than maintaining an outright ban on convergence, Congress should permit mergers so long as "the proposed consolidation . . . would be . . . in the public interest."[65] They also constantly found support from elements in the U.S. military which saw amalgamation as vital to national security interests—much along the same lines that had compelled the formation of Cable and Wireless. To their credit, Congress and the Federal Trade Commission stayed the course, with the latter arguing that the structure of the U.S. communication industries was already so Byzantine as to prevent effective regulation. The fact that Western Union continued to oppose the combination of cable and wireless firms also helped stem the tide, although it had no such compunctions at home, where it acquired the Postal Telegraph Company (a subsidiary of ITT) in 1943.[66] Matters followed this well-worn path into the 1940s.

CONCLUDING REFLECTIONS: TOWARD A CRITIQUE OF THE POLITICAL ECONOMY OF COMMUNICATION AND EMPIRE

The most enduring impact of the Cable and Wireless merger was not the standards of economic performance or technological prowess that it set but the links between communication, consolidation, and security that it tightened. While Cable and Wireless became a model that others sought to emulate, the conditions of its own creation masked the fact that many basic issues had never been adequately addressed. Upon closer inspection, one thing that stands out is the poor quality of the assessment that went into its creation.

While the evidence presented suggested a large impact on rates and revenue from competition between cable and wireless systems, this evidence was so general that meaningful analysis was, and still is, difficult to make. Consequently, the impact of competition remained hazy, and the cable companies' economic conditions remained much as they had been in the past—a tightly held secret. In addition, the British hearings never asked why only markets served by British cable concerns were wilting while those served by U.S. (and other) firms were expanding. Rather than tough analysis, the British cable and wireless merger was filtered through three ideological screens: technological determinism, nationalism and imperialism, and empire security.

Technological Determinism

The complexity of mergers and "inter-media" competition was limited by a technological determinist discourse that laid much of the blame for the British cable organizations' woes at the feet of new technology. That same discourse ignored the fact that the British communication organizations, from Marconi to the Eastern Company to the Pacific Cable Board, and the national and imperial government agencies involved were working with a faulty model of media evolution that assumed that new media simply supplanted the old. Such views were, to put it mildly, unsupported by historical research, nor, in fact, did they conform to the views of the U.S. companies or "communication experts." In fact, even some British experts, from Charles Bright at an early point and subsequently Frank J. Brown, former assistant secretary of the GPO and chronicler of world communication in the late 1920s, rejected this simplistic view of media evolution. These experts saw the "new media" of wireless as augmenting the existing cable infrastructure, with each communications media finding its own areas of preeminence over time.[67]

Technological determinism also screened out the fact that different institutions handled technological change very differently. This is crucial because instead of suggesting that fixed technological conditions eliminate choices, a more sociologically oriented perspective highlights how organizations, institutions, and people choose between the available options to shape the evolution of technology, markets, and policies.

This is not to say that technological choices do not have consequences. In this chapter we have seen that they most assuredly do. Choices at the outset of the 1920s made by the Eastern Associated Company objectively wed it to an obsolete network infrastructure. Blaming competing technology for the dire situation of the Eastern Company diverted attention from the fact that its earlier decisions had tied it to network standards that were transcended within two years. The proceedings, recommendations, and reports of the Imperial Communication Conferences, however, did not raise such issues. Similar questions can be raised about the wisdom of choices made in the 1920s by the British government and Pacific Cable Board to expand the latter's system by buying old cables that had been laid by the Direct U.S. Company in 1874 and by the Halifax and Bermudas Company and Direct West India Company in 1890 and 1898, respectively. At least one participant to the Imperial Communications Conferences—the Empire Press Union—raised the idea that corporate lethargy and obsolete technology and practices at the Eastern Company were a considerable problem that had only been improved, albeit not enough, by competition from wireless. Beyond these furtive glimpses, however, such issues were absent. This is crucial, because one of the tasks of ideology and power is to remove important issues from the realm of serious scrutiny. The fact that technological determinism underpinned a new "policy norm"—that is, that wireless technology should be institutionally subordinated to cable communications—made it all the more influential as ideology. In fact, ideology had been institutionalized.

Nationalism and Imperialism

The second ideological screen that was institutionalized at this juncture was that of nationalism and imperialism as a warrant for consolidation. To be sure, this was not the first time that such ideas were present and, indeed, as we have shown throughout this book, such concerns were always present. The main difference, however, is that the creation of Cable and Wireless signaled a resurgence of the ideology of nationalism. U.S. companies, most notably the All-America Company and Commercial Company, also turned hard toward a more nationalistic outlook during the 1920s and adopted more hawkish views of their relationship to U.S. foreign policy objectives. The change was striking with respect to the latter as it veered sharply from its identity as a cosmopolitan global corporation. And as it grew closer to the All-America Company, it took on the attributes that had previously defined the latter company: patriotic language, greater representation of the U.S. foreign policy elite on its board of directors, and so on. Fuel was added to the nationalist fire when both corporations were swallowed by ITT, ever comfortable as it was to meet U.S. foreign policy aims and the needs of state power whatever they might be.

Nationalism was pitted against nationalism in an escalating spiral as the Eastern Company, followed by Marconi, obscured its strained economic and technological conditions by a strident rhetoric of national rivalry whose target was first and foremost ITT. Serious scholarship today—and probably then— would have little room for the conspiratorial undertones that laced their presentation of an impending American take-over of world communication. However, surprisingly, that discourse set the tone for British policy discussions and much subsequent scholarship on global communication history ever since. As Daniel Headrick notes, in a preface to a quote from a Marconi executive in the publication *United Empire*, "The British response to American merger rumours was properly horrified."[68] Yet, rather than justifying the caricature of American communications power that Marconi painted, it is more important to assess Marconi's claims about the supposed phalanx of U.S. communication companies, backed by J. P. Morgan and in league with the Germans, poised to take over the communications of the British Empire. Rather than taking this sophomoric portrait of bad foreign corporations, suspect bankers, and evil Germans seriously, a better analysis might indicate that there was no unified phalanx of U.S. communication companies, but there were durable cleavages between them. It might also highlight long-standing cleavages in U.S. politics and the well-known restrictions of the Radio Act as barriers to the nightmare scenario mobilized by the British global communication players. Again, this is not to say that nationalistic factors were insignificant. They certainly were not. Indeed, so enamored with nationalism and patriotism had leading lights in the U.S. global communications elite become—especially David Sarnoff and Owen Young of RCA and Lieutenant Colonel Sosthenes Behn of ITT—that they leaned on them heavily to pitch the merits of their own merger, going so far as to invite the participation of the U.S. government in the firm that would emerge. This too was significant, because it reflected a rapprochement between RCA, the U.S. military establishment, and the view that RCA was an appropriate "tool" of U.S. foreign policy. The fact was that nationalism had become a superheated ideology and was being pressed into service by many of the major players by 1929 to such an extent that it did represent a fundamental break with the past. In fact, it is at this point in time that the "struggle for control" model of media history becomes most obvious. The problem is, however, that too many historians have stretched this moment back into the past and thus obscured the more interdependent and cooperative character of the global media system, at least prior to World War I.

Communication, Security, and Empire

The third ideological screen mobilized by the growing forces of consoli-dation—in Britain, France, the United States, Germany, and Fascist Italy—tied the defense of nations and empires to strict control over all means of com-munication and content. Given the nature of social and rhetorical phenom-ena, such discourses likely magnified and accelerated the processes they pur-ported to describe. As ideology and in the British context, the discourse of nationalism, security, and empire also masked the fact that the empire was coming under greater strain.[69] Social changes within the Dominions meant that second-generation emigrants born abroad no longer had deep connec-tions to the "motherland." Such factors galvanized officials like never before in support of imperial cable and wireless services, with real tangible benefits in terms of improved social communication, most notably by the Pacific Cable Board system in the 1920s. However, such actions always had something of a "rearguard" character about them. Indeed, growing nationalism in the Do-minions meant that they were more inclined to develop their own communi-cations capabilities.

Throughout the 1920s British imperial news had also become a much harder sell. Seen through the lens of cultural theorist Raymond Williams, empire was becoming a residual cultural formation rather than the dominant one it had been in the past.[70] Considered from the view of popular culture, it was clear that people and, increasingly but not uniformly, journalists and news organizations were tired of the turgid speeches and long-winded "white papers" that flowed at cheap rates from the metropole to the periphery. Pithily put, audiences wanted more sports, less politics. In the "zones of formal empire," as the Victorian radical Edward Carpenter put it, the "semen of democracy" was spawning even more intense nationalist and anti-imperialist movements—which, incidentally, found favor among British, European, and American anti-imperialist movements that had risen in prominence and in-fluence after the mid-1910s.[71]

In essence, cultural changes were way out in front of the political changes that would lead to the dismantling of the British Empire after World War II. That the "cultural glue" model of news and empire was in retreat was evident in the fact that almost all British imperial news services during the 1920s were carried at a loss and propped up by ever more generous subsidies. The main beneficiary of such largesse was Reuters, which, like Marconi and the Eastern Company, became a leading corporate protagonist behind the "empire for-ever" view of things.[72] Yet, members of the British press complained bitterly about "doped news" and how hidden subsidies were compromising the quality of the British press. The Canadian Press, Canada's national commercial news agency, was so concerned that it put its own editors in London during the 1920s to sift through the Reuters service before it was forwarded on to members of the Canadian news media.

Britain, of course, was not alone in such matters. A 1922 Parliamentary Inquiry in France also revealed "an unhealthy relationship between the government, the banks and the press" in the wake of allegations of corrupt links between French news organizations and the government as well as of bribes received by the French press from Serbia, Argentina, and Brazil in return for favorable coverage of economic conditions in these countries.[73] Likewise in the United States, news played a role in attempts to consolidate the bonds of the U.S. Empire and to carve out an expanded role for U.S. interests in China. Although most American observers eschewed the idea that their news organizations had been transformed into propaganda agencies, Walter Lippmann and Charles Merz illuminated grave deficiencies in the foreign news coverage of these news organizations. In "A Test of the News," Lippmann and Merz studied the reporting by the *New York Times* on the Russian Revolution between 1917 and 1920, concluding that the paper's coverage was severely compromised by poor journalistic standards, excessive reliance on official and anonymous sources, the biases of journalists and editors, a lack of resources, and a fuzzy line between news and editorials.[74] According to Lippmann and Merz, stereotypes, rumor, biased expectations, and facts and interpretations supplied by officials and foreign policy elites overwhelmed reliable, objective, and independent news in the *Times*'s foreign events coverage. Moreover, because the newspaper played an agenda-setting role, these problems rippled across the U.S. news media. The American news media, in this assessment, did not function as objective analysts of world events but as conveyor belts between "official interpretations" and the public.

The problem had become so severe that the League of Nations convened several conferences to investigate the issues between 1927 and 1933. While the League had largely been neutered by this time, the deficiencies the conferences chronicled showed how bad the problems affecting the production and distribution of foreign news had become. Consisting of newspaper owners, representatives of news agencies, and journalists from sixteen organizations, the Conferences of Press Experts drew attention to the fact that the lines between news and propaganda were perilously thin and were leading to the incitement of hatred and violence among nations. The Conferences of Press Experts also pointed to the inadequacy of communication facilities and other restrictions on the free flow of information. In sum, the following recommendations, first announced in 1927 and repeated in their basic form at each of the subsequent conferences in 1932 and 1933, were offered to improve the quality of international news: (1) provide cheaper rates, (2) permit news organizations to use code, (3) provide better transportation and postal rates for newspapers within countries, (4) allow open access to news sources and copyright for news once published, (5) professionalize journalism and adopt a code of ethics, (6) eliminate censorship during peacetime, (7) curb news that incited hatred and violence, and (8) foster trust in the news media as vehicles for the cultivation and representation of public opinion.[75]

It was a tall order and registered the persistent hopes of those for whom liberal globalism still held much appeal. This appeal was, unfortunately, now largely irrelevant. That the conferences were but a side show was indicated by the fact that the big three global news agencies—Reuters, Associated Press, and Havas—met just before and after the first conference to revise the global news cartel. Yet, by that time the global news cartel was on its last legs. It was revised again in 1932 to expand the role of Associated Press and finally put out of its misery in 1934.

Kent Cooper, the head of Associated Press at this time, celebrated the demise of the cartel in his 1942 book, *Barriers Down*, and concocted a historical account of the news agencies laden with the sanctimonious rhetoric of American exceptionalism, democracy, and free speech. However, Cooper's criticism did little to indicate how the cartel had at least been a unidimensional means of holding the rapidly disintegrating global order together and that neither the news cartel nor those like himself had done much to promote a thicker version of globalism based on freedom of the press, democracy, cultural understanding, and cooperation—at least in any way that sought to institutionalize such norms. Instead, that honor fell to Walter Rogers, the Institute of International Law, John Dewey, Walter Lippmann (at least during his "liberal phase" in the 1910s and early 1920s), John Henniker Heaton, Kang Youwei, and others of their kind. While the whole notion of liberal globalism might be deeply problematic, and in the hands of Wilson and the U.S. administration it often was, it seemed a far better alternative to the kind of world order that was shaping up in the early 1930s.

As markets and culture receded as bonds of global interdependence, military considerations surged to the fore. But stepping back from that precipice for a moment, some final observations on competition between cables and wireless communications reveal additional factors that helped to propel these changes. Serious consideration of the United States experience shows that the competition between media, coupled with surging global economic activity in the middle of the 1920s, instigated a breakneck pace of innovation that prevented the collapse of cable companies and underpinned a decade of strong economic performance. Moreover, the U.S. experience also showed that by the late 1920s global links were a modest source of revenue for Western Union (about 6 to 7 percent), RCA (6 percent), and ITT (20 percent).[76] Instead, the crucial source of revenue and control did not stem from global markets at all, but from domestic communication networks. This was absolutely vital and a key determinant of success in global markets. The importance of controlling local networks was a lesson that Cable and Wireless learned well and followed in the 1930s as it, too, turned to acquiring urban and national telephone systems in Turkey, Cyprus, and the British colonies in Asia, the Caribbean, Africa, and the Middle East.

Yet, just as Cable and Wireless came around to the American point of view with respect to the relative balance between local and global communications, so, too, did the American communication industries continue to assimilate elements of the British experience. In particular, the link between communication and security continued to press forward relentlessly, growing ever bolder over time. Indeed, by the 1940s and under the pressure of World War II, that position had surged to the foreground. Congressional resistance remained, but it became harder to sustain as RCA and ITT, supported by key elements of the U.S. military, used the economic crisis of the 1930s and the destruction of World War II to push hard for changes that would otherwise have been impossible during less volatile times.[77] The Federal Communications Commission also began promoting consolidation after 1940. Even the State Department briefly adopted this position, although it reverted to its previous stance quite quickly and by 1945 was opposing amalgamation on the grounds, among others, that creating a unified U.S. global communications firm would be a lot "easier to put together than it is to take apart."[78] The push toward consolidation was also curbed by the sustained opposition of Western Union and, by this time, AT&T. This in itself was revealing, indicating that cleavages within the upper echelons of the U.S. communication industries still persisted and, crucially, that inter-elite cleavages were decisive in the formation of U.S. approaches to global communication policy. Nonetheless, RCA and ITT had gained strong support from the FCC and, more importantly, from the U.S. Navy for its vision of a massive communications conglomerate and tightly restricted control over information flows into and out of the United States. One of the clearest expressions of such views is contained in the following statement by Secretary of the Navy James Forrestal in 1945 and the startled response it evinced:

> MR. FORRESTAL: . . . Modern communications are the warp and woof of international society, and, therefore, are a matter of sovereign interest rather than private interest. Our diplomatic and military affairs are so vitally dependent upon the comprehensiveness, efficiency, reliability, and security of international communications that the continuation of private competition in such communications can no more be rationalized than could the administration by private enterprise of the diplomatic and military affairs themselves. In other words, they are so closely intertwined it is impossible to separate them.
>
> SENATOR REED: You do not really mean that, do you?
>
> MR. FORRESTAL: Yes sir. I have reached that conclusion reluctantly.[79]

While consolidation of the private communication firms continued to be resisted, the U.S. military had stealthily built up its own system of global communication that, as Forrestal boasted, was now larger than that of "the private industries combined."[80] The change represented the rise of the military-information-communication complex in the United States, much along the lines that had developed in Britain and France but not so overwhelmingly so as to constitute the authoritarian model of communication that had been adopted in Germany and Italy. That the pattern of development followed by the United States in the 1930s and 1940s followed that staked out by the British discourse of communication and empire was stated often to justify the former's similar path. In sum, the United States had moved far from the ideals that had animated its approach to global communication policy after World War I. Whether they were mere rhetoric—which we believe was certainly not so—at the very least the stark contrast between the words of the 1920s and those of the 1940s revealed that massive historical changes had occurred in the global political economy of communication.

Conclusions: The Moving Forces
of the Early Global Media

> A CEO as reviled as [Rupert] Murdoch might take comfort in know-
> ing he is just one man in a long line of cable barons. To be sure, these
> captains of the media industry ruled over a very different kingdom a
> century ago. But they cut a similar figure in the public imagination: the
> controlling oligarch holding information captive.—L. PETERSON,
> "The Moguls are the Medium"

One of the key tasks of this book has been to show the depth and duration of globalism between 1860 and 1930. Toward those ends, we have demonstrated that a combination of a technological revolution, the expansion of markets, international law, imperialism, and a discourse of modernization from the mid-nineteenth century on, especially after about 1870, were globalizing mechanisms in their own right and the basis upon which the activities of "global actors" depended. The technological revolution allowed for the near instantaneous worldwide movement of information and news through undersea cable and wireless linked inland to telegraph systems and the press. Time and again, however, we have also shown that this revolution did not facilitate long-distance social communication for the general public, at least not to the same extent as cheap mass mail service, better road facilities, railways, and fast steamships. Nonetheless, many of the "global media reformers" did believe that the new technologies could turn their dreams of almost instantaneous global communications into reality for the many, rather than the few, and strove relentlessly to do so. And in so doing, their visions anticipated a time in which citizens had greater access to the means of long-distance social communication, along the lines that the World Wide Web serves in our own time.

Other authors, past and present, such as Herbert Feis and Geoffrey Jones, also lend support to our claim that the period covered in the preceding pages was one in which a strong form of globalism emerged. For both of these authors, business enterprises and investment across national borders, independent of monopolistic rights or the protection of territorial imperialism, began early in the nineteenth century but only really started to take off near the end of the century, either in 1870, according to Feis, or a decade later in Jones's view. Indeed, Jones refers to the period between 1880 and 1929 as "the first global economy." Over this period, 1913 was the high point when multinational corporations and finance reached an importance in the global economy that was not attained again until the 1990s. And while Jones sets the date at 1880 for the start of this period, we think that our evidence regarding the rise of multinational firms and global media cartels requires that we push that date back at least a decade. Otherwise, the timelines of our respective analyses fit quite neatly.

The factors which stimulated the initial growth of multinational firms were, not surprisingly, in large measure, those that stimulated easier distance communications. In the early nineteenth century, geographical distance imposed great problems in managing international enterprises. Messages could take months to reach their destinations and months before a reply was received. But as Jones notes:

> Nevertheless, the reasons why multinational investment occurred at all was that the once overwhelming obstacles imposed by geography were being eroded by telegraph, railroads and steamships. Moreover, . . . governments imposed few restrictions on foreign-owned firms. People could move freely across borders without passports or work visas. Capital, though not trade, could flow freely across borders. The City of London served as a global information hub. Colonialism and informal imperialism imposed Western institutions and legal systems widely. "Regime changes" could be enforced on countries which opposed globalization. If firms could build the administrative structures, there were few exogenous obstacles to their organization in a coordinated or even "global" fashion.[1]

While Jones devotes only a few pages to the actual analysis of the multinational communications corporations, the above quotation fortuitously contains a succinct summary of many themes upon which the axle of our analysis turns: international investment, government policies, imperialism, international news, and information. All combine in a broad sweep which we term "thick globalism." Jones is also important insofar as he not only maps the rise of globalization but also its reversal. This is a key point because in our analysis it is not only politics, strategic interests, and imperialism—themes covered well by existing studies—but also the view of rapid communications as a double-edged sword that defined the times. That is, rapid communications and voluminous flows of information simultaneously furthered inter-dependence while diffusing economic instability. Indeed, for contemporaries this was crucial to understanding the global financial crises of 1873, 1890, and 1929 as well as justification for putting "developing countries" under the supervision of financial experts and the control of powerful nations, that is, a form of imperialism, either informal or formal. The spread of such crises contains interesting parallels to our own time. Indeed, three years for each crisis to spread through the global system is about the same time that it took for recent events on the periphery of the world economy—the Asian Financial Crisis of 1998—to reverberate back and wipe out the speculative excesses of the so-called new economy in 2001. But these are more than just economic phe-nomena; they are cultural as well, as anybody who has read *Wired* knows, and as shown by a cursory glance at the rhetoric that flourished around the new economy—the dot.com-, Internet-, anything-that-has-to-do-with-new-technology economy. Our point here is that the link between communication, financial crises, and imperialism have been vastly under-researched in both the communications literature as well as among economic historians, al-though the latter generally have much to say about the severity of economic crises in the nineteenth century.

Such themes also give a sense of the dynamic yet precarious nature of globalizing processes. That is crucial because it is such conditions that allow not just for the rise of globalizing forces but also their reversal. And in the period we have covered, this could be seen by way of a backlash against globalism beginning in 1914, which was somewhat reversed in the early 1920s but which ultimately led to globalism's collapse altogether around 1930. The crisis of 1929 slammed the brakes on the further development of global com-munication markets. Investment in the global communications infrastructure that had skyrocketed in the 1920s evaporated. The revenues and profits of all the major global communications companies plummeted. Stock prices that had soared on lusty talk of a communications and consumer revolution tanked, with some, such as RCA, regaining their previous highs only in the mid-1960s. Information that had flown speedily and relatively unobstructed between nations shrivelled.[2] More broadly, countries abandoned the "gold standard" one by one, and restrictive policies blocked the flow of investment, goods, information, and immigrants between nations. A decade that had begun with a wave of democratization ended as counter-revolutions and authoritarianism pushed back the tide. Reflecting on the outcomes, Karl Polanyi noted that attempts to restore free markets had failed, while "free governments had been sacrificed."[3] In short, globalism collapsed, and similar processes would not be revived until the 1970s and 1980s.[4]

Within this overall context, we would briefly like to emphasize six points which stand out in our analysis. The first is the multinationalism of the communications firms and the willingness of governments to use "foreign" cables, not least in the pursuit of imperial goals. Likewise, Jones notes that "although London was exceptional in the degree of cosmopolitanism in the nineteenth century, the lack of concern about the nationality of ownership was general. This stance did not shift with the intensification of nationalistic rivalries in Europe before World War I."[5] Although the war brought massive expropriations of foreign assets by the belligerent nations, it is apparent from our analysis that this extraordinary measure of earlier openness was only one element in a broader willingness by governments to pursue laissez-faire goals and by corporations to expand into foreign markets. These companies typically formed alliances and networks with similar companies of other national origins when they crossed national borders. Interestingly, in the broader debates among economic and foreign policy historians, our analysis strongly supports Michael Hogan's focus on cooperation in Anglo-American economic diplomacy in the ten years between 1918 and 1928.[6] Indeed, we suggest that such cooperation actually extended back deeper in time and involved a greater range of actors than even Hogan suggests. In short, at least with respect to the "global media" and for other industries, as Jones indicates, the era was characterized by the "*internationalization* of control" rather than the "*struggle* for control," in contrast to the views of most media historians and a line of writing in other disciplines that stretches back at least as far as Ludwell Denny's work in 1930.[7]

In the case of the cable and wireless companies, the substantial amount of capital investment needed and the complexity of the technology tended to require a substantial organizational structure. On one hand, this begot efforts to adopt common technical, legal, and political standards, mostly through the ITU (International Telecommunications Union), diplomacy, treaties, and licenses but also through, albeit with limited success, the League of Nations after World War I. On the other hand, private regimes of cooperation were just as important, with cartels created in each sector—cables, telegraphs, wireless, and news—a clear sign that the media industries preferred cooperation to competition. And along the same lines, there were overlapping directorships, shared investment, price fixing, and strenuous efforts to "bring into the fold" any firm which appeared to pose a threat to stable markets (the British treatment of the German Cable Company and actions by Eastern Company affiliates in the Asian and South American markets are good cases in point). Of course, companies and different cartels did compete, but it was really only the war and its aftermath which saw the rise of that politicoeconomic nationalism which led some governments—often influenced by lobby interests—to consider foreign-owned corporations a potential danger to security and perhaps economic welfare. For example, the American Navy's antagonism and the constant harping by All-America Cable on the dangers of foreign competition led to American Marconi's enforced melding into RCA. But, in line with the functioning of "capitalist imperialism," governments and globally oriented corporations often did not see eye to eye.[8] The latter's "processes of capitalist accumulation" had an independent momentum which led them on occasion, as did Western Union in the Miami incident, to move in directions contrary to the prevailing logics of political power and diplomacy.

Second, while Japan's use of its cable and wireless interests in the pursuit of territorial imperialism in Korea, Manchuria, and China was an extreme case of state and commercial interests working together in a particularly vicious form of imperialism, such are the complex modalities of this phenomenon that one can learn much from Frank Ninkovich's definition of imperialism: "It exists when an important aspect of a nation's life is under the effective control of an outside power." However, Ninkovich continues, this control may vary from complete sovereign domination via colonialism to control over important aspects of government usually associated with sovereignty, as in many protectorates. And, crucially, control can be informal as well as formal: "It may be exercised through the workings of private social forces without overt political control from the outside."[9] We have been concerned to stress the importance of such "private social forces," not least in the form of the flows—capital, commerce, expertise, technology—which tended to create varying degrees of asymmetry between those giving and those receiving them. From this perspective, one focuses less on government interests than on the extraordinary dynamism with which the major multinational corporate and financial interests spread across the globe in the fifty years before 1914. True, our examples drawn from the cable and later the wireless companies show that such companies were often exploitative and arrogant and corrupt, exacting huge concessions and charges as their representatives ventured not just into established settler-colonies but also into recently formed republics anxious to enhance their global images and into huge, unstable political entities such as China, where the Beijing government's standing was often in contention. This was a mode of economic imperialism, to be sure, and those who opposed it by force could, as did the unfortunate Chinese who supported the Boxer Rebellion, find themselves brutally suppressed by Western military power. However, the Western companies dangled the carrot of Western "civilization" before the indigenous elites of the colonies and nations which they entered. Thus, for example, in Latin America, the cable companies became part of a "modernizing" process which saw the development of railways, telegraphs, tramlines, port facilities, better newspapers, and speedy communication with the outside world.

Our third point derives from the above comments. The common view of formal and informal imperialism as a process of one-way exploitation (notably of racially different peoples) by imperial states and corporations is unduly simplistic. Thus, we have seen the importance of reform elements among certain business, political, and intellectual classes in many countries, including China, the Ottoman Empire, Persia, and South America, which hoped to use foreign assistance cautiously in a bid to gain a stronger sense of national self-worth and, in some cases, democratic institutions. Again, the growth of national and international news via telegraph, cable, and wireless helped to disseminate a sense of national consciousness, not least in the Middle East, and thus itself added one more nail into the coffin of imperial power, be it European or Ottoman. In this sense, our book goes some small way to answer the concern of historian Linda Colley, that is, the need to study the impact, on each other, of the citizens of the imperial nations and the peoples with whom they interacted in a formal or informal controlling relationship. The most important message of Colley's book *Captives* is that the various peoples ruled within the British Empire were rarely passive victims of foreign domination; on the contrary, their actions and reactions had a profound impact on British imperial policy and public consciousness.[10] Extend this to our work and beyond formal empire, and one can see the frequency and importance of such actions and reactions: the development of the China Telegraph Administration which ultimately became a partner in the Asian cable cartel; the swiftness with which the Japanese government learned of the dangers inherent in long-term concessions to European companies; the ultimately successful attempts by the Brazilian state to exert sovereign power in order to control the terms of cable concessions. And within the British Empire, who were some of the most severe critics of the cable companies and felt among the most exploited by them? They were the politicians, officials, and spokespeople for business and press interests in the so-called "white Dominions." The Eastern Associated

Telegraph Company and its affiliates did not care much about the racial origins or imperial loyalty of those over whom it exerted monopoly control. It was an equal opportunity exploiter.

So, in 2006, we are in the midst of a second phase of globalism which has critics in plenty, not least because of the tendency to see this phenomenon as merely the globalization of capitalism rather than a thicker, more humane approach to global order. We also have a large number of studies of imperialism, past and present, with the current role of the United States and global capital much in mind. So what, if anything, has changed? A better question might be, what has not changed? Several points of continuity stand out for us, though there are undoubtedly others. Now as in the past, it is necessary to understand globalization and imperialism as two sides of a coin, but not simply with the latter being interpreted as "the highest stage of finance," as Lenin put it, nor as cloaked in the image of Darth Vader. Instead, imperialism can probably be better interpreted as a process that occurs when there is a "crisis" of globalization and as responses to such crises are pitched in a language that fits snugly within the discourse of modernization, liberalism, and democracy. That, after all, is how "good people" can make sense of such a phenomenon and helps to explain the "new imperialism" of the late nineteenth century and the liberal imperialism now splashed across the media screens of our twenty-first century.

Again, a fourth point, the advent of electronic communications media was constantly shaped by the dialectic between the free flow of information versus its control. That dialectic was at play notably in Britain's use of free trade to maintain London's position as the hub of world communication and as a means for enabling surveillance and censorship. The issue of government information and propaganda was also one of the legacies of World War I, epitomized by the British government's control of Reuters. And a similar logic could be seen in Wilson's attempts to create a more open global media system after the war while intensively using propaganda to shape the public mind, both at home and abroad. After the war, however, the unwillingness of many journalists to handle what they considered "tainted news" was an accolade to the profession. And today a similar logic continues, whether in the form of establishing media outlets in zones of conflict, as with the U.S.-created Radio Sawa and Al Hurra television network in the Middle East or the slipping of hundreds of news segments made by U.S. federal agencies into television news programs as a routine part of the Bush administration's publicity machine. While such practices and the use of government-sponsored plants placed into news conferences to ask "politic" questions have drawn an occasional editorial slap on the wrist, it seems that such practices have garnered more sarcastic ire from *Doonesbury* than from the U.S. media.[11]

To continue our points, as Laura Peterson noted in a commentary on some of our work, there is a continuity between the control over the flow of information held by the early cable and media barons and that held by the owners and controllers of the contemporary global media. Whether contemporary media barons such as Rupert Murdoch wield greater control than their predecessors is an open question, but the desire for influence continues, and the breadth of concerns raised by media concentration today would certainly ring a bell with the global media reformers of the past. Another element of continuity rests in technology and the organizations that surround and shape its development, control, and use. The global cable system did not become obsolete with the advent of wireless and long-distance telephone service. Indeed, the majority of our global communication still depends on a vast net of fiber-optic cables that spans the globe. And while the character of cartels is no longer the same, these networks are still owned and managed by highly complex "strategic alliances" consisting of the largest private and government-owned telecommunications operators in the world.

Finally, the global communication policies of the American governments in the late nineteenth and early twentieth centuries and in the U.S. Cable License Act of 1921 contain the precursors of the country's policies today. Indeed, the free flow of information doctrine can be traced back to that time, not, as many have suggested, to just after World War II. The United States also continues to leverage access to its own markets in order to prize open foreign markets and also to foist network security functions onto foreign telecommunications providers by having cables landing in the country reviewed by the National Security Agency. In the contemporary age of the security-conscious imperial state, cable landing licenses remain a powerful tool for achieving hegemony in markets and embedding the security of "imperial" interests and the nation-state deep in the fabric of communication networks.[12] Thus, policies established in the nineteenth century continue to shape the global media in our own time. So, then, *plus ça change . . .*

Notes

INTRODUCTION

1   Jones, *Multinationals and Global Capitalism*; O'Rourke and Williamson, *Globaliza-tion and History*; Feis, *Europe*; Polanyi, *The Great Transformation*.
2   Boyd-Barrett and Rantanen, *The Globalization of News*, 2.
3   Ibid.; Desbordes, "Western Empires and News Flows in the 19th Century."
4   Chandler, *Scale and Scope*.
5   Jones, *Multinationals and Global Capitalism*, 282.
6   Hills, *The Struggle for Control of Global Communications*, 179–80.
7   Hogan, *Informal Entente*.
8   Wilson, *Submarine Telegraphic Cables in Their International Relations*, 7–8.
9   NAC, Fleming, RG3, vol. 627.
10   Harvey, *The New Imperialism*, 26.
11   Gallagher and Robinson, "The Imperialism of Free Trade," 1.
12   Harvey, *The New Imperialism*, 26.
13   Ibid., 37.
14   Comor, *The Global Political Economy of Communications*, chap. 1; Cox, "Global Perestroika," 30–43.
15   Lukes, *Power*.
16   Harvey, *New Imperialism*, 29.
17   As paraphrased in Sklar, *The Corporate Restructuring of American Capitalism*, 81.
18   The following countries adopted constitutional governments during the time frame of our study, some including a "bill of rights" modeled along liberal lines and drafted with the aid of European legal scholars: Romania and Tunisia (1861), Egypt (1866), Serbia in the 1860s, Persia (1906), the Ottoman Empire (1876), China (1911) as well as several Latin American countries (circa 1890s); Findley, *Bureaucratic Reform in the Ottoman Empire*; Kedourie, *Politics in the Middle East*; Marichal, *A Century of Debt Crisis in Latin America*; Lee and Goldman, *An Intellectual History of Modern China*, 13–96.
19   Rosenberg, *Financial Missionaries to the World*; Ninkovich, *The United States and Imperialism*.
20   Kindleberger, *Manias, Panics, and Crashes*.
21   Ninkovich, *The United States and Imperialism*; Ambrosius, *Wilsonianism*, 24.
22   Giddings, *Democracy and Empire*; Reinsch, *Intellectual and Political Currents in the Far East*.
23   Ignatieff, *Empire Lite*; Cooper, "The New Liberal Imperialism."

24   Koskenniemi, *The Gentle Civilizer of Nations*, 12–17; Clark, *International Communi-cations*, 127–32.

1   BUILDING THE COMMUNICATION INFRASTRUCTURE

1   Kieve, *The Electric Telegraph*, 16–17; our translation.
2   Codding and Rutkowski, *The International Telecommunications Union in a Changing World*, 5–7; Koskenniemi, *The Gentle Civilizer of Nations*, 4–19.
3   Figures expressed in U.S. dollars with exchange rate (1914) of $1 = £.21, 5.8 francs, or 4.2 marks. Feis, *Europe*, 14, 51, 71.
4   Kieve, *The Electric Telegraph*, 52–52, 88–89.
5   Blondheim, *News Over the Wires*, 41–45; Thompson, *Wiring a Continent*, 217–25.
6   Thompson, *Wiring a Continent*, 310–30.
7   *DUS V. AATC*, 62.
8   U.S. Senate, *Report on Magnetic Telegraph Companies*, 1858, 2–5.
9   Kieve, *The Electric Telegraph*, 106–10; Thompson, *Wiring a Continent*, 331–35.
10   Kieve, *The Electric Telegraph*, 49–72; Thompson, *Wiring a Continent*, 254.
11   Quoted in Cookson, "Ruinous Competition," 95.
12   Britain, *Report of Joint Committee of Inquiry*, xiv–x.
13   Kieve, *The Electric Telegraph*, 112; Bright, *The Story of the Atlantic Cable*, 179. These companies were ultimately combined in 1872 to create the Eastern Telegraph Company. This company, in turn, was the nucleus of the Eastern and Associated Telegraph Companies, in effect a holding company consisting of the Telegraph Construction and Maintenance Company and the Eastern Telegraph Company as well as the numerous regional firms that were created or acquired by Pender and his partners as they extended their cable system to the Far East, South America, and Africa. The original company names were usually maintained, and in the case of the Telegraph Construction and Maintenance Company, the name of the formerly independent firms were subsequently used to acquire concessions and create cable and telegraph systems in South America and Africa, as the next two chapters show. Key figures in these affiliates included Charles T. Bright and Matthew Gray. These points are crucial because many writers have mistakenly identified Bright, Gray, and the India Rubber Gutta Percha Telegraph Company, for instance, as rivals to the Eastern Telegraph Company. Two final points of nomenclature: First, we use the Eastern and Associated Telegraph Companies and Eastern Telegraph Company interchangeably, given that many authors use the latter as synonymous with the former. Second, we often refer to the Eastern and Associated Telegraph Companies in the singular, i.e., as the Eastern Associated Telegraph Company, for ease of exposition.
14   Carter, *Cyrus Field*, 272.
15   Wardl, *Telegraphic Communication with India*, 4.
16   Wilson, *Submarine Telegraphic Cables in Their International Relations*, 3–4.
17   Wardl, *Telegraphic Communication with India*, 1–3; HOC, August 14, 1860, 1252–54.
18   Wardl, *Telegraphic Communication with India*, 8–16.
19   HOC, August 14, 1860, 1252.

20   Wardl, *Telegraphic Communication with India*, 5–11.

21   Ibid., 12–16; Britain, *Report of Joint Committee of Inquiry*, xiv–xx.

22   December 26, 1862, as quoted in Wardl, *Telegraphic Communication with India*, 21.

23   Ibid., 24–30.

24   Findley, *Bureaucratic Reform in the Ottoman Empire*; Kedourie, *Politics in the Middle East*; and Feis, *Europe*, 332–43.

25   Britain, *Telegraph Line—Persia*, 21; Britain, *Telegraph between India and Ottoman Territory*, 5.

26   Britain, *Telegraph Line—Persia*, 1; Wardl, *Telegraphic Communication with India*, 33.

27   Britain, *Telegraph Line—Persia*, 1–11.

28   Ibid., 2; Kedourie, *Politics in the Middle East*, 83.

29   Britain, *Telegraph Line—Persia*, 2.

30   Ibid., 13–15.

31   Ibid., 17.

32   Ibid., 21; Wardl, *Telegraphic Communication with India*, 36–39.

33   Britain, British-Persian Telegraph Convention, renewal 1866.

34   Wardl, *Telegraphic Communication with India*, 38.

35   Britain, *Indo-European Telegraph*, 1839–1931; Wardl, *Telegraphic Communication with India*, 38–39.

36   Britain, British-Persian Telegraph Convention, renewal 1873; Wardl, *Telegraphic Communication with India*, 40–45.

37   Britain, British-Persian Telegraph Convention, renewal 1873. Wardl, *Telegraphic Communication with India*, 40–45; Anderson to Andrews, October 8, 1870.

38   Read, *Reuters*, 49–50.

39   Britain, British-Persian Telegraph Convention, renewal 1892, art. x–xii; Britain, British-Persian Telegraph Convention, renewal 1873.

40   Britain, *Memorial Calcutta and Bombay Telegraph Communications*, 18.

41   Britain, *Memorial Calcutta and Bombay Telegraph Communications*, 1867–68, 13; also see Britain, *Memorial by Bankers and Merchants of City of London*.

42   Britain, *Memorial Calcutta and Bombay Telegraph Communications*, 13.

43   Ibid., 15.

44   Ibid., emphasis added.

45   Wardl, *Telegraphic Communication with India*, 45–49.

46   See Britain, The Telegraph Purchase Act, 1868; and Britain, The Telegraph Act, 1869.

47   PRO, *Miscellaneous Correspondence between Treasury Office and Various HMG Officials*, 1870; "Ocean Pacific Cable," *New York Times*, May 16, 1870, 5(2).

48   Lange, *Partner og rivaler*.

49   Cowderoy, *Direct Telegraphic Communication with Europe and the East*.

50   Britain, *Telegraph communication (East India)*, 19.

51   Ibid.

52   Sprye, January 15, 1866, 24.

53   Ibid., 23.

54   Ibid.

55   The British documents refer to the Tulumgong; however, contemporary reference to

this indigenous tribe who lived in Sumatra (now in Indonesia) is the Tulungagung. Britain, *Telegraph Communication (East India)*, 16–21; Cowderoy, *Direct Telegraphic Communication with Europe and the East*, 3–8.

56   Britain, *Telegraph Communication (East India)*, 21.

57   Correspondence between Prince Gong and Thomas F. Wade, May 7, 1870, CWA, *Chinese Concessions to Eastern Extension Cos.*, 12.

58   Correspondence between Prince Gong and General Raasloff, January 11, 1875, ibid., 9–13.

59   Baark, *Lightning Wires*, 70–71.

2   FROM GILDED AGE TO PROGRESSIVE ERA

1   Headrick, *The Invisible Weapon*, 29.

2   O'Rourke and Williamson, *Globalization and History*; Cameron and Bovykin, *International Banking*.

3   Feis, *Europe*, 5.

4   Figures expressed in U.S. dollars with exchange rate (1914) of $1 = £.21, 5.8 francs, or 4.2 marks. These rates are used consistently throughout the book. Feis, *Europe*, 14, 51, 71.

5   O'Rourke and Williamson, *Globalization and History*, 209.

6   CFB, 1873, 22; Feis, *Europe*, 27, 54, 71; O'Rourke and Williamson, *Globalization and History*, 211, 229.

7   Feis, *Europe*, 21–23, 51; Hobson, *Imperialism*, 38–49.

8   Feis, *Europe*, 11–14, 44–51, 71; Kindleberger, *Manias, Panics, and Crashes*; Marichal, *A Century of Debt Crisis in Latin America*, 4.

9   Britain, Pacific Cable Committee, 1899. Emphasis added.

10   Quoted in Sklar, *The Corporate Restructuring of American Capitalism*, 63–64.

11   Ibid., 82; Rosenberg, *Financial Missionaries to the World*, 31–35.

12   U.S. Secretary of State, Hamilton Fish to foreign governments, October and November, 1869, reproduced in U.S., *Digest of International Law*, 1:475–76.

13   Quoted in Clark, *International Communications*, 140–41. See Koskenniemi, *The Gentle Civilizer of Nations*, 105, on the importance of Renault to formalization of international law. Renault won the Nobel Peace Prize in 1907, and Asser won, with another person, in 1911.

14   FRUS, Davis to Morton, 1883, 253–91; FRUS, Schuyler to Fish, 1875, 1:1071–74.

15   U.S., *Cable-Landing Licenses*, 1921, 371–79, for a summary of concessions and the "security clause" therein.

16   Clark, *International Communications*, 127–32; Koskenniemi, *The Gentle Civilizer of Nations*, 12–13, 17, 105–6, 274–78.

17   Quoted in Clark, *International Communications*, 145.

18   Participation in such conferences was not limited just to Britain, the United States, France, and Germany but also included Denmark, Spain, Portugal, the Ottoman Empire, Japan, China, Argentina, Brazil, and, indeed, altogether some twenty to thirty countries.

19 Sharp, "International News Communications," 9; Cookson, "Ruinous Competition," 97.

20 CFB, 1873, 28.

21 PRO, French Atlantic Cable Co., 1868, 2–31; Cookson, "Ruinous Competition," 96–100.

22 Harlow, *Old Wires and New Waves*, 299.

23 Quoted in U.S., *Cable-Landing Licenses*, 1921, 7.

24 *DUS v.* AATC, 62.

25 *DUS v.* AATC, 57–67; *Times* (London), February 1, 1870, 5(c); PRO, French Atlantic Cable Co., 1868; GTTC, *Minutes*, 212.

26 GTTC, *Minutes*, Memorandum and Articles of Association, July 11, 1873, 211; GTTC, *Minutes*, January 1, 1874.

27 See, for example, statements in the 1890s which contemplate getting into "telephone, wireless, grammaphone [*sic*], or any other method now or hereafter known, discovered, or invented." Quoted from GTTC, *Minutes*, 1873–1950.

28 GTTC, *Minutes*, 1873–1950.

29 GTTC, *Minutes*, 1880–1892, 2.

30 PRO, DUS Articles of Association, 1873; U.S., Message of President, 1887, 4–10.

31 *DUS v.* AATC, 3–10; Canada, Marine Telegraph Bill, 1–24; Canada, *Debates*, March 23, 1875, 870–73.

32 Monck was British governor of North America from 1861 to 1867 and Canada's first governor general (1867–68).

33 *DUS v.* AATC; Canada, Marine Telegraph Bill, 1–24; Canada, *Debates*, March 23, 1875, 870–73. GTTC, *Minutes*, 1872–1880.

34 GTTC, *Minutes*, 1872–1880.

35 CWA, Anglo-American Company contracts, 1880.

36 The two companies had been formed in 1879 and 1881, respectively.

37 CWA, Anglo-American Company contracts, 1882a, 21.

38 Headrick, *The Invisible Weapon*, 35.

39 Anglo-American Company Contracts, 1882b, 3–4. CWA, Anglo-American Company contracts, 1882a, 20.

40 U.S., Message of President, 1887, 38–59.

41 Statement by J. H. Carson, Britain, Pacific Cable Committee, 1899, 120.

42 GTTC, *Minutes*, 1880–1892, 212.

43 Pacific Cable Committee, 1899, 81; Bright, *The Story of the Atlantic Cable*, 212–16; Blondheim, *News Over the Wires*, 151–64.

44 Ward statement to Pacific Cable Board, 1899, 81; NAC, "Memo on Imperial Conference," April 3, 1911, 4–5; U.S., *Cable-Landing Licenses*, 1921, 209.

45 U.S., *Cable-Landing Licenses*, 1921, 202–10.

46 Scrymser, *Personal Reminiscences of James A. Scrymser in Times of Peace and War*, 68–74.

47 Quoted in U.S., *Cable-Landing Licenses*, 1921, 54.

48 CWA, West India Company Agreements, 4–5; Ahvenainen, *The History of the Caribbean Telegraphs before the First World War*, 18–31; Scrymser, *Personal Reminiscences of*

James A. Scrymser in Times of Peace and War, 15, 69–71; U.S., *Cable-Landing Licenses*, 1921, 85.

49   Western Union, *Annual Reports*, 1876–1906; Field, "American Imperialism," 662; Hogan, *Informal Entente*, 49–66.

50   International Ocean Telegraph Company et al., 1869; Scrymser, *Personal Reminiscences of James A. Scrymser in Times of Peace and War*, 74; Ahvenainen, *The History of the Caribbean Telegraphs before the First World War*, 24.

51   Ahvenainen, *The History of the Caribbean Telegraphs before the First World War*, 28. Parenthetically, note that the companies being discussed here were quite small, and none were among the tier one or two players outlined earlier in this chapter.

52   Clause 4 of the agreement, International Ocean Telegraph Company et al., 1869.

53   Britain, *Royal Commission*, 1910, 43.

54   Britain, *Royal Commission*, 1910, 44–45; Britain, *Evidence*, 1902, 116–17; Ahvenainen, *The History of the Caribbean Telegraphs before the First World War*, 31–43.

55   Ahvenainen, *The History of the Caribbean Telegraphs before the First World War*, 28; also see NAC, RG3, vol. 1002, file 21–5–10. RG Pacific Cable Board (1914). *Notes on West Indies Communication*, Papers Circ. no. 223.

56   WTC, Brazil, box 6; Berthold, *History of the Telephone and Telegraph in Brazil*, 12; U.S., *Cable-Landing License*, 1921, 39–91.

57   The claim that the India Rubber and Gutta Percha Telegraph Company was an arm of the Telegraph Construction and Maintenance Company is important because we assume that companies set up by it, notably the West Coast of America Company, are creatures of the Pender group of companies. The assumption is based on the analysis presented in chapter 1, namely, that the Gutta Percha Telegraph Company was one of two firms (with Glass Elliot) that were merged in 1864 to create the Telegraph Construction and Maintenance Co. See our chapter 1, n. 13, and Ahvenainen, *The History of the Caribbean Telegraphs before the First World War*; also Berthold, *History of the Telephone and Telegraph in Brazil*; *History of the Telephone and Telegraph in Uruguay*; *History of the Telephone and Telegraph in Argentina*, 19–21.

58   Ahvenainen, *The European Cable Companies in South America*, 45–57.

59   Ibid., 52.

60   U.S., *Cable-Landing Licenses*, 1921, 91. Emphasis added.

61   WTC, Brazil, box 6; Berthold, *History of the Telephone and Telegraph in Uruguay*, 22; *History of the Telephone and Telegraph in Argentina*, 1.

62   Ahvenainen, *The European Cable Companies in South America*, 64–67.

63   Hogan, *Informal Entente*.

64   Ahvenainen, *The European Cable Companies in South America*, 64–70, 94.

65   Ibid., 36–39; Berthold, *History of the Telephone and Telegraph in Uruguay*, *History of the Telephone and Telegraph in Brazil*, and *History of the Telephone and Telegraph in Argentina*; and Rippy, "Notes on Early Telephone Companies of Latin America."

66   Two more companies, the Oriental Telegraph Company and the Canadian-owned River Plate Telegraph and Telephone Company, were created in 1871 and 1887, respectively, a point which we develop further below. See Berthold, *History of the Telephone and Telegraph in Uruguay*, 3–4, 18, and Ahvenainen, *The European Cable Companies in South America*, 158, 189.

67 Stewart, "Notes on an Early Attempt to Establish Cable Communication between North and South America"; Berthold, *History of the Telephone and Telegraph in Chile.*

68 E. M. Archimbald, British Consul General, New York, to Simon Stevens, February 12, 1873, reproduced in Stewart, "Notes on an Early Attempt to Establish Cable Communication between North and South America," 123.

69 On skates, fine china, speculative manias, and the global economic crisis of 1873, see Kindleberger, *Manias, Panics, and Crashes*; Marichal, *A Century of Debt Crisis in Latin America*, 98–125.

70 All-America Cable Company, *A Half Century of Cable Service to the Americas*, 14–15.

71 Reprinted at July 28, 1872, in *FRUS*, 1872, vol. 1, 42–43.

72 About $1.75 per word from Buenos Aires to London in 1890. See Brown, *The Cable and Wireless Communications of the World*, 128.

73 WTC, Brazil, box 1, July 12, 1885.

74 London *Times*, April 5, 1875, 6b; CFB, 1873, 25, 42, 58.

75 Kindleberger, *Manias, Panics, and Crashes*; Marichal, *A Century of Debt Crisis in Latin America*, 98–12.

76 CFB, 1873, 1874, and 1878.

77 Ahvenainen, *The European Cable Companies in South America*, 137–40.

78 Berthold, *History of the Telephone and Telegraph in Chile*, 30; Johnson, *Pioneer Telegraphy in Chile*, 112–20.

79 See Britain, *Inter-Dept. Comm. Report*, 1902, appendix G, 75–76; Britton and Ahvenainen, "Showdown in South America," 16–17.

80 In addition to Berthold, *History of the Telephone and Telegraph in Argentina*; *History of the Telephone and Telegraph in Brazil*; WTC, Brazil, boxes 1 and 2.

81 Berthold, *History of the Telephone and Telegraph in Brazil*, 15–18.

82 Ahvenainen, *The European Cable Companies in South America*, 170–78, 229–32; Marichal, *A Century of Debt Crisis in Latin America*, 140–64; Pletcher, *The Diplomacy of Trade and Investment*, 311–13.

83 In contemporary parlance, it was a build-own-transfer model of development. Berthold, *History of the Telephone and Telegraph in Uruguay*, 10; Britain, *Inter-Dept. Comm. Report*, 1902, 74.

84 Berthold, *History of the Telephone and Telegraph in Uruguay*, 14

85 Britton and Ahvenainen, "Showdown in South America," 18; Berthold, *History of the Telephone and Telegraph in Chile*, 17.

86 As reproduced in Scrymser, *Personal Reminiscences of James A. Scrymser in Times of Peace and War*, 90. Note the anticipated connection between Scrymser's central and south American cables and future connections to the Far East.

87 All-America Cable Company, *A Half Century of Cable Service*, 20, 49–69; Central and South American Telegraph Company, *Via Galveston*; Western Union, *Annual Reports*, 1876, 1890; U.S., *Cable-Landing Licenses*, 1921, 376–79; Field, "American Imperialism," 662.

88 Scrymser, *Personal Reminiscences of James A. Scrymser in Times of Peace and War*, 95–100; Field, "American Imperialism," 662–65.

89 U.S., *Cable-Landing Licenses*, 1921, 375–79; Berthold, *History of the Telephone and Telegraph in Chile*, 32.

90   Britton and Ahvenainen, "Showdown in South America," 20–21.

91   All-America Cable Company, *A Half Century of Cable Service*, 22; Central and South American Telegraph Company, *Via Galveston*, 9.

92   Berthold, *History of the Telephone and Telegraph in Uruguay*, 3–4, 18, and Ahvenainen, *The European Cable Companies in South America*, 158, 189.

93   WTC, Brazil, box 6, 110; U.S., *Cable-Landing Licenses*, 1921, 89–91.

94   Britton and Ahvenainen, "Showdown in South America," 20–22; CWA, West Coast of America Telegraph Company, *Report*, 1890; Britain, *Inter-Dept. Comm. Report*, 1902, appendix G, 76–77.

95   Britain, *Inter-Dept. Comm. Report*, 1902, appendix G, 69–71, and Ahvenainen, *The European Cable Companies in South America*, 166–70.

96   PRO BT 31/5102/34363, South American Cable Company, Ltd 1891.

97   Berthold, *History of the Telephone and Telegraph in Brazil*, 21.

98   Britain, *Inter-Dept. Comm. Report*, 1902, appendix G, 69–71.

99   U.S., *Cable-Landing Licenses*, 1921, 89–92.

100 Britain, *Evidence*, 1902, 116; Ahvenainen, *The History of the Caribbean Telegraphs before the First World War*, 57–64, 88–94.

101 The point, as we state elsewhere, is to reduce the complexity of the "alphabet soup." The nomenclature also fits with that of other companies, such as the Anglo-American Company, and was in fact employed in the major U.S. *Cable Landing Licenses Hearings* of 1921.

102 Britain, *Evidence*, 1902, 130; Ahvenainen, *The History of the Caribbean Telegraphs before the First World War*, 102–3.

103 Ahvenainen, *The History of the Caribbean Telegraphs before the First World War*, 102–3.

104 U.S., *Cable-Landing Licenses*, 1921, 89–92; Britain, *Evidence*, 1902, 130.

105 CFB, 1887, 48–49; U.S., *Cable-Landing Licenses*, 1921, 89–92; Ahvenainen, *The History of the Caribbean Telegraphs before the First World War*, 91–98.

106 Britain, *Evidence*, 1902, 115–20; Britain, *Royal Commission*, 43; Ahvenainen, *The History of the Caribbean Telegraphs before the First World War*, 143.

107 Britain, *Inter-Dept. Comm. Report*, 1902, 20; Britain, *Evidence*, 1902, 115–16, 151; Britain, *Royal Commission*, 1910, 134–43.

108 On these general changes in Germany and the Felten and Guilleaume's status in the German domestic economy, see Feis, *Europe*, 62–77; Chandler, *Scale and Scope*, 66–69.

109 We will refer to these companies collectively as the German Cable Company.

110 Britain, *Evidence*, 1902, 150.

111 Ibid., 48–53.

112 NAC, Marquess of Salisbury to Count Hatzfeld, January 3, 1900. Robert Cecil, the Marquess of Salisbury, or Lord Salisbury, was the British Prime Minister, for the third time, between 1895 and 1902, so the shift in policy was coming from the highest levels of the British government.

113 U.S., *Cable-Landing Licenses*, 1921, 202–10.

3   INDO-EUROPEAN MARKETS AND AFRICA

1   Wardl, *Joint Purse*, 1.

2   Britain, British-Persian Telegraph Convention, renewal 1873; Britain, British-Persian Telegraph Convention, renewal 1892.

3   Hay to Duke of Argyll, February 22, 1871; Wardl, *Joint Purse*, 4, 10–16.

4   Anderson to Andrews, October 8, 1870.

5   Hay to Duke of Argyll, February 22, 1871.

6   Ibid.

7   Ibid.; Melvill to Anderson, March 31, 1871.

8   The rate is for a twenty-word message. Hay to Duke of Argyll, February 22, 1871; Wardl, *Joint Purse*, 8.

9   Wardl, *Joint Purse*, 1.

10   Ibid., 12–16.

11   Wardl, *Joint Purse*, 2; Hay to Duke of Argyll, February 22, 1871, 37.

12   Feis, *Europe*, 313–43.

13   It must be stressed that these were ideals, but nonetheless, the fact that they were institutionalized at all revealed a fundamental cultural transformation and attempts to wrestle with questions of modernity, Islam, democracy, and liberalism. On these subjects in the Ottoman Empire and Persia, see, for example, Feis, *Europe*, 310–40; Cole, *Colonialism and Revolution in the Middle East*; Findley, *Bureaucratic Reform in the Ottoman Empire*; Kedourie, *Politics in the Middle East.*

14   Barty-King, *Girdle Round the Earth*, 60–62.

15   Wardl, *Joint Purse*, 1.

16   Ibid.; Barty-King, *Girdle Round the Earth*, 62.

17   Wardl, *Joint Purse*, 4.

18   Ibid., 12–16; Anderson to Champlain, December 4, 1879; Anderson to Parry, January 24, 1879.

19   Wardl, *Joint Purse*, 8. Emphasis added.

20   Ibid., 10.

21   Ibid., 12–16.

22   Cole, *Colonialism and Revolution in the Middle East*, 110.

23   Quote from ibid., 112. As for telegraph revenues, CFB, 1878, 24–25, notes that in 1875 the state-owned Egyptian telegraph system was the fourth largest source of revenue after the Contributions Fonciers, Contributions Personelles, and Customs departments.

24   Boehmer, *Empire, the National and the Post Colonial*, 9; Cole, *Colonialism and Revolution in the Middle East*, 112. Even the conservative Israeli scholar Kedourie (*Politics in the Middle East*, esp. 61, 80–83) points to these constitutional developments in Egypt and the Ottoman Empire more broadly as liberal, secular, and cosmopolitan in orientation.

25   Emphasis added. CFB, 1879, 35–36.

26   See Feis, *Europe: The World's Banker, 1870–1914*, 390–95.

27   Barty-King, *Girdle Round the Earth*, 73.

28 See Denison-Pender's comments to the Balfour Committee, in Britain, *Evidence*, 1902, 26, regarding some of the terms of the "lease" governing its operations in Egypt.

29 Anderson to Parry, January 24, 1879; Wardl, *Joint Purse*, 4; Britain, *Evidence*, 1902, 18; Barty-King, *Girdle Round the Earth*, 66–67.

30 Anderson to Parry, January 24, 1879; Wardl, *Joint Purse*, 4; Britain, *Evidence*, 1902, 18.

31 Britain, *Evidence*, 1902, 21.

32 Wardl, *Joint Purse*, 2; Britain, *Inter-Dept. Comm. Report*, 1902, 5–6.

33 Britain, *Inter-Dept. Comm. 1st Report*, 1902, 10–12.

34 Except for the northern coastline and a few islands off the west coast of Africa, which were used as a hub en route from Europe to South America.

35 Quote is from Marichal, *A Century of Debt Crisis in Latin America*, 152. On the diamond and gold mining boom on the east coast of Africa from the early 1880s, see Hobsbawm, *The Age of Empire*, 62–63, and Ferguson, *Empire*, 223–27. However, in almost every measure of integration into the global economy, relative to South America, Asia, and the Ottoman Empire, Africa fared badly. Thus, in crude terms, the problem in Africa (South Africa excepted), then as today, was not an excess of foreign investment and "overexploitation" but underinvestment and a *lack* of integration, which thereby resulted in imperialism becoming the dominant mode of development. Thus, while investment in gold and diamond mining and a few other areas did become significant, it was sector and location specific, while, as Ferguson states, "hardly anything was invested in the new Africa acquisitions" (243). See Feis, *Europe*, 21–23, 51; Hobson, *Imperialism*, 38–49.

36 CFB, 1891, 50–90, and 1892, 17.

37 Koskenneimi, *The Gentle Civilizer of Nations*, 106–10, 121–27. In addition to these countries, the following were in attendance: Austria-Hungary, Belgium, Denmark, Italy, the Netherlands, Portugal, Russia, Spain, Sweden, and Turkey. Ferguson, *Empire*, 236.

38 Both quotes are from the *General Acts* as reproduced in Ferguson, *Empire*, 236.

39 See Koskenniemi, *The Gentle Civilizer of Nations*.

40 Ferguson, *Empire*, 236.

41 Ibid., 223.

42 Britain, *Inter-Dept. Comm. Report*, 1902, 63.

43 Ferguson, *Empire*, 223–27.

44 Britain, *Inter-Dept. Comm. Report*, 1902, 36–38. The issue of subsidies is discussed and fully cited immediately below.

45 Total cable subsidies from Britain and British colonies were about $13.9 million, with approximately $7.5 million earmarked for Africa. For a peek ahead regarding the data on French, German, and Portuguese "matching subsidies" to the Eastern Associated Company, see Britain, *Inter-Dept. Comm. Report*, 1902, 59, and the notes to the tables in appendix G to that report, 72–73. Further explanation is developed below.

46 PRO, South American Cable Company, 1891. Coggeshall, *An Annotated History of Submarine Cables and Overseas Radiotelegraphs*, 85, also refers to the mystery over the Spanish company.

47   Headrick, *The Invisible Weapon*, 64–65.

48   PRO, South American Cable Company, 1891; also see Bright, *The Story of the Atlantic Cable*, 168; King, *Girdle Round the Earth*, 96–97. See this volume, chapter 1, n.13, for support of our claim regarding Gutta Percha's connection to the Telegraph Construction and Maintenance Company.

49   Barty-King, *Girdle Round the Earth*, 395; *Inter-Dept. Comm., Report*, 72–73, regarding opening of both companies for business in 1886 and their position within the Eastern Associated Company.

50   Headrick, *The Invisible Weapon*, 65; Britain, *Inter-Dept. Comm. Report*, 1902, appendix G, 72–73, and appendix J, 80.

51   Ferguson, *Empire*, 237.

52   Ibid., 225–26.

53   See, for example, Britain, *Inter-Dept. Comm. Report*, 1902, appendix E (subsidies), 55–56.

54   Here we use this nomenclature to refer to both of Pender's companies on the west coast of Africa: the African Direct Telegraph Company and the West African Telegraph Company. As for the unpaid subsidies to these companies, see ibid., appendix G, 73.

4   ELECTRONIC KINGDOM AND WIRED CITIES

1   Dayer, *Bankers and Diplomats in China*, 5; Feis, *Europe*, 433–45; Furth, "Intellectual Change," 3–47.

2   Baark's analysis is more nuanced and focuses a great deal on the role of modernizers in China.

3   *FRUS*, Gong to Avery, January 12, 1875, 262–63.

4   *FRUS*, Avery to Fish, December 20, 1874, 223–37, 238.

5   *FRUS*, Seward to Cadqalader, August 27, 1874, 335.

6   *FRUS*, Fish to Lindencrone, January 22, 1874, 370–71; *FRUS*, Fish to Avery, March 4, 1875, 274–75.

7   *FRUS*, Avery to Fish, December 20, 1874, 238.

8   *FRUS*, Lindencrone to Fish, January 7, 1874, 378.

9   *FRUS*, Avery to Fish, December 20, 1874, 224–26.

10   *FRUS*, Seward to Davis, June 23, 1874, 378; Baark, *Lightning Wires*, 136.

11   *FRUS*, Seward to Davis, June 23, 1874, 323; *FRUS*, Seward to Cadqalader, August 27, 1874, 337; *FRUS*, Avery to Fish, December 20, 1874, 223–27.

12   *FRUS*, Avery to Fish, December 20, 1874, 223–27.

13   *FRUS*, Williams to Secretary of State Fish, February 9, 1874, 246–47; also see Baark, *Lightning Wires*, 83, on this point.

14   *FRUS*, Lindencrone to Fish, January 7, 1874, 378; *FRUS*, Williams to Secretary of State Fish, February 9, 1874, 246–47; *FRUS*, Avery to Fish, December 20, 1874, 223–27.

15   Vittinghoff, "Unity vs. Uniformity," 92–93; Wagner, "The Role of the Foreign Community in the Chinese Public Sphere," 432. The following two paragraphs are based

on these two sources, with additional background context from Furth, "Intellectual Change"; Ninkovich, *The United States and Imperialism*; and Feis, *Europe.*

16   *FRUS*, Avery to Fish, December 20, 1874, 225–26; *FRUS*, Gong to Avery, January 12, 1875, 262–63.

17   *FRUS*, Avery to Fish, December 20, 1874, 237–41.

18   Baark, *Lightning Wires*, 117–19, 154; Headrick, *The Invisible Weapon*, 63–65.

19   Correspondence between Prince Gong and General Raasloff, January 11, 1875, CWA, *Chinese Concessions to Eastern Extension Cos.*, 9–13; *FRUS*, Avery to Fish, January 27, 1875, 260.

20   *FRUS*, Lindencrone to Fish, January 7, 1874, 379.

21   *FRUS*, Avery to Fish, January 27, 1875, 261.

22   *FRUS*, Avery to Fish, December 20, 1874, especially 241; *FRUS*, Avery to Fish, March 18, 1875, 275–78.

23   Emphasis added. *FRUS*, Avery to Fish, March 19, 1875, 278.

24   *FRUS*, Avery to Fish, December 20, 1874, 238.

25   Ibid., 237–41.

26   Ibid., 241.

27   Ibid., 239–41.

28   *FRUS*, Avery to Fish, May 14, 1875, 335–36.

29   Emphasis added. *FRUS*, Avery to Fish, December 20, 1874, 239.

30   *FRUS*, Avery to Fish, May 14, 1875, 331; *FRUS*, Seward to Mr. Davis, July 21, 1875, 326.

31   *FRUS*, Avery to Fish, May 14, 1875, 355.

32   *FRUS*, Seward to Cadqalader, August 27, 1874, 336.

33   *FRUS*, Avery to Fish, May 14, 1875, 329.

34   *FRUS*, Avery to Steward, February 3, 1875, 265. Baark claims that the approach merely restated the Wade-Gong agreements of 1870 (*Lightning Wires*, 123), but it is clear from the foreign correspondence cited here that C. F. Tietgen and "foreign officials" were emphatic that it was a radical break with the past.

35   *FRUS*, Seward to Cadqalader, August 27, 1874, 338.

36   The company had raised its rates more than threefold within five years of its inception, from an original rate of roughly $12.50 for twenty words in 1870, to $30 in 1873, and to $40 in 1875. CWA, *China Agreements*, May 13, 1870; CWA, *China Agreements*, February 11, 1873; CWA, *China Agreements*, September 10, 1875, 40–41. For opposition to rate hikes, see Ahvenainen, *The Far Eastern Telegraphs*; on Russian influences and cleavages between the company and others, see Baark, *Lightning Wires*, 117–21, and Lange, *Partner og Rivaler.*

37   *FRUS*, Avery to Fish, May 14, 1875, 330.

38   Quoted from official Chinese sources in Baark, *Lightning Wires*, 152, n. 110.

39   *FRUS*, Young to Frelinghuysen, November 28, 1882, 142–53.

40   On the position of these figures in Chinese society and politics in general, see Furth, "Intellectual Change." Relative to China's national communication system, see Baark, *Lightning Wires*, generally and 110.

41   The beginning of the company's operations is noted in Baark, *Lightning Wires*, 88.

42   Emphasis added. *FRUS*, Young to Frelinghuysen, November 28, 1882, 143–44.

43   Ibid., 147.

44   Baark, *Lightning Wires*, 159–62.

45   Recall that these events were occurring at the same time that relations between Pender and Tietgen were also deteriorating in South America (see this volume, chapter 2). *FRUS*, Young to Frelinghuysen, November 28, 1882, 148.

46   These comments and the last section of this chapter are based on the following: CWA, *China Agreements*, January 12, 1883; CWA, *China Agreements*, May 7, 1883; CWA, *China Agreements*, January 1, 1884; CWA, *China Agreements*, February 27, 1884; CWA, *China Agreements*, October 29, 1886.

47   CWA, *China Agreements*, October 29, 1886, 1.

48   CWA, *China Agreements*, January 12, 1883; CWA, *China Agreements*, May 7, 1883.

49   Barty-King, *Girdle Round the Earth*, 96.

50   CWA, *China Agreements*, January 12, 1883, articles 13–16.

51   CWA, *China Agreements*, February 11, 1882.

52   CWA, *China Agreements*, May 7, 1883, article 5; CWA, *China Agreements*, August 10, 1887.

53   CWA, *China Agreements*, January 12, 1883, article 20. As for the period covering China's obligations, the first formal mention was a deadline of May 1903 in CWA, *China Agreements*, August 10, 1887, article 18, and extended to December 31, 1910, by CWA, *China Agreements*, Chinese Telegraph Convention, July 11, 1896, article 16. A new date of December 31, 1930, is stated in CWA, *Japan and China Agreements*, December 22, 1913.

54   Britain, *Hansard*, July 23, 1900, 219, regarding troops.

55   Quoted in Dayer, *Bankers and Diplomats in China*, 12.

5   THE POLITICS OF GLOBAL MEDIA REFORM I

1   Standage, *The Victorian Internet*, vii.

2   Anderson to Parry, January 24, 1879.

3   Kieve, *The Electric Telegraph*, 236.

4   Parsons, *The Telegraph Monopoly*, 1899, 37.

5   U.S., Department of Labor, *Investigation of Western Union*, 1909, 21.

6   Bright, *Telegraphy, Aeronautics and War*, 1918, 139.

7   Coates and Finn, *A Retrospective Technology Assessment*, 87.

8   Ahvenainen, *The Far Eastern Telegraphs*, 205; Britain, *Inter-Dept. Comm. Report*, 1902, 36–37.

9   Gramling, AP: *The Story of News*, 74.

10   Desbordes, "Western Empires and News Flows in the 19th Century," 4.

11   Ibid., 3–5; Rantanen, "The Struggle for Control of Domestic News Markets," 37–41.

12   Johnson, *Pioneer Telegraphy in Chile*, 110.

13   U.S. Senate, *Navy Pacific Cable*, 1900, 44.

14   Desbordes, "Western Empires and News Flows in the 19th Century"; Rantanen, "The Struggle for Control of Domestic News Markets."

15   Britain, *Dominions Royal Commission*, 1914, 43.

16   Ibid., 54.

17 "The Great International Telegraph Enterprises," *New York Times*, July 31, 1866, 4.

18 "A Guinea a Word," *New York Times*, August 1, 1866, 4.

19 "The Atlantic Telegraph: Its Monopoly by England," *New York Times*, November 24, 1866, 4.

20 "Ocean Telegraph: The Empire and the Cable Companies," *Times* (London), May 24, 1912, 15–16.

21 NAC, Heaton to Fleming, March 11, 1886.

22 See Sassoon's obituary in the *Times* (London), May 25, 1912, 11. Ironically, he died on the same day (May 24) that the *Times* devoted a large supplement on telegraph communication and Empire unity. Sassoon was Conservative-Unionist member for Hythe, Kent, from 1899 to 1912, and a member of a wealthy, well-known Jewish family. From 1900, he chaired the Imperial Telegraph Committee of the British House of Commons.

23 See, for example, Squier's detailed statement to U.S. Senate, *Navy Pacific Cable*, 1990.

24 Twain, *Following the Equator*.

25 Britain, *Evidence*, 1902, 89.

26 Britain, *Inter-Dept. Comm. Report*, 1902, appendix G, 59, 72–73.

27 NAC, Fleming to Lemieux, April 16, 1909.

28 NAC, "Explanatory Note," and NAC, "The Empire Cables"; NAC, Ottawa City Board of Trade, "An Address to His Excellency, Earl Grey," 1907.

29 See Britain, The Telegraph Purchase Act, 1868.

30 Heaton, "Postal and Telegraphic Reforms," 17.

31 U.S., H.R. *Pacific Cable*, 1900, 34.

32 Ibid., 91.

33 Boyce, "Imperial Dreams and National Realities," 39–70.

34 Correspondence between T. V. Lister, Foreign Office, to the Under Secretary of State, Colonial Office, June 7, 1892. NAC, RG25, A1, vol. 53.

35 NAC, Jones to Macdonald, 2339 Order in Council; NAC, Fleming to Langevin, April 8, 1886.

36 NAC, Anderson to Pender, July 20, 1880.

37 Britain, Pacific Cable Committee, 1899, 89. See also NAC, Smith to Mulock, October 5, 1904. Staniforth Smith became a strong, and knowledgeable Australian supporter of the Pacific Cable.

38 NAC, Pacific Cable Committee, 1899, 23–24. Evidence to the committee was given three years before, in 1986.

39 Ibid., 488.

40 A collection of Colonial Conference papers is reproduced in Ollivier, *The Colonial and Imperial Conferences from 1887 to 1937*, 33.

41 Britain, Pacific Cable Committee, 1899, 95; see also similar arguments of the Treasury and the Government of India in PRO, Mercer Memo, April 29, 1897.

42 Livingston, *The Wired Nation Continent*, chap. 4; New Zealand, Copyright Telegrams Committee Report, 1896.

43 Richeson, *Sandford Fleming and the Establishment of a Pacific Cable*, 315–16.

44 CWA, *Report of the Twenty-Seventh Ordinary General Meeting*, April 27, 1887, 3.

45 Ibid.

46 Boyce, "Imperial Dreams and National Realities," 48–58; also NAC, Fleming to Tarte, July 1, 1899.

47 Britain, Pacific Cable Committee, 1899, 99.

48 Ibid., 6.

49 Part of the debate is recorded in NAC, Tupper-Pender Debate, 1894.

50 Ibid.

51 Britain, Pacific Cable Committee, 1899.

52 CWA, Eastern Telegraph Company, 1899; NAC, Fleming to Tarte, July 1, 1899.

53 NAC, Fleming to Tarte, July 1, 1899.

54 PRO, Mercer Memo, April 29, 1897. As another memo from the Treasury to the Colonial Office states, "It would be preferable not to present the Report and Proceedings of the Pacific Cable Committee to Parliament until Her Majesty's Government has arrived at a decision upon . . . the undertaking." PRO, Hamilton Memo, April 29, 1897.

55 PRO, Hamilton Memo, April 29, 1897.

56 Kennedy, "Imperial Cable Communications and Strategy," 738; Britain, *Inter-Dept. Comm. Report*, 1902, 32.

57 Britain, Pacific Cable Committee, 1899, 112.

58 CWA, *Eastern Telegraph, Pacific Agreements*, July 26, 1904; CWA, *Eastern Telegraph, Pacific Agreements*, September 12, 1905.

59 Ahvenainen, *The Far Eastern Telegraphs*, 163–70.

60 Pacific Cable Company of New York, *Pacific Cable*.

61 U.S., *Pacific Cable Bills*, 14.

62 Ibid., 46–47.

63 Ibid., 5.

64 CWA, Memo on Commercial Pacific Company, 1934, 1.

65 CWA, *Pacific Agreements*, September 25, 1905.

66 CWA, Memo re American Pacific Combination, July 1, 1905.

67 "Ocean Telegraph: The Empire and the Cable Companies," *Times* (London), May 24, 1912, 15–16.

68 CWA, *Eastern Telegraph, Pacific Agreements*, July 26, 1904; CWA, *Eastern Telegraph, Pacific Agreements*, September 12, 1905; CWA, Memo on Commercial Pacific Company, 1934.

69 Hoare, "The Era of Unequal Treaties."

70 Checkland, *Britain's Encounter with Meiji Japan*, 51–52, 75–76.

71 "Telegraphic Communication with the Far East—Agreement with China and Japan 1907," July 29, 1908.

72 Yang, *Submarine Cables and the Emerging Japanese Empire*.

73 Ibid.

74 Gordon, *A Modern History of Japan*, 119.

75 Iriye, "Japan's Policies towards the U.S.," 418.

76 "Telegraphic Communication with the Far East—Agreement with China and Japan 1907," July 29, 1908, 1–2.

77   On telegraph use, see Duus, *The Cambridge History of Japan*, 399. On newspapers, Gordon, *A Modern History of Japan*, 79.

78   "Ocean Telegraph: The Empire and the Cable Companies," *Times* (London), May 24, 1912, 15–16.

79   Britain, *Inter-Dept. Comm. Report*, 1902, 14.

80   Ibid., 35.

6   THE POLITICS OF GLOBAL MEDIA REFORM II

1   Emphasis added. Lippmann, *Liberty and the News*, 43; Lippmann and Merz, "A Test of the News."

2   Rodgers, *Atlantic Crossings*, 52.

3   Sklar, *The Corporate Restructuring of American Capitalism*, 99–100.

4   O'Rourke and Williamson, *Globalization and History*.

5   Note that there were several kinds of message services, with ordinary, government, and press being the main before several new categories designed to enhance "social communication" were introduced in 1913–14, as discussed below; more were introduced after World War I. Pacific Cable Board, *Via Pacific*, 19.

6   NAC, "Report of the Chairman of the Pacific Cable Board," 1909.

7   Pacific Cable Board, *The All Red Route Via Pacific*, 1924.

8   Boyce, "Imperial Dreams and National Realities," 66–68, and NAC, Memo on Pacific Cable, May 1, 1905.

9   Australia, Report of the Select Committee, 287–401. Rantanen, "The Struggle for Control of Domestic News Markets," 40–42. The Board of Railway Commissioners in Canada reached similar conclusions with respect to a complex four-way alliance between the Canadian Pacific Telegraph Company and the Western Union as the network providers and Associated Press and Canadian Press as the news agencies. As in Australia, conditions were imposed as a result of the committee report that led to the effective dismantling of this situation. See Winseck, *Reconvergence*, 103–9.

10   NAC, Ross to Laurier, August 4, 1909. Here Ross notes that many erstwhile supporters of the state cable had abandoned it as having "through no fault of its own, a notoriety for inaccuracy."

11   NAC, Ross to Deputy Minister of Trade and Commerce, August 1907, 2.

12   Ibid.

13   NAC, Fleming, *Cheap Telegraph Rates*, February 28,1902.

14   "Ocean Telegraph: The Empire and the Cable Companies," *Times* (London), May 24, 1912, 16.

15   NAC, "Survey of Disabilities," December 31, 1908.

16   Ibid.

17   Britain, *Inter-Dept. Comm. Report*, 1902, 25.

18   NAC, "Memo on Imperial Conference," April 3, 1911, 7–8.

19   *Punch*, July 16, 1909.

20   NAC, Lemieux to Earl Grey, January 13, 1910.

21   From editor of New Zealand Press Association, November 3, 1908, contained in Heaton, *Universal Penny-a Word Telegrams*, 1908, 52 (originally published in the

Financial Review, 1908, vol. and no. unknown. A copy can be found in NAC, MG29, vol. 25, BI); also "An Extraordinary Monopoly," January 15, 1909, 9–10.

22 Donald, *The Imperial Press Conference in Canada*, 121.

23 Heaton, "How to Smash the Cable Ring," 1907, and "The Cable Telegraph Systems of the World," 1907. The latter contains a list of the world's cables and his general philosophy which linked cheap rates to communications and international peace.

24 "Cable Rate Abuses," *New York Times*, November 29, 1908.

25 Ibid.

26 Pike, "National Interests and Imperial Yearnings," 1998, 22–48.

27 NAC, "Memo on Imperial Conference," 1911.

28 The following source contains a compendium of original documents: Ollivier, *The Colonial and Imperial Conferences from 1887 to 1937*, 319.

29 Canada, *An Act to control the Rates and Facilities of Ocean Cable Companies*, May 4, 1910, specifically clause 5.

30 "4 Cent Cable Soon to Cross Atlantic," *New York Times*, December 8, 1908.

31 NAC, Dispatch from Lord Crewe to Earl Dudley, October 14, 1910.

32 "Atlantic Cable Amalgamation," *Times* (London), April 14, 1911.

33 Oslin, *The Story of Telecommunications*, 262.

34 Ibid., 259.

35 "Hits back at Col. Clowry," *New York Times*, November 13, 1910, 10.

36 Bright, *Imperial Telegraphic Communication*, xvii.

37 See, notably, the following articles in the *Times* (London): "Atlantic Cable Amalgamation," April 14, 1911; "Cable Companies' Combination," September 15, 1911; and "Shareholders' Sanction," September 30, 1911.

38 Coggeshall, *An Annotated History of Submarine Cables and Overseas Radiotelegraphs*, 33.

39 Ibid.

40 Sklar, *The Corporate Restructuring of American Capitalism*, 187, 221.

41 "Letters by Cable Is the Plan Now," *New York Times*, September 15, 1911, 6.

42 Coggeshall, *An Annotated History of Submarine Cables and Overseas Radiotelegraphs*, 34.

43 "Shareholders' Sanction," *Times* (London), September 30, 1911.

44 Britain, *Royal Commission*, 1912, 274.

45 Britain, *Royal Commission*, 1914.

46 Britain, *Royal Commission*, 1918, 335.

47 "Cable Reductions Start Rumors of Rate War," *New York Times*, January 7, 1912, 8.

48 Britain, *Royal Commission*, 1917, 37.

49 Sharp, "International News Communication," 16.

50 "International Communications," *New York Times*, June 13, 1912, 10.

51 Ibid.

52 Ibid.

53 John Henniker Heaton was not the only major advocate for cable reform to die close to the initial eruption of the Guns of August. Edward Sassoon preceded him in 1912, and Sandford Fleming died in 1915.

54 Britain, *Royal Commission*, 1914, 48. We are indebted to discussions with Professor

Peter Putnis of the University of Canberra, Australia, for the point regarding Indian diasporic communities and long-distance communication in the Pacific region. He notes that this was already reflected in cheaper postal rates between India, Australia, and, for example, Fiji. These points are now, in fact, being taken up by the historical study of diasporic communities and the formation of Chinese, Indian, and other business networks as part of the social infrastructure of globalization.

55   Sklar, *The Corporate Restructuring of American Capitalism*, 205, nn. 35, 299.

56   PRO, *Treasury Minutes*, 1916.

57   Headrick, *The Invisible Weapon*, 169.

58   U.S., *Cable-Landing Licenses*, 131; Britain, *Correspondence*, 1921.

59   See U.S., *International Communications*, 1923, 8.

60   Such individuals consisted of, among others, people with long and intimate connections to the U.S. "foreign policy establishment," including John Hay, Elihu Root Sr., Elihu Root Jr., former President William Taft, and Frank Kellogg. See especially Knock, "Wilsonian Concepts and International Realities at the End of War," 111–29. Also R. Johnson, *The Peace Progressives and American Foreign Relations*, chap. 3.

61   Britton and Ahvenainen, "Showdown in South America," 20.

62   Ahvenainen, *The European Cable Companies in South America*, 190; Berthold, *History of the Telephone and Telegraph in Uruguay*, 3, 18.

63   Central and South American Telegraph Company, *Via Galveston*, 9; Britain, *Inter-Dept. Comm. Report*, 1902, 36–39; Ahvenainen, *The European Cable Companies in South America*, 385.

64   FRUS, Trail to Baron de Cotegipe, February 28, 1887. FRUS, Bayard to Jarvis, September 5, 1887. Also see Ahvenainen, *The European Cable Companies in South America*, 292–99.

65   Pletcher, *The Diplomacy of Trade and Investment*, 387–95; Feis, *Europe*, 23, 56; Marichal, *A Century of Debt Crisis in Latin America*, 183; Britton and Ahvenainen, "Showdown in South America," 4–6, 23–25.

66   Marichal, *A Century of Debt Crisis in Latin America*, 156–59; CFB, *Annual General Report*, 1891,75. Also see CFB, *Annual General Reports* for the next two years and compare them with those of the last five years of the 1880s; Pletcher, *The Diplomacy of Trade and Investment*, chaps. 11 and 12.

67   Quoted in Ninkovich, *The United States and Imperialism*, 14.

68   Desbordes, "Western Empires and News Flows in the 19th Century," 3–6; Gallagher and Robinson, "The Imperialism of Free Trade," 1–15; Marichal, *A Century of Debt Crisis in Latin America*, 147–57.

69   Desbordes, "Western Empires and News Flows in the 19th Century," 10.

70   Read, *Reuters*, 86.

71   Desbordes, "Western Empires and News Flows in the 19th Century," 7.

72   Ahvenainen, *The European Cable Companies in South America*, 288.

73   Quoted in Triana, *The Pan-American Financial Conference of 1915*, 6.

74   U.S., Noyes, Associated Press, 5.

75   Associated Press was established in 1900 as a news cooperative, the United Press by E. W. Scripps in 1907, and the smaller International News Service by William Randolph

Hearst in 1909. These dates reflect their modern versions, while predecessors of the first two went back into the latter half of the nineteenth century, with the immediate predecessor to Associated Press of New York being Associated Press of Illinois, formed in 1893 and signing its first agreements with the other "global news agencies" in that year. Associated Press, *M. E. S. in Memoriam*, 1921.

76   U.S., Noyes, Associated Press, 4; Read, *Reuters*, 86.

77   Central and South American Telegraph Company, *Via Galveston*, 1–5.

78   Pletcher, *The Diplomacy of Trade and Investment*, refers extensively to American newspaper coverage of these events in Latin America and the role of the U.S. government and public opinion therein. He also identifies two good sources for representative samplings of U.S. press coverage along these lines: *Public Opinion* and *Literary Digest*. See, especially, chaps. 6, 11, and 12.

79   John Britton, U.S. foreign policy historian at Francis Marion University in South Carolina, brought these examples to our attention in personal correspondence with the authors by way of informal review of the present manuscript (July 2004), and the following two quotes are from that correspondence. Additional sources covering these events are listed in the following note.

80   See, especially, Rosenberg, *Financial Missionaries to the World*, chap. 5, and Ninkovich, *The United States and Imperialism*, 13–17; but also Pletcher, *The Diplomacy of Trade and Investment*, 315–21, and R. Johnson, *The Peace Progressives and American Foreign Relations*.

81   Britain, *Inter-Dept. Comm. Report*, 1902, appendix G, 73.

82   O'Rourke and Williamson, *Globalization and History*.

83   The original reference to the All-America Company in the quote has been changed to reflect the fact that the company was operating under its old name, the Central and South American Company, during the time referred to in the quote and was not formally renamed until 1920. Quote from U.S., *Cable-Landing Licenses*, 192–93, 202–10.

84   PRO, South American Cable Company, 1891; Ahvenainen, *The European Cable Companies in South America*, 282–83; Headrick and Griset, "Submarine Telegraph Cables," 565.

85   Ahvenainen, *The European Cable Companies in South America*, 379; Britain, *Inter-Dept. Comm., Report*, 36–39; at U.S., *Cable-Landing Licenses*, 202–10.

86   Ahvenainen, *The European Cable Companies in South America*, 282–84.

87   Ibid., 57, 318–21.

88   Much of the following discussion of the German company, unless cited otherwise, draws heavily on Ahvenainen, who has used German, French, and South American sources. With respect to the above proposals, see Ahvenainen, *The European Cable Companies in South America*, 286–87, 316–18.

89   U.S., *Cable-Landing Licenses*, 202–10.

90   Ahvenainen, *The European Cable Companies in South America*, 286–91.

91   Ibid., 316.

92   Ibid., 292–96.

93   Ibid., 301–10.

94 This attitude did not apply to all Latin America governments, and foreign diplomats often saw those of, say, Venezuela, Bolivia, Ecuador, Peru, or the Caribbean and Central America as congenitally incapable of achieving political stability or sound management of their national economies. We will discuss these general claims in the context of specific examples below, but for the time being, we use statements made by American and British officials in *FRUS*, Wilson to the Secretary of State, April 20, 1909, 61. For more general analyses along these lines, see Rosenberg, *Financial Missionaries to the World*, 40–47; Munro, *Intervention and Dollar Diplomacy in the Caribbean*, 1964.

95 Ahvenainen, *The European Cable Companies in South America*, 293–99.

96 Comparative figures based on Britain, *Inter-Dept. Comm. Report*, 1902, 31.

97 Ahvenainen, *The European Cable Companies in South America*, 310–11.

98 Ibid., 311.

99 Ibid., 339–40.

100 U.S., *Memo*, 28–29.

101 WTC, Brazil, box 6.

102 Robertson to chairman Western Telegraph Company, January 26, 1910, ibid.

103 Robertson to Steer Hodson, December 12, 1911, ibid.

104 Ibid.; Ahvenainen, *The European Cable Companies in South America*, chap. 12.

105 Robertson to Steer Hodson, March 10, 1910, WTC, Brazil, box 6.

106 U.S. *Cable-Landing Licenses*, 92; Robertson to Steer Hodson, March 10, 1910, WTC, Brazil, box 6; Ahvenainen, *The European Cable Companies in South America*, 365.

107 PRO, South American Cables, Kelly to Curzon, January 19, 1921; Ahvenainen, *The European Cable Companies in South America*, 323.

108 *FRUS*, Charles Wilson to Secretary of State, April 20, 1909, 61. Robertson to chairman of the Western Telegraph Board in London, January 26, 1910, WTC, Brazil, box 6.

109 U.S., *Memo*, 35–45.

110 Berthold, *History of the Telephone and Telegraph in Uruguay*, 16.

111 WTC, Brazil, box 6, 415.

112 *FRUS*, U.S. Chargé d'Affaires in Brazil to the Secretary of State, March 20, 1917, 45.

113 Ahvenainen, *The European Cable Companies in South America*, 311–13.

114 Ibid., 371–73; Berthold, *History of the Telephone and Telegraph in Uruguay*, 18.

115 Ahvenainen, *The European Cable Companies in South America*, 300, 374–75.

116 Yang, "Submarine Cables and the Emerging Japanese Empire."

117 The following analysis relies on CWA, *Japan and China Agreements*, August 23, 1913; and CWA, *Japan and China Agreements, Joint Purse*, August 23, 1913; also see CWA, *Eastern Telegraph, Pacific Agreements*, September 12, 1905, and CWA, Commercial Pacific Cable Company Memo, November 19, 1906. Secondary sources are from Yang, "Submarine Cables."

118 Ninkovich, *Modernity and Power*, 26; Dayer, *Bankers and Diplomats in China*, 69; Feis, *Europe*, 444–45.

119 MacMillan, *Paris 1919*, 329.

120 "Shanghai Cable (1914) and a Communications Policy" in Yang, "Submarine Cables."

121   Hobsbawm, *The Age of Empire, 1875–1914*, 281.

122   "Conclusions" in Yang, "Submarine Cables."

123   Figures from Headrick, *The Invisible Weapon*, 29.

7   WIRELESS, WAR, AND NETWORKS

1   Lewis and Pearlman, *Media and Power*, 35.

2   Dunlap, *Marconi*, 252.

3   "The Phenomenal Growth of Wireless Telegraphy" in *Wireless*, 1909.

4   Headrick, *The Invisible Weapon*, 1991; Aitken, *The Continuous Wave*; Hills, *The Struggle for Control of Global Communications*; Douglas, *Inventing American Broadcasting*; Griset, *Enterprise, Technologie et Souveraineté*.

5   Emphasis added. Britain, *Radiotelegraphic Convention Report*, 1907, vii–viii.

6   Douglas, *Inventing American Broadcasting*, 216.

7   Britain, *Radiotelegraphic Convention Report*, 1907, xxi.

8   Henniker Heaton's evidence to Britain, *Radiotelegraphic Convention Report*, 1907.

9   British Parliamentary Debates, vol. 167, 1st sess., December 18, 1906, 1324.

10   Donaldson, *The Marconi Scandal*, 1982.

11   "Imperial Wireless Chain Litigation," *Wireless World*, June 1918, 113–14.

12   Headrick, *The Invisible Weapon*, notably 125–27.

13   Aitken, *The Continuous Wave*, 356.

14   Chandler, *Scale and Scope*, 69

15   Hannah, "Visible and Invisible Hands in Great Britain," 67.

16   Douglas, *Inventing American Broadcasting*, 80.

17   Streeter, *Selling the Air*, 72–73.

18   Aitken, *The Continuous Wave*, 82.

19   Streeter, *Selling the Air*, 79.

20   Aitken, *The Continuous Wave*, 133–37, 282.

21   Douglas, *Inventing American Broadcasting*, 177–78, 183–85.

22   As quoted ibid., 115.

23   Streeter, *Selling the Air*, 78.

24   Winston, *A History of Communication Technology*, chap. 1.

25   Griset, "L'Etat et les Télécommunications Internationales au Debut de XXème Siècle en France," 181–207.

26   Headrick, *The Invisible Weapon*, 124–27; Bullard, "The Naval Radio Service."

27   Bullard, "The Naval Radio Service," 10, 16.

28   Ibid., 20–22.

29   NARA, Addenda to American Publishers Committee on Cable and Radio Communications, 1920, 1.

30   See Aitken, *The Continuous Wave*, 133–37.

31   Dunlap, *Marconi*, 154.

32   "First Wireless Press Message Across Atlantic," *New York Times*, October 18, 1907.

33   "The New Achievements," *New York Times*, January 29, 1912, 2.

34   "Airman Sends and Receives Wireless," *New York Times*, April 11, 1912, 3.

35 "The Triumph of the Wireless Telegraph," *New York Times*, May 6, 1912, 8.

36 Marconi Wireless Company Ltd., *Report of Directors*, July 24, 1919.

37 Aitken, *The Continuous Wave*, 134.

38 Our translation. Griset, *Enterprise, Technologie et Souveraineté*, 203.

39 Hills, *The Struggle for Control of Global Communications*, 164–65.

40 The main source for this information is U.S., *Memo*, notably 35–43.

41 Britain, *Royal Commission*, 1910, 44.

42 Britain, *Minutes*, 1910, 123–24.

43 See Fessenden's evidence to Britain, *Minutes*, 1910, 16.

44 Britain, *Royal Commission*, 1910, 45.

45 Ahvenainen, *The History of Caribbean Telegraphs before the First World War*, 183.

46 U.S. *Memo*, 1920, 32–45.

47 Bright, "Cable versus Wireless Telegraphy," notably 1077, 1085–86.

48 Griset, "L'Etat et les Télécommunications Internationales au Debut de XXème Siècle en France," 203.

49 Aitken, *The Continuous Wave*, 259.

50 The Tuckerton station was not handed over by the Germans to the French firm for which it had been built. However, it and the Sayville stations were taken over, by the U.S. Navy, in 1915, albeit communications with Germany naturally continued. See later in this chapter.

51 See Griset, *Enterprise, Technologie et Souveraineté*, 284.

52 Hills, *The Struggle for Control of Global Communications*, 167–68.

53 Cooper, *Barriers Down*, 5–76.

54 See Hills, *The Struggle for Control of Global Communications*, 168.

55 Griset, *Enterprise, Technologie et Souveraineté*, 283.

56 Ibid., notably 291–99.

57 Britain, *Wireless Report*, 1924, 42.

58 As cited ibid., 42–43; Griset, *Enterprise, Technologie et Souveraineté*, 322–25. The latter gives an account of the complex arrangements leading to the creation of TSF which need not concern us here.

59 U.S., *Cable-Landing Licenses*, 45–51, 106–7, 175.

60 Cooper, *Barriers Down*, 75. The following paragraph is based mainly on U.S., *Cable-Landing Licenses*, particularly the testimony of the managing directors of Associated Press and United Press as well as Walter Rogers.

61 Emphasis added. Desbordes, "Western Empires and News Flows in the 19th Century," 8–9.

62 U.S., *Cable-Landing Licenses*, 51.

63 Cooper, *Barriers Down*, 116, and Rogers at U.S., *Cable-Landing Licenses*, 50–51.

64 U.S., *Cable-Landing Licenses*, 50, 174–77, 282–84.

65 Bright, "Telegraphs in War-Time," 876.

66 Ibid., 875.

67 R. Jones, *A Life at Reuters*, 160–61.

68 Douglas, *Inventing American Broadcasting*, 275.

69 Read, *Reuters*, 167.

70   PRO, Treasury Secretary (1917–22), *Reuters Ltd., Agreement,* and important corre-
     spondence therein: Roderick M. Jones correspondence with Alexander Lawrence,
     Treasury Solicitor's Office, June 4, 1919; Lawrence correspondence with S. Geselee,
     Foreign Office, April 20, 1919; Lawrence correspondence with C. H. Montgomery,
     Foreign Office, re New Public Policy Letter with Reuters, October 31, 1918; Mark F.
     Nathier, Letter to the Directors of the Reuters Telegraph Company, Ltd., November
     21, 1916; Robert Cecil, Foreign Office, correspondence with Roderick Jones and Mark
     Napier outlining arrangements with Reuters, December 8, 1916.

71   Ibid.

72   PRO, Cecil to Jones and Napier, December 8, 1916.

73   R. Jones, *A Life at Reuters,* 212; Read, *Reuters,* 158.

74   PRO, Lord Burnham Statement to the Newspaper Proprietors Association, Decem-
     ber 1916.

75   See PRO, Clements Memorandum, August 16, 1920.

76   MacMillan, *Paris 1919,* 179.

77   Quoted in Douglas, *Inventing American Broadcasting,* 279.

78   Ibid.

79   U.S., *Cable-Landing Licenses,* 179.

80   Ibid., 280.

81   Ibid., 177–78.

82   "Government Control of Radio Communication," attachment to Marconi Wireless
     Company of America, *Annual Report, 1918.* Evidence of Edward McNally before
     the House of Representatives Committee on Merchant Marine and Fisheries, De-
     cember 17, 1918, 11.

83   See Donaldson, *Marconi,* 247–48.

84   Douglas, *Inventing American Broadcasting,* 278.

85   Ibid., 279; Sarnoff, *Looking Ahead,* 11–13.

86   U.S., *Navy in Peace,* 9.

87   Sobel, RCA, 25.

88   Marconi Wireless Company of America, *Annual Report, 1918.*

89   Aitken, *The Continuous Wave,* 330–34, 370–71.

90   U.S., *Report,* 1923, 16.

91   Griset, *Enterprise, Technologie et Souveraineté,* 308–9; U.S., H.R., *Radio Stations,* July
     1919.

92   Streeter, *Selling the Air,* 69.

93   Quotation in Bilby, *The General,* 45.

94   Streeter, *Selling the Air,* 82–83.

95   U.S., *Report,* 1923, 22.

96   Ibid., 34–35.

97   Streeter, *Selling the Air,* 83.

98   Hills, *The Struggle for Control of Global Communications,* 192; Aitken, *The Continu-
     ous Wave,* 356–57.

99   Aitken, *The Continuous Wave,* 428.

8   THICK AND THIN GLOBALISM

1   Australia, Report of the Select Committee on Press Cable Services; Rantanen, "The Struggle for Control of Domestic News Markets," 37–41.

2   Cassidy, *dot.con*, 70.

3   Rosenberg, *Financial Missionaries to the World*, 122–50; R. Johnson, *The Peace Progressives and American Foreign Relations*, 3–5.

4   Ninkovich, *Modernity and Power*, 29–42; Ambrosius, *Wilsonianism*, 24.

5   Dewey, "Imperialism Is Easy." Dewey was vice president of the Anti-Imperial League from 1910 to 1920.

6   Headrick, *The Invisible Weapon*, 174.

7   Ninkovich, *Modernity and Power*, iv–v, 41–62; Ambrosius, *Wilsonianism*, 27–35.

8   The archives of the United States, Canada, and Britain that we rely on most also contain extensive records of the views of other nations that were involved in negotiating issues of global communication policy.

9   Ackerman, *Report of the Graduate School of Journalism*, 8.

10   NARA, Rogers, *An American International and Inter-Colonial Communications Program*, 1918, 3.

11   Ibid., 22.

12   NARA, Power, Memo to Lansing, January 15, 1919, 9.

13   NARA, Rogers, Transpacific Communication, July 30, 1921; NARA, Power, Memo to Lansing, January 15, 1919, 6.

14   NARA, Rogers, Outline of Communications Program, September 11, 1918, 6; NARA, Power, Memo to Lansing, January 15, 1919, 36–39; NARA, Brown to Rogers, January 20, 1920.

15   PRO, Treasury, *Memoranda of the Imperial Communications Committee re the Washington Conference*, 1920.

16   NARA, Mackay to Colby, August 19, 1920, 5–6.

17   Headrick and Griset, "Submarine Telegraph Cables," 571.

18   NARA, Mackay to Colby, August 19, 1920; PRO, South American Cables, Nye to Chief Censor, March 28, 1918.

19   NARA, Mackay to Colby, August 19, 1920.

20   NARA, To Lansing, from unknown, September 14, 1920; NARA, Colby to American Commissioners, Berlin, November 18, 1920; Coggeshall, *An Annotated History of Submarine Cables and Overseas Radiotelegraphs*, 85–90.

21   Mackay Companies, *Annual Report for 1918*, 2–4, *Annual Report for 1919*, 7; "Burleson Attacked in Highest Court," *New York Times*, February 21, 1919, 20; "Cable Company Asks Court to Annul Seizure," *New York Times*, December 5, 1918, 1.

22   NARA, Cable-Using Newspapers and Press Associations of America Memo, October 6, 1920, 3–5; NARA, Addenda to American Publishers Committee on Cable and Radio Communications, 1920; NARA, National Foreign Trade Council, November 26, 1919; NARA, American Manufacturers Export Association, 1920; NARA, Merchants Association of New York, January 8, 1920.

23 For the disclosure, see U.S., *Cable-Landing Licenses*, 1921, 269–70; NARA, Rogers, Transpacific Communication, July 30, 1921, 19.

24 NARA, Rogers, Memo on Pacific Communication, October 9, 1918, 2; NARA, Rogers, Transpacific Communication, July 30, 1921, 19; NARA, Rogers, *An American International and Inter-Colonial Communications Program*, 1918, 1; NARA, Power, Memo to Lansing, January 15, 1919, 24–28.

25 U.S., FTC, *Report on the Radio Industry*, 1923.

26 NARA, Rogers, Transpacific Communication, July 30, 1921, 20.

27 NARA, Rogers, Memo on International Communications Conference, June 8, 1921, 20, 28. The Universal Electrical Communications Union was an organization proposed to unify the International Telegraph Union and the International Radiotelegraph Union (formed 1903). While this bid was premature, it did mobilize efforts which led to the amalgamation of these areas under the International Telecommunications Union in 1927.

28 Reproduced in NARA, Rogers, Memo on International Communications Conference, June 8, 1921, 29–30.

29 NARA, Rogers, Subcommittee of the Interdepartmental Committee on Communications, July 30, 1921; NARA, Rogers, Transpacific Communication, July 30, 1921, 4.

30 U.S., Conference on International Communication, 1921, 2.

31 NARA, Alexander to Long, 1920, 1–3; NARA, State Department Conference with Prince de Bearn, July 16, 1920, 3.

32 The analysis in this and the following paragraphs draws heavily on Ninkovich, *Modernity and Power*, 41–62.

33 Emphasis added. NARA, Power, Memo to Lansing, January 15, 1919, 1.

34 The Council of Ten consisted of the leaders, each with their foreign secretary, of the United States (Woodrow Wilson), Britain (David Lloyd George), France (Georges Clemenceau), and Italy (Vittorio Orlando), with two more representatives from Japan quickly added at Britain's insistence. The group excluded smaller countries and neutrals and in March 1919 was reduced to the Council of Four after the Japanese representatives and the foreign ministers were excluded from subsequent meetings conducted between Wilson, Lloyd George, Clemenceau, and Orlando. MacMillan, *Paris 1919*, 53.

35 NARA, British Memo to Lansing, August 18, 1919; PRO, Imperial Communications Committee, July 5, 1919; PRO, Committee of Imperial Defense, 1920.

36 Japan probably seemed to think that the "international control"–"spheres of influence" conundrum had been taken care of by the Lansing-Ishii agreements of 1916, which appeared to offer U.S. recognition of Japan's "special interests" in China in return for its acceptance of the United States' own sphere of interest in the Philippines, Guam, and Hawaii. MacMillan, *Paris 1919*, 329.

37 U.S., Conference on International Communication, 1921, 2–4.

38 NARA, To Lodge, from Unknown, October 4, 1919. Opposition in the United States was based more on the kind of globalism that the League or Wilson embodied. Opponents included conservative internationalists as well as anti-imperialists, who saw the Mandate System created under the League, in particular, as a new form of

empire that was at odds with American democratic traditions. Then, of course, there were the isolationists. Thus, it was the combination of conservative international-ism, anti-imperialism, *and* isolationism that opposed Wilson and the League. See R. Johnson, *The Peace Progressives and American Foreign Relations*, 75–100, 200–12.

39   NARA, Cable-Using Newspapers and Press Associations of America Memo, October 6, 1920; NARA, Addenda to American Publishers Committee on Cable and Radio Communications, 1920; NARA, National Foreign Trade Council, November 26, 1919; NARA, American Manufacturers Export Association, 1920; NARA, Merchants Asso-ciation of New York, January 8, 1920.

40   NARA, Cable-Using Newspapers and Press Associations of America Memo, October 6, 1920.

41   NARA, Long, Memo to Lansing, November 25, 1918.

42   NARA, Power, Memo to Lansing, January 15, 1919, 30.

43   Ibid., 13–14.

44   PRO, International Conference Memo, May 27, 1921.

45   Ibid., 1.

46   Ibid.

47   NARA, British Memo to Lansing, August 18, 1919; PRO, Imperial Communications Committee, July 5, 1919; PRO, Committee of Imperial Defense, 1920; NARA, State Department Conference with Prince de Bearn, July 16, 1920, 3.

48   NARA, Alexander to Long, 1920, 1–2.

49   Headrick, *The Invisible Weapon*, 213.

9   COMMUNICATION AND INFORMAL EMPIRES

1   Feis, *Europe*, 23, 56.

2   Marichal, *A Century of Debt Crisis in Latin America*, 183.

3   U.S., *Memo*, 1920, 3–4; U.S., *Inter-American Committee on Electrical Communica-tions*, 1924, 7–13, 135–40.

4   Munro, *Intervention and Dollar Diplomacy in the Caribbean*, 530–46; Marichal, *A Century of Debt Crisis in Latin America*, 189.

5   U.S., *Memo*, 8, 41–42.

6   Rogers, in U.S., *Cable-Landing Licenses*, 51.

7   Keylor, "Versailles and International Diplomacy," 481–83; Widenor, "The Construc-tion of the American Interpretation," 550–54.

8   Rogers, in U.S., *Cable-Landing Licenses*, 50–51; Rogers, "Tinted and Tainted News," *Saturday Evening Post*, July 21, 1917.

9   Cooper, *Barriers Down*, 70–77; Hawkins, in U.S., *Cable-Landing Licenses*, 176; Martin, in U.S., *Cable-Landing Licenses*, 282–84; Rogers, in U.S., *Cable-Landing Licenses*, 50.

10   Hawkins, in U.S., *Cable-Landing Licenses*, 177.

11   Ibid., 176–77; Martin, ibid., 283.

12   NARA, Power, Memo to Lansing, January 15, 1919, 17; NARA, Cable-Using Newspapers and Press Associations of America Memo, October 6, 1920, 3–5.

13   The decision is reproduced in U.S., *Cable-Landing Licenses*, 55–56.

14 PRO, South American Cables, Kelly to Curzon, January 19, 1921. U.S., *Cable-Landing Licenses*, 83, 116. The All-America Company had full access to Buenos Aires and the rest of Argentina and Uruguay, thus our estimate that the firm had access to about 75 to 85 percent of the east coast of South America.

15 U.S., *Memo*, 1920, 29; PRO, South American Cables, Brown to Under Secretary of State, September 15, 1921.

16 PRO, South American Cables, Kelly to Curzon, January 19, 1921.

17 Western Union, *Annual Report for 1919*, 1920, 10; Western Union, *Annual Report for 1920*, 1921, 22–23; U.S., *Cable-Landing Licenses*, 98–102.

18 Mackay Companies, *Annual Report for 1917*, 1918, 6.

19 U.S., *Cable-Landing Licenses*, 102.

20 PRO, South American Cables, Tower to Balfour, February 15, 1919; PRO, South American Cables, Denison-Pender to Foreign Office, November 7, 1918.

21 PRO, South American Cables, Sperling to Foreign Office, July 18, 1921.

22 PRO, South American Cables, Paget to Curzon, May 8, 1920.

23 See U.S., *Cable-Landing Licenses*, 59–71, for the correspondence between the three companies with respect to the proposed cartel. Also see Western Union's *Annual Reports* for 1919 and 1920, 10, 18, respectively.

24 U.S., *Cable-Landing Licenses*, 233.

25 Ibid., 140.

26 Ibid., 53.

27 Ibid., 39.

28 Newcomb Carlton, quoted ibid., 107.

29 Quoted ibid., 102.

30 See ibid., 19–21, 109–13. Also see the statement of the Western Union's position in its 1920 *Annual Report*, 19–27.

31 Verbatim from U.S., *Cable-Landing Licenses*, 9; also see 230–33.

32 "Warns of British Control of Cables," *New York Times*, August 16, 1922, 8; "Carlton Answers Rogers on Cables," *New York Times*, August 17, 1922, 17; "Defends Hughes on Cable Policy," *New York Times*, August 18, 1922, 12.

33 PRO, Dossier 1120 Telegraphs and Cables, 1922; PRO, South American Cables, Brazil Extract.

34 U.S., *Regulation of International Services*, 1933, 10.

35 PRO, GPO, Purchase of Cable, 1920–23.

36 Britain, *Wireless Report*, 1924, 39–43; U.S., *Report*, 1923, 58–59.

37 U.S., *Report*, 1923, 61.

38 Emphasis added. Stanford Hooper, Navy, correspondence with Secretary of the Navy Denby, December 22, 1921. Quoted in Aitken, *The Continuous Wave*, 491–92.

39 Acting Secretary of the Navy Roosevelt to the Secretary of State, October 3, 1923. Quoted ibid., 492.

40 Aitken, *The Continuous Wave*, 491–92.

41 U.S., *Report*, 1923, 61.

42 Britain, *Summary of Evidence*, 1928, 3.

43 U.S., *Report*, 1923, 61.

44   Emphasis added. Correspondence to James R. Sheffield, Navy Department, December 16, 1921, ibid., 63–64.

45   Ninkovich, *The United States and Imperialism*, 27, 32.

46   Quoted in Dayer, *Bankers and Diplomats in China*, 27.

47   Ibid., 57–60.

48   Furth, "Intellectual Change," 33.

49   Ibid., 88; Ninkovich, *The United States and Imperialism*, 168–69.

50   Vittinghoff, "Unity vs. Uniformity," 7; PRO, Telegraphs and Cables—China Related, Hurd, July 12, 1921; PRO, Cables to China, Wellesley to Treasury, January 17, 1921.

51   Z. C. Tang, *Shanghai Shi*, 335–37; X. Tang, *Global Space and the Nationalist Discourse of Modernity*, 57–58; Furth, "Intellectual Change," 13; Reinsch, *Intellectual and Political Currents in the Far East*.

52   See Liang, *Yinbingshi Heji-Wenji*, 57. Parenthetically, Kang was not intoxicated by 100-proof utopian ideals, but stated that his vision of *da tong* (the great community) represented a distant potential that could be derived from the ever increasing interdependence visible in his own time, to which he pointed to the telegraph, the Universal Postal Union, and international law as his touchstones. He also did not imagine that the "state" would be superseded, but believed that political power would be organized through a world parliament. See Hsiao, *A Modern China and a New World*, 456–60.

53   Vittinghoff, "Unity vs. Uniformity"; Wagner, "The Role of the Foreign Community in the Chinese Public Sphere."

54   NARA, Rogers, Transpacific Communication, July 30, 1921, 6, 26; PRO, Cables to China, Wellesley to Treasury, January 17, 1921.

55   PRO, Cables to China, Wellesley to Treasury, January 17, 1921, 478.

56   NARA, Balfour to Lloyd George, February 4, 1922; NARA, Rogers, Memo for Secretary, December 8, 1921, 1–3.

57   NARA, Balfour to Lloyd George, February 4, 1922, 3.

58   NARA, Rogers, Memo for Secretary, November 28, 1921, 2; NARA, Rogers to Harrison, June 7, 1922.

59   Cooper, *Barriers Down*, 42, 99, 120–28, 153, 173–99; Read, *Reuters*, 181–82.

60   U.S., *Report*, 1923, 5; also Owen Young to Charles E. Hughes, Secretary of State, January 9, 1922, ibid., 66–67.

61   Eastern Telegraph Company, *Report*, 1923, 4.

62   PRO, Cables to China, Denison-Pender to Wellesley, April 8, 1921, 424; PRO, Cables to China, Wellesley to Treasury, January 17, 1921, 445.

63   Figures from PRO, Cables to China, Loans, 1925; PRO, Cables to China, Wellesley to Treasury, January 17, 1921, 445–46; also see U.S., *Report*, 1923, 60. For the International Western Electric Company deal and its cancellation, see NARA, Rogers to Root, December 22, 1921.

64   PRO, Telegraphs and Cables, Foreign Office Cypher to Balfour, December 22, 1921.

65   PRO, Cables to China, Denison-Pender to Wellesley, April 8, 1921.

66   NARA, Rogers et al., *Report*, 1922, 4.

67   PRO, Telegraphs and Cables, Foreign Office Cypher to Balfour, December 22, 1921, 1.

68   U.S., *Report*, 1923, 120; NARA, Rogers et al. *Report*, 1922, 1.

69   NARA, Balfour to Lloyd George, February 4, 1922, 4–5.

70   PRO, Cables to China, Loans, 1925.

71   Western Union, *Annual Report for 1927*, 1928, 8.

72   AT&T sold its international operations after the Department of Justice began raising objections that it was using its domestic monopoly to cross-subsidize its presence in global markets. U.S., *Preliminary Report on Communication*, 1934, 78–83; U.S., *Report on Communication Companies*, 1935, 3721–25.

10   EURO-AMERICAN COMMUNICATION MARKET AND MERGERS

1   NARA, To Lansing, from unknown, September 14, 1920; NARA, Rogers, Memo on International Communications Conference, June 8, 1921, 16; Mackay Companies, *Annual Report*, 1922, 6.

2   Eastern Telegraph Company, *Report*, 1923, 7.

3   Emphasis added. Western Union, *Annual Report for 1926*, 7.

4   U.S., *Report*, 1923, 34–35.

5   U.S., Conference on International Communication, 1921, 339–49.

6   Rogers testimony in U.S., *Cable-Landing Licenses*, 46.

7   By the mid-1930s, Press Wireless Service operated an extensive network of news distribution, in cooperation with a variety of state-owned and commercial wireless networks in ninety countries. For the evolution of the service, see U.S., *Report*, 1923, 6, 53; U.S., *Report on Communication Companies*, 1935, 152.

8   Britain, *Wireless Report*, 1924, 3.

9   Ibid., 1, 23–27, 39–40.

10   Ibid., 4.

11   Ibid., 13.

12   Ibid., 16–42.

13   Ferguson, *Empire*, 292–356.

14   Britain, *Wireless Report*, 13–18.

15   Ibid., 43.

16   NARA, Rogers, Memo re Owen Young, 1921, 1; Mackay Companies, *Annual Report for 1922*, 5; U.S., *Report on Communication Companies*, 1935, 80.

17   Aitken, *The Continuous Wave*, 481.

18   The rates were roughly one-half, one-third, and one-quarter the regular rate for plain language messages. See Eastern Telegraph Company, *Annual Report*, 1923, 6; Mackay Companies, *Annual Report for 1922*, 14.

19   U.S., *Report on Communication Companies*, 1935, 454.

20   The experimental development of shortwave occurred between 1920 and 1924, and the technology was not commercialized until 1927, although service from Britain to Canada began in November 1926, just as the Pacific Cable Board's new high-speed cable was opening for service. This new innovation was absolutely decisive and laid the foundations for the development of a new generation of wireless services, includ-

ing international shortwave broadcasting, television, facsimile, and so forth. It was also superior to cables for international telephone service and served as the medium for this service, at least across the Atlantic, until it was superseded by a new generation of cables in the mid-1950s. Shortwave beam wireless services were five times as fast as the old long-wave technology (250 words per minute versus 50), and the concentrated signal made it more reliable, harder to intercept, and cheaper to operate. The front-end capital costs of setting up a beam wireless system were also claimed to be about 10 percent of those associated with a similar cable network. However, shortwave, or beam wireless as it was often called, was still slower than cables (with the Western Union's Permalloy transatlantic cables operating at speeds of 1,200 words per minute in 1924 and over 2,000 in 1930); its operating costs were more expensive; it was less secret, reliable, and secure; and it would take a very long time to reach all of the places that the global cable system had managed to reach during the past six decades. Given all of these factors, debates over which of the two media were superior raged for years. One point in this is absolutely key: the answers to the debates lay not in the technology alone, nor even primarily, but in how systems, policies, and social practices that had been built up in the past adjusted to, as well as managed, the introduction of this new technology. The technical discussion of beam wireless is based on Aitken, *The Continuous Wave*, 512; Brown, *The Cable and Wireless Communications of the World*, 90–94.

21   Mackay Companies, *Annual Report for 1925*, 2–3.

22   U.S., *Report on Communication Companies*, 1935, 3826.

23   Mackay Companies, *Annual Report for 1926*, 2.

24   There were questionable accounting practices that cast some doubt on these figures. U.S., *Report on Communication Companies*, 1935, 133.

25   Mackay Companies, *Annual Report for 1922*, 4, *Annual Report for 1923*, 3.

26   Mackay Companies, *Annual Report for 1923*, 9, *Annual Report for 1925*, 6.

27   Mackay Companies, *Annual Report for 1921*, 7–8.

28   Ibid., 8; Western Union, *Annual Report for 1927*, 7; Coggeshall, *An Annotated History of Submarine Cables and Radiotelegraphs*, 85–87.

29   Western Union, *Annual Reports* for 1927–30; Mackay Companies, *Annual Reports* for 1925–27; Coggeshall, *An Annotated History of Submarine Cables and Overseas Radiotelegraphs*, 85–90.

30   Britain, *Imperial Wireless and Cable Conferences*, 1928; Barty-King, *Girdle Round the Earth*, 201; Coggeshall, *An Annotated History of Submarine Cables and Overseas Radiotelegraphs*, 85–90.

31   Mackay Companies, *Annual Report for 1926*, 6–8.

32   Western Union, *Annual Reports* for 1927, 6.

33   The above figures are based on U.S., *Preliminary Report on Communication*, 1934, 76, 133, 143–49; U.S., *Report on Communication Companies*, 1935, 3729, 3754, 3977–80.

34   Quote from Sobel, *I.T.T.*, 41.

35   Based on tally of acquisitions from U.S., *Report on Communication Companies*, 1935, 3758–77.

36   Sobel, *I.T.T.*, 44; U.S., *Preliminary Report on Communication*, 1934, 78–83 and U.S., *Report on Communication Companies*, 1935, 3721–25.

37   "U.S. Bid for the Telegraphs," *Daily Express*, March 31, 1928.

38   Ibid.

39   The system was carrying 10 million words in 1922–23, up from 3 million before war, and continued to grow thereafter to roughly 11.3 million by 1925–26. See Britain, *Imperial Economic Conference*, 1924; Brown, *The Cable and Wireless Communications of the World*, 137. Trends after that are discussed below.

40   Britain, *Imperial Economic Conference*, 1924, 369.

41   Boyce, "Canada and the Pacific Cable Controversy, 1923–28."

42   Ibid., 84–85.

43   NAC, Headlam, Circular no. 89, November 8, 1927.

44   Brown, *The Cable and Wireless Communications of the World*, 137.

45   The Eastern Telegraph Company further muddled the issue by calculating its losses against what it *would* have made had it not been forced to adopt cheaper rates, although this strategy was often resisted by British policy makers. However, it does show just how far the company was willing to go to create the impression of a crisis.

46   Britain, *Imperial Wireless and Cable Conferences*, 1928, paras. 10–20.

47   NAC, Murphy to Skelton, July 7, 1927.

48   These were the five policy options considered. Britain, *Imperial Wireless and Cable Conferences*, 1928, paras. 25–30.

49   Britain, *Summary of Evidence*, 1928, fifth meeting January 31, 1928, 18.

50   The following section, unless otherwise noted, is based on Britain, Imperial Wireless and Cable Conference, *Cable and Wireless Agreements*, 1929. We will refer to the company as "Cable and Wireless" throughout because over time that was how the new company came to be known after the title was formally adopted in 1936.

51   Britain, *Imperial Wireless and Cable Conferences*, 1928, para. 35.

52   CWA, Imperial and International Communications Ltd., *First Report*, 1929, 6.

53   Britain, *Imperial Wireless and Cable Conferences*, 1928, schedules 2 and 3, clauses 3 and 4, and para. 4.

54   Ibid., schedule 3, clauses 1 and 2.

55   "Post Office is Neglecting Empire Beam Wireless. Favouritism towards the U.S.," *Daily Mail*, January 1, 1930.

56   Britain, *Imperial Wireless and Cable Conferences*, 1928, para. 17, point H.

57   Ibid., paras. 3, 9, 11, and 15.

58   Ibid., paras. 18, 19, 20–23, and schedule 3, clauses 6(2), 9, and 10.

59   "Cable Stock Weak," *Daily Mail*, January 31, 1930.

60   Hills, *The Struggle for Control of Global Communications*, 231.

61   Headrick, *The Invisible Weapon*, 212.

62   Ibid., 208; Western Union, *Annual Report for 1927*, 7–8.

63   "International T and T Acquired RCA Communications, Inc.," *New York Times*, March 29, 1928; U.S., *Hearing on S. 6*, 1929.

64   U.S., *Preliminary Report on Communication*, 1934, xxiv–xxv.

65   Ibid., xxix.

66 Ibid.; U.S., *International Communications*, 1945, 117–18.

67 Brown, *The Cable and Wireless Communications of the World*, 97–100. For an excellent treatment of technological determinism, see Winston, *A History of Communication Technology*.

68 Headrick, *The Invisible Weapon*, 209.

69 The point is lamented and ably made by the conservative historian and sympathetic observer of British imperialism, Ferguson, in *Empire*, chap. 6.

70 Williams, *The Sociology of Culture*.

71 Boehmer, *Empire, the National and the Post Colonial*, 17.

72 As cited in chap. 7, n. 70, this volume. For an apologetic view of Reuters's views, see those of its in-house historian, Read, *Reuters*, 158–68.

73 Feis, *Europe*, 157–58.

74 Lippmann and Merz, "A Test of the News."

75 League of Nations, *The League of Nations and the Press*, 1928, 46–47; NAC, *Press and International Relations Conference*.

76 ITT kept its foreign telegraph, wireless, and telephone systems. The 20 percent figure refers to its long-distance networks.

77 Quoting 1943 State Department memo in U.S., *International Communications*, 1945, 231.

78 Ibid., 223.

79 Ibid., 10.

80 Ibid., 12.

CONCLUSIONS

1   Jones, *Multinationals and Global Capitalism*, 167.

2   U.S., *Preliminary Report on Communication*, 1934, 4; U.S., *Report on Communication Companies*, 1935, 3826–33.

3   Polanyi, *The Great Transformation*, 233–39.

4   Jones, *Multinationals and Global Capitalism*, 20–21, 31.

5   Ibid., 282.

6   Hogan, *Informal Entente*.

7   Denny, *America Conquers Britain*.

8   Harvey, *The New Imperialism*.

9   Ninkovich, *The United States and Imperialism*, 5.

10 Colley, *Captives*, notably the epilogue.

11 Barstow and Stein, "Under Bush, a New Age of Prepackaged TV News," *New York Times*, March 13, 2005.

12 References to these points can be found in the Federal Communications Commission documents, Cable-Landing Licenses, August 13, 1999, DA 99-1636, file SCL-98-005, and Proceedings of the Undersea Cable Public Forum before the Federal Communications Commission, Washington, D.C., November 8, 1999, http://www.fcc.gov.

Bibliography

PRIMARY SOURCE MATERIALS

**British Reports, Minutes of Evidence, Conferences,
Correspondence, and Other Materials**

Anderson to Andrews, October 8, 1870.

Copy of Correspondence between the Secretary of State for India. Letter from James Anderson, Managing Director of British to India Submarine Telegraph Co., to Under Secretary of State and W. Andrews, Managing Secretary of Indo-European Telegraph Company (Ltd.), October 8, 1870. British Telecommunications Archives (hereafter BT), post 82/91.

Anderson to Champlain, December 4, 1879.

Letter from James Anderson, Managing Director of Eastern Telegraph Company (hereafter ETC) to Major Champlain, Indo-European Telegraph Department, December 4, 1879. British Library: L/PWD/7/182.

Anderson to Parry, January 24, 1879.

Letter from James Anderson to C. H. B. Parry, General Post Office, January 24, 1879. British Library: L/PWD/7/182.

Britain, British-Persian Telegraph Convention, renewal 1866.

Convention between the British and Persian Governments for Prolonging Conventions between H. M. and Shah of Persia relative to Telegraphic Communication between Europe and India. British Parliamentary Papers, 1866 (3687) LXXVI 521 mf 72.594.

Britain, British-Persian Telegraph Convention, renewal 1873.

Convention between the British and Persian Governments for Prolonging Conventions between H. M. and Shah of Persia relative to Telegraphic Communication between Europe and India. British Parliamentary Papers, 1872 1873 (C.7967) LXXV 685 mf 79.682.

Britain, British-Persian Telegraph Convention, renewal 1892.

Agreement between the British and Persian Governments for Prolonging Conventions between H. M. and Shah of Persia relative to Telegraphic Communication between Europe and India, Tehran. British Parliamentary Papers, January 1892 XCV.727, articles X–XII.

Britain, *Colonial Conference*, 1894.

Colonial Conference (cmnd 7753). London, August 1894.

Britain, *Correspondence*, 1921.

Correspondence Respecting Alleged Delay by British Authorities of Telegrams to and from the United States (cmnd 1230). London: HMSO, 1921.

Britain, *Evidence*, 1902.

Lord Balfour of Burleigh. *Minutes of Evidence Taken before the Inter-Departmental Committee Appointed to Inquire into the Telegraphic System of Communication between Different Parts of the Empire.* London: HMSO, 1902.

Britain, *Hansard,* July 23, 1900.

Britain, House of Commons, *Parliamentary Debates—Hansard,* July 23, 1900.

Britain, *Imperial Economic Conference,* 1924.

Imperial Economic Conference of Great Britain, the Dominions, India and the Colonies and Protectorates (cmnd 2009). London: HMSO, 1924.

Britain, Imperial Wireless and Cable Conference, *Cable and Wireless Agreements,* 1929.

Imperial Wireless and Cable Conference, *Cable and Wireless Agreements,* 1929. PRO BT 33/2500.

Britain, *Imperial Wireless and Cable Conferences,* 1928.

Imperial Wireless and Cable Conferences—Proceedings (1). Empire Press Union, Memorandum, January 20, 1928. BT Post 33/2500.

Britain, *Indo-European Telegraph.*

India Office, Public Works Department Records, *Indo-European Telegraph,* 1839–1931 (British Library).

Britain, *Inter-Dept. Comm. 1st Report,* 1902.

Lord Balfour of Burleigh. *Preliminary Report of the Inter-Departmental Committee on Cable Communications.* London: HMSO, 1902.

Britain, *Inter-Dept. Comm. Report,* 1902.

Lord Balfour of Burleigh. *Second Report of the Inter-Departmental Committee on Cable Communications.* Cd 1056. London: HMSO, 1902.

Britain, *Memorial by Bankers and Merchants of City of London.*

Memorial by Bankers and Merchants of City of London Recommending Laying of Submarine Telegraph between Suez and India by Red Sea. Proposal by Submarine Telegraph Construction and Maintenance Company. British Parliamentary Papers, 1867–68, L.227 mf 74.595.

Britain, *Memorial Calcutta and Bombay Telegraph Communications.*

Memorial to Governor General from Commercial Communities of Calcutta and Bombay, on Telegraph Communications. British Parliamentary Papers, 1867–68 (269) L22 mf 74.595.

Britain, *Minutes,* 1910.

Minutes of Evidence taken in London before the Royal Commission on Trade Relations between Canada and the West Indies. Part 4, appendix 19, 1910.

Britain, Pacific Cable Act, 1927.

Pacific Cable Act, 1927, *Reports and Accounts of the Pacific Cable Board for the Year Ended 31 March, 1928.* HMSO: London, 1928.

Britain, Pacific Cable Committee, 1899.

Government Post Office, Pacific Cable Committee. *Report, Minutes and Proceedings, etc.* London: Darlington and Sons, 1899.

Britain, *Radiotelegraphic Convention Report,* 1907.

Report of the Select Committee on Radiotelegraphic Convention together with the Proceedings of the Committee, Minutes of Evidence and Appendix. London: HMSO, 1907.

Britain, *Report of Joint Committee of Inquiry.*

Report of Joint Committee of Inquiry into the Construction of Submarine Telegraph Cables. House of Commons General Papers, 1861.

Britain, *Royal Commission*, 1910.

Report of the Royal Commission on Trade Relations between Canada and the West Indies (Cmd 5369). London: HMSO, 1910.

Britain, *Royal Commission*, 1912.

Dominions Royal Commission, Minutes of Evidence taken in London, October and November, 1912, Part II. December 1912 (Cmd 5369). London: HMSO, 1912.

Britain, *Royal Commission*, 1914.

Dominions Royal Commission, Minutes of Evidence taken in London in November 1913, etc. London: HMSO, 1914.

Britain, *Royal Commission*, 1917.

Dominions Royal Commission, Fifth Interim Report (Canada), 1917.

Britain, *Royal Commission*, 1918.

Dominions Royal Commission, Final Report. London: HMSO, 1918.

Britain, *Summary of Evidence*, 1928.

Imperial Wireless and Cable Conference, *Report: The Summary of Evidence of the Imperial Wireless and Cable Conference* (cmnd 3163). London: HMSO, 1928. Also PRO BT 33/2500.

Britain, The Telegraph Purchase Act, 1868.

The Telegraph Purchase Act, 1868, *Statutes of the United Kingdom*, 31–32 Vict. chap. 60. London: George E. Eyre and William Spottiswoode.

Britain, The Telegraph Act, 1869.

The Telegraph Act, *Public General Statutes*, 32–33 Vict. ch. 73, London: George E. Eyre and William Spottiswoode.

Britain, *Telegraph between India and Ottoman Territory*.

Telegraphic Communication between India and Ottoman Territory, Constantinople. British Parliamentary Papers, September 1864 1865 (3431) LVII 487 mf 17.442.

Britain, *Telegraph Communication (East India)*.

Telegraph Communication (East India), Correspondence and Papers Relative to Establishment of Telegraphic Communication between India, Singapore, China and Australia, since February 21, 1865. Presented to House of Commons, May 3, 1868. British Parliamentary Papers, 1867–68 (269) L.183 mf 74.594.

Britain, *Telegraph Line—Persia*.

Correspondence Respecting the Construction of Telegraph Line through Persia. British Parliamentary Papers, 1864, LXVI.207.

Britain, *Wireless Report*, 1924.

Report of the Imperial Wireless Telegraph Committee (cmnd 2060). London: HMSO, 1924.

Hay to Duke of Argyll, February 22, 1871.

Copy of Correspondence between the Secretary of State for India. Letter from Lord William Montagu Hay to his Grace and Duke of Argyll, K.T., Her Majesty's Secretary of States for India in Council, February 22, 1871. BT, post 82/91.

HOC, August 14, 1860.

Telegraph Communication with India—Observations. Britain, House of Commons, August 14, 1860, Vol. 160, 5th session.

Melvill to Anderson, March 31, 1871.

Copy of Correspondence between the Secretary of State for India. Letter from the Assistant Under Secretary of State for India, J. C. Melvill, to Sir James Anderson, Managing Director of the Eastern Extension Telegraph Company, March 31, 1871. BT, post 82/91.

Sprye, January 15, 1866.

Copy of a Letter from Captain Richard Sprye to the Secretary of State for India regarding Extension of the Indo-European Telegraph from Pegu to Hong Kong and Chinese Open Ports, January 15, 1866. Printed by the House of Commons, June 14, 1866. BT, post 83/43.

"Telegraphic Communication with the Far East—Agreement with China and Japan 1907," July 29, 1908.

Untitled article in *Commercial Intelligence*, July 29, 1908 (published). BT, post 30/1601, file 17.

Cable and Wireless Archives (Porthcurno, UK): Agreements, Correspondence, and Other Documents

CWA, Anglo-American Company contracts, 1880.

Agreement between the Anglo-American Telegraph Co, the Direct U.S. Cable Co., and Compagnie Française du Télégraphe de Paris à New York, September 24, 1880. In L. B. Carlslake, 1904, *Anglo-American Telegraph Company, Ltd, Contracts, etc.* Not indexed.

CWA, Anglo-American Company contracts, 1882a.

Agreement between the Anglo-American Telegraph Co, the Direct U.S. Cable Co., Compagnie Française du Télégraphe de Paris à New York (the European Companies) and the American Telegraph and Cable Company, May 12, 1882. In L. B. Carlslake, 1904, *Anglo-American Telegraph Company, Ltd, Contracts, etc.* Not indexed.

CWA, Anglo-American Company contracts, 1882b.

Agreement between the Anglo-American Telegraph Co, the Direct U.S. Cable Co., Compagnie Française du Télégraphe de Paris à New York and the American Telegraph and Cable Company, May 12, 1882.

CWA, *China Agreements*, May 13, 1870.

China Agreements. Agreement between the China Telegraph Company and the Great Northern Telegraph Company, May 13, 1870. DOC/EEACTC/1/68.

CWA, *China Agreements*, February 11, 1873.

China Agreements. Agreement between the Eastern Extension and Australasia and China Telegraph Company and the Great Northern Telegraph Company, February 11, 1873. DOC/EEACTC/1/68.

CWA, *China Agreements*, September 10, 1875.

China Agreements. Agreement between the China Submarine Telegraph Company and the Great Northern Telegraph Company (Appendix), September 10, 1875. DOC/EEACTC/1/68.

CWA, *China Agreements*, February 11, 1882.

China Agreements. Agreement between the Great Northern Telegraph Company, Eastern Extension, Australasia and China Telegraph Company, Ltd., and the Oriental Telephone Company and the China and Japan Telephone Company, February 11, 1882. DOC/EEACTC/1/68.

CWA, *China Agreements*, January 12, 1883.

> *China Agreements.* Joint instructions for the stations in Hong Kong and China of the Eastern Extension, Australasia and China Telegraph Company, Ltd., and the Great Northern Telegraph Company, January 12, 1883. DOC/EEACTC/1/68.

CWA, *China Agreements*, May 7, 1883.

> *China Agreements.* Regulation of Telegraphic Communication between Hongkong, drawn up by the Central Agency of the Chinese Telegraph Administration and the British Eastern Extension, Australasia and China Telegraph Company, Ltd., May 7, 1883. DOC/EEACTC/1/68.

CWA, *China Agreements*, January 1, 1884.

> *China Agreements.* Agreement between the Eastern Telegraph Company, Ltd., and the Eastern Extension, Australasia and China Telegraph Company, Ltd., re. the relationship between Indo-European joint purse and Asian joint purse, January 1, 1884. DOC/EEACTC/1/68.

CWA, *China Agreements*, February 27, 1884.

> *China Agreements.* Agreement between the Great Northern Telegraph Company, Eastern Extension, Australasia and China Telegraph Company, Ltd., and the Oriental Telephone Company and the China and Japan Telephone Company, February 27, 1884. DOC/EEACTC/1/68.

CWA, *China Agreements*, October 29, 1886.

> *China Agreements.* Joint instructions relating to the "Heads of Agreement" arranged at Copenhagen in August 1886. October 29, 1886. DOC/EEACTC/1/68.

CWA, *China Agreements*, August 10, 1887.

> *China Agreements.* Agreement between the Imperial Chinese Telegraph Administration, of the one part, and the Great Northern Telegraph Company and the Eastern Extension, Australasia and China Telegraph Company, Ltd., of the other part, August 10, 1887. DOC/EEACTC/1/68.

CWA, *China Agreements*, July 11, 1896.

> *China Agreements*, Chinese Telegraph Convention, July 11, 1896. DOC/EEACTC/1/68.

CWA, *Chinese Concessions to Eastern Extension Cos.*

> *Chinese Concessions to Eastern Extension Cos.* Correspondence between Prince Gong, Minister of Zhongli Yamen, and foreign diplomats. DOC/EEACTC/1/68.

CWA, Commercial Pacific Cable Company Memo, November 19, 1906.

> Commercial Pacific Cable Company Confidential memo regarding Supplemental Agreements made with Japan, November 19, 1906. DOC/CPCC/12/4.

CWA, Eastern Telegraph Company, 1899.

> Report of a Deputation from the Eastern Telegraph Company, Limited, and the Eastern Extension, Australasia and China Telegraph Company to the Chancellor of the Exchequer and the Secretary of State for the Colonies on the All British Pacific Cable, June 29, 1899. DOC/ETC/1/260.

CWA, *Eastern Telegraph, Pacific Agreements*, July 26, 1904.

> *Eastern Telegraph, Pacific Agreements, etc.* Agreements between Eastern Telegraph Company and various others, July 26, 1904. DOC/CW/12/4.

CWA, *Eastern Telegraph, Pacific Agreements*, September 12, 1905.

> *Eastern Telegraph, Pacific Agreements, etc.* Mackay Agreement with Japanese, September 12, 1905. DOC/CW/12/4.

CWA, Imperial and International Communications Ltd., *First Report*, 1929.

> Imperial and International Communications Ltd., *First Report*, April 1928 to December 1929. DOC/1/1 and I.C/1/9/2.

CWA, *Japan and China Agreements*, August 23, 1913.

> *Japan and China Agreements,* Concession, dated August 23, 1913, granted to the Great Northern Telegraph Company, of Denmark by the Imperial Japanese Government. DOC/EEACTC/1/751.

CWA, *Japan and China Agreements, Joint Purse*, August 23, 1913.

> *Japan and China Agreements*, Japan-China Traffic Joint Purse, August 23, 1913. DOC/EEACTC/1/75/6.

CWA, *Japan and China Agreements*, December 22, 1913.

> *Japan and China Agreements*, Chinese Telegraph Convention (ratified by the Chinese, Danish, and British governments, December 31, 1913), December 22, 1913. DOC/EEACTC/1/75/1.

CWA, Memo on Commercial Pacific Company, 1934.

> Memorandum on Commercial Pacific Company, 1934. DOC/CU/7/15.

CWA, Memo re American Pacific Combination, July 1, 1905.

> Memorandum respecting American Pacific Combination, July 1, 1905. Records Related to the Commercial Cable Company, box 1. DOC/CW/7/1.

CWA, *Pacific Agreements*, September 25, 1905.

> *Pacific Agreements, etc.* Book of compiled agreements, September 25, 1905. DOC/CW/12/4.

CWA, *Report of the Twenty-Seventh Ordinary General Meeting*, April 27, 1887.

> *Report of the Twenty-Seventh Ordinary General Meeting of the Eastern Extension, Australasia, and China Telegraph Company, Limited*, April 27, 1887. DOC/EEACT/1/84.

CWA, West Coast of America Telegraph Company, *Report*, 1890.

> West Coast of America Telegraph Company, *Report of the Proceedings at the Extraordinary General Meeting*, September 1, 1890.

CWA, West India Company Agreements, 1869.

> International Ocean Telegraph Company and West India and Panama Telegraph Company, Ltd., July 24, 1869.

Foreign Relations of the United States: Correspondence (United States Department of State, *Papers Relating to the Foreign Relations of the United States, Transmitted to Congress, with the Annual Message of the President*, . . . [280 vols.; Washington, D.C., Government Printing Office, 1861–1960; New York: Kraut Reprint, 1969])

FRUS, Avery to Fish, December 20, 1874.

> Correspondence between Benjamin P. Avery, U.S. Legation, Beijing, to Secretary of State Hamilton Fish, December 20, 1874 (2 entries). China. *1874*, Vol. 1, 223–27, 237–41.

FRUS, Avery to Fish, January 27, 1875.

Correspondence between Benjamin P. Avery, U.S. Legation, Beijing, to Secretary of State Hamilton Fish, no. 142, January 27, 1875. China. *1875*, Vol. 1.

FRUS, Avery to Fish, March 18, 1875.

Correspondence from Benjamin P. Avery, U.S. Legation, Beijing, to Secretary of State Hamilton Fish, no. 147, March 18, 1875. China. *1875*, Vol. 1.

FRUS, Avery to Fish, March 19, 1875.

Correspondence from Benjamin P. Avery, U.S. Legation, Beijing, to Secretary of State Hamilton Fish, with enclosures from local U.S. consular officials—Mr. Delano—in Fuzhou, no. 148, March 19, 1875. China. *1875*, Vol. 1.

FRUS, Avery to Fish, May 14, 1875.

Correspondence from Benjamin P. Avery, U.S. Legation, Beijing, to Secretary of State Hamilton Fish, with enclosures from various foreign ministers to one another and from Avery to Head of Danish Ministerial Delegation to Beijing, General Raasloff, no. 167, May 14, 1875. China. *1875*, Vol. 1.

FRUS, Avery to Steward, February 3, 1875.

Benjamin P. Avery, U.S. Legation, Beijing, to Mr. W. Seward, U.S. Consul General, Beijing, no. 142, February 3, 1875. China. *1875*, Vol. 1.

FRUS, Bayard to Jarvis, September 5, 1887.

T. F. Bayard, U.S. Secretary of State, to Mr. Jarvis, Dept. of State, no. 60, September 5, 1887. *1887*.

FRUS, Davis to Morton, 1883.

Correspondence from Acting Secretary of State John Davis to Levy P. Morton, U.S. Embassy, Paris, 1883. *1883*.

FRUS, Fish to Avery, March 4, 1875.

Correspondence from Secretary of State Hamilton Fish to Benjamin P. Avery, U.S. Legation, Beijing, no. 146, March 4, 1875. China. *1875*, Vol. 1.

FRUS, Fish to Lindencrone, January 22, 1874.

Correspondence from Secretary of State Hamilton Fish to J. Hegerman Lindencrone, Legation of Denmark, no. 194, January 22, 1874. China. *1874*, Vol. 1.

FRUS, Gong to Avery, January 12, 1875.

Correspondence from Prince Gong, Zongli Yamen, to Benjamin P. Avery, U.S. Legation, Beijing, no. 142, January 12, 1875. China. *1875*, Vol. 1.

FRUS, Lindencrone to Fish, January 7, 1874.

Correspondence from J. Hegerman Lindencrone, Legation of Denmark, to Secretary of State Hamilton Fish, with numerous enclosures, no. 193, January 7, 1874. China. *1874*, Vol. 1.

FRUS, Schuyler to Fish, 1875.

Correspondence from Eugene Schuyler to Secretary of State Hamilton Fish, 1875. Russia. *1875*, Vol. 1.

FRUS, Seward to Cadqalader, August 27, 1874.

Correspondence from Mr. George F. Seward, U.S. Consul General, Beijing, to Mr. Cadqalader, with numerous enclosures, no. 165, August 27, 1874. China. *1874*, Vol. 1.

FRUS, Seward to Davis, June 23, 1874.

> Correspondence between Mr. George F. Seward, U.S. Consul General, Beijing, to Mr. Davis, no. 160, June 23, 1874. China. *1874*, Vol. 1.

FRUS, Seward to Davis, July 21, 1875.

> Correspondence from Mr. George F. Seward, U.S. Consul General, Beijing, to Mr. Davis, no. 162, July 21, 1875. China. *1875*, Vol. 1.

FRUS, Trail to Baron de Cotegipe, February 28, 1887.

> Charles B. Trail, Legation of the United States, Rio de Janeiro, to Baron de Cotegipe, Ministry of Foreign Affairs, no. 74, February 28, 1887. *1887*.

FRUS, U.S. Chargé d'Affaires in Brazil to the Secretary of State, March 20, 1917.

> U.S. Chargé d'Affaires in Brazil to the Secretary of State, March 20, 1917. *1917*.

FRUS, Wilson to the Secretary of State, April 20, 1909.

> Message from Charles Wilson, U.S. Chargé d'Affaires, Buenos Aires, to the Secretary of State, April 20, 1909, in "Concessions Granted by the Argentine Republic." *1910*.

FRUS, Young to Frelinghuysen, November 28, 1882.

> Correspondence from J. Russell Young, U.S. Legation, Beijing, to Secretary of State Frelinghuysen, with enclosures of correspondence between Young and Prince Gong, Zongli Yamen, and from Great Northern Company agreement regarding construction of telegraph lines in China, no. 68, November 28, 1882. China. *1883*, Vol. 1.

Miscellaneous

Australia, Report of the Select Committee on Press Cable Services.

> Australia, Report of the Select Committee on Press Cable Services, Commonwealth Parliamentary Papers. *Journal of the Senate*, Vol. 1, 1909.

Canada, *An Act to control the Rates and Facilities of Ocean Cable Companies*, May 4, 1910.

> *An Act to control the Rates and Facilities of Ocean Cable Companies, and to Amend the Railway Act with Respect to Telegraphs and Telephones and the Jurisdiction of the Board of Railway Commissioners.* Assented May 4, 1910. Statutes of Canada 1910.

Canada, *Debates*, March 23, 1875.

> Canada, House of Commons, *Parliamentary Debates*. March 23, 1875, 870–73.

Canada, Marine Telegraph Bill.

> Canada. Marine Telegraph Bill: Schedule of Documents. *Sessional Papers* 38 Victoria (no. 20), A. 1875.

CFB.

> Council of Foreign Bondholders, *Annual General Reports*, 1873–87.

DUS V. AATC.

> *Direct U.S., Appellants, and the Anglo American Telegraph Company, et al., Respondents.* Record of Proceedings. Privy Council on Appeal from the Supreme Court of Newfoundland, 1874. Library of Congress.

Eastern Telegraph Company, *Report*, 1923.

> Eastern Telegraph Company, *Report of the Proceedings of the Ninety-Second General Meeting*, July 24, 1923.

GTTC, *Minutes.*

> Globe Telegraph and Trust Company. *Board Meeting Minute Books*, 1872–1950. City of London Libraries, Guildhall, MS 24.

League of Nations, *The League of Nations and the Press*, 1928.

League of Nations, *The League of Nations and the Press. International Press Exhibition, Cologne, May to October 1928*. Geneva: League of Nations (Information Section), 1928.

Mackay Companies.

Mackay Companies, *Annual Reports*, 1918–28. New York, 1919–29.

Marconi Wireless Company of America, *Annual Report, 1918*.

Marconi Wireless Company of America, *Annual Report, 1918*. New York, 1919.

Marconi Wireless Company, Ltd. *Report of Directors*, July 24, 1919.

Marconi Wireless Company, Ltd. *Report of Directors*, July 24, 1919. Reprinted in *Wireless World*, September 1919.

New Zealand, Copyright Telegrams Committee Report, 1896.

Appendix to the Journals of the House of Representatives of New Zealand, Vol. 4. Wellington, 1896.

Western Union.

Western Union, *Annual Reports*, 1876–1906.

Wireless, 1909.

Wireless, 1909 (issue no. and date unknown). McNicol Collection on Telecommunications, Jordan Library, Queen's University at Kingston.

National Archives of Canada: Reports, Correspondence, and Other Documents

NAC, Anderson to Pender, July 20, 1880.

Correspondence between J. Anderson and J. Pender, July 20, 1880. RG25, A1, Vol. 53.

NAC, Dispatch from Lord Crewe to Earl Dudley, October 14, 1910.

Dispatch from Lord Crewe, Secretary of State for the Colonies, to Earl Dudley, Governor General of Australia, October 14, 1910. RG25, series A3, Vol. 1106, file 1910–979.

NAC, "The Empire Cables."

"The Empire Cables." RG3 Vol. 627, folder 65162–4.

NAC, "Explanatory Note."

"Explanatory Note." RG3 Vol. 627, folder 65162–4.

NAC, Fleming, *Cheap Telegraph Rates*, February 28,1902.

Sanford Fleming, Address to the Annual Meeting of the Canadian Press Association, February 28, 1902. RG 6 A–1, Vol. 108, file 1145.

NAC, Fleming to Langevin, April 8, 1886.

Correspondence from Sandford Fleming to Hector Langevin, Minister of Public Works, April 8, 1886. RG25, A1, Vol. 53.

NAC, Fleming to Lemieux, April 16, 1909.

Letter from Sandford Fleming to Canadian Postmaster General Rodolphe Lemieux, April 16, 1909. RG3, Vol. 626.

NAC, Fleming to Tarte, July 1, 1899.

Fleming to Tarte, July 1, 1899. RG3, Vol. 627, folder 65162.

NAC, [Fleming], *To the Citizens of the Empire*, 1907.

Board of Trade of the City of Ottawa (written by Sandford Fleming). *To the Citizens of the Empire*. Ottawa, 1907. RG3 Vol. 627, folder 65162–4.

NAC, Headlam, Circular no. 89, November 8, 1927.

M. F. Headlam, Acting Chair of Pacific Cable Board, Circular no. 89, November 8, 1927. *House of Commons for Canada: Documents Relating to the Transactions of the Pacific Cable Board (1923–29)*. RG3, Vol. 631, file 65162–13.

NAC, Heaton and Fleming.

Correspondence of Henniker Heaton and Sandford Fleming. MG29, Vol. 22, B1.

NAC, Jones to Macdonald, 2339 Order in Council.

Correspondence between Owen Jones, Secretary of Pacific Telegraph Company, to Prime Minister Sir John A. Macdonald, P.C., 2339 Order in Council. RG25, A1, Vol. 53.

NAC, Lemieux to Earl Grey, January 13, 1910.

Lemieux to Governor General Earl Grey, January 13, 1910. RG3, Vol. 627, file 65162–4.

NAC, Marquess of Salisbury to Count Hatzfeld, January 3, 1900.

Correspondence from Marquess of Salisbury to Count Hatzfeld, January 3, 1900. CO 42 873 (mf B789).

NAC, "Memo on Imperial Conference," April 3, 1911.

Canadian Post Office Department. "Memorandum on the Imperial Conference on the Atlantic Cables," April 3, 1911, author unknown. RG3, Vol. 626, file 61670.

NAC, Memo on Pacific Cable, May 1, 1905.

Canada Post Office Department. Memorandum on the Pacific Cable, May 1, 1905 (probably written by senior administrator William Smith). RG3, Vol. 638, file 65162.

NAC, Murphy to Skelton, July 7, 1927.

History of Imperial Cables, Charles Murphy, Postmaster General, to C. D. Skelton, Under Secretary of State for External Affairs, Canada, July 7, 1927. RG3, Vol. 1001, file 21–5–6.

NAC, Ottawa City Board of Trade, "An Address to His Excellency, Earl Grey," 1907.

Board of Trade of the City of Ottawa, "An Address to His Excellency, Earl Grey, Governor General of Canada, with His Excellency's Reply and other Documents Bearing on the Proposed Imperial Cable Service to Girdle the Globe." Ottawa, 1907.

NAC, Pacific Cable Committee, 1899.

Pacific Cable Committee. *Report, Minutes and Proceedings, etc.* London: HMSO, 1899. RG3, Vol. 627, folder 65162.

NAC, Press and International Relations Conference.

Press and International Relations Conference (item 14), Cooperation of the Press in the Organization of Peace, n.d. (contents review 1925–33 period). RG25, Vol. 2371, file s/233/1.

NAC, "Report of the Chairman of the Pacific Cable Board," 1909.

"Report of the Chairman of the Pacific Cable Board," 1909. RG14E1, Vol. 2025.

NAC, Ross to Deputy Minister of Trade and Commerce, August 1907.

Memorandum from D. W. Ross to Deputy Minister of Trade and Commerce regarding Pacific Cable, August 1907. RG3, Vol. 1002, file 21–5–11.

NAC, Ross to Laurier, August 4, 1909.

Memorandum from D. W. Ross to Prime Minister Sir Wilfrid Laurier, August 4, 1909. RG3, Vol. 1002, file 21–5–11.

NAC, Smith to Mulock, October 5, 1904.

Correspondence from Senator Staniforth Smith (Australia) to William Mulock, October 5, 1904. RG3, Vol. 680, file 112139.

NAC, "Survey of Disabilities," December 31, 1908.

Canadian Post Office Department. "Survey of Disabilities under which the Pacific Cable Labours at Present," December 31, 1908, author unknown. MG29, Vol. 99, folder 22.

NAC, Tupper-Pender Debate, 1894.

Tupper-Pender Debate, Britain, *Colonial Conference*, 1894. RG6 A1, vol. 108, folder 1145.

National Archives and Records Administration (United States): Reports, Correspondence, and Other Documents

NARA, Addenda to American Publishers Committee on Cable and Radio Communications, 1920.

Addenda to the General Memorandum of the American Publishers Committee on Cable and Radio Communications, October 29, 1920. RG43, entry 75, box 3.

NARA, Alexander to Long, 1920.

J. Alexander, Secretary, Department. of Commerce, Memo to Breckinbridge Long, Third Assistant Secretary of State, July 8, 1920. RG59, 574 D1/74.

NARA, American Manufacturers Export Association, 1920.

American Manufacturers Export Association, Correspondence with Robert Lansing, Secretary of State, January 7, 1920. RG59, 574 D1/14.

NARA, Balfour to Lloyd George, February 4, 1922.

Arthur James Balfour, Correspondence with David Lloyd George, First Lord of the Treasury Foreign Office, February 4, 1922. RG59, 574 D1, box 4.

NARA, British Memo to Lansing, August 18, 1919.

Unknown author, Memorandum from British Government to Robert Lansing, Secretary of State, regarding Proposed World Communication Conference, August 18, 1919. RG59, 574 D1/4.

NARA, Brown to Rogers, January 20, 1920.

James W. Brown, Editor and Publisher, Correspondence with Walter Rogers, January 20, 1920. RG43, entry 75, box 2.

NARA, Cable-Using Newspapers and Press Associations of America Memo, October 6, 1920.

Memorandum of Cable-Using Newspapers and Press Associations of America, October 6, 1920. RG43, entry 75, box 3.

NARA, Colby to Alexander, July 17, 1920.

Bainbridge Colby, Department of State, to J. Alexander, Secretary of Commerce, July 17, 1920. RG59, 574 D1/84a.

NARA, Colby to American Commissioners, Berlin, November 18, 1920.

Bainbridge Colby, Secretary of State, Telegram to American Commissioners, Berlin, November 18, 1920. RG59, 574 D1.

NARA, To Lansing, from unknown, September 14, 1920.

Unknown author, Correspondence to Robert Lansing, Secretary of State, September 14, 1920. RG59, 574 D1.

NARA, To Lodge, from Unknown, October 4, 1919.

Unknown author, Memorandum to Senator Henry Cabot Lodge, October 4, 1919. RG59, 574 D1/7.

NARA, Long, Memo to Lansing, November 25, 1918.

Breckinbridge Long, Third Assistant Secretary of State, Memorandum to Secretary of State Robert Lansing regarding international telegraphic communication, November 25, 1918, 3. RG59, 574 D1.

NARA, Mackay to Colby, August 19, 1920.

Clarence Mackay, President, Commercial Cable Company, Correspondence with Bainbridge Colby, Secretary of State, August 19, 1920. RG59, 574 D1.

NARA, Merchants Association of New York, January 8, 1920.

Merchants Association of New York, Correspondence with Alvee A. Adee, Second Assistant Secretary of State, January 8, 1920. RG59, 574 D1/15.

NARA, National Foreign Trade Council, November 26, 1919.

National Foreign Trade Council, Correspondence with William Phillips, Assistant Secretary of State, November 26, 1919. RG59, 574 D1/12.

NARA, Power, Memo to Lansing, January 15, 1919.

Ernest E. Power, Third Assistant Secretary of State, Memorandum to Secretary of State Robert Lansing regarding International Telegraphic Communication, January 15, 1919. RG59, 574 D11 box 5558.

NARA, Rogers, *An American International and Inter-Colonial Communications Program,* 1918.

Walter Rogers, *An American International and Inter-Colonial Communications Program,* n.d. (most likely 1918). RG59, 574 D1, box 4.

NARA, Rogers et al., *Report,* 1922.

W. S. Rogers, F. J. Brown, M. Girardanu, K. Yoshino, and S. Inada. *Report of Experts.* Enclosure to Dispatch no. 22, February 4, 1922. RG59, 574 D1, box 4.

NARA, Rogers to Harrison, June 7, 1922.

Memorandum of Walter S. Rogers to Mr. Harrison, June 7, 1922. RG59, 574 D11, box 5558.

NARA, Rogers, Memo on International Communications Conference, June 8, 1921.

Walter Rogers, Memorandum to Under-Secretary of State regarding International Communications Conference, June 8, 1921. RG59, 574 D11, box 5558.

NARA, Rogers, Memo on Pacific Communication, October 9, 1918.

Walter Rogers, Memorandum of a meeting of the Interdepartmental Committee on Pacific Communication, October 9, 1918. RG59, 574 D1, box 4.

NARA, Rogers, Memo re Owen Young, 1921.

Walter Rogers, Memorandum regarding Owen Young, 1921. RG59, 574 D1, box 4.

NARA, Rogers, Memo for Secretary, November 28, 1921.

Walter Rogers, Memorandum for the Secretary, November 28, 1921. RG59, 574 D1, box 4.

NARA, Rogers, Memo for Secretary, December 8, 1921.

Walter Rogers, Memorandum for the Secretary, December 8, 1921. RG59, 574 D1, box 4.

NARA, Rogers, Memo for Secretary, March 21, 1922.

Walter Rogers, Memorandum for the Secretary, March 21, 1922. RG59, 574 D1, box 4.

NARA, Rogers, Outline of Communications Program, September 11, 1918.

Walter Rogers, Outline of Communications Program, September 11, 1918. RG59, 574 D1, box 4, file 188.

NARA, Rogers to Root, December 22, 1921.

Walter Rogers, Memorandum for Senator Root, December 22, 1921. RG59, 574 D1, box 4.

NARA, Rogers, Subcommittee of the Interdepartmental Committee on Communications, July 30, 1921.

Walter Rogers to Subcommittee of the Interdepartmental Committee on Communications, July 30, 1921. RG43, entry 75, box 3.

NARA, Rogers, Transpacific Communication, July 30, 1921.

Walter Rogers, Transpacific Communication, July 30, 1921. RG59, 574 D11, box 5558.

NARA, State Department Conference with Prince de Bearn, July 16, 1920. U.S. State Department, Conference with the Charge d'Affaires of the French Embassy, Prince de Bearn, July 16, 1920. RG59, 574 D1/4.

Public Records Office (Britain): Company Charters, Reports, Correspondence, and Other Documents

PRO, Cables to China, Denison-Pender to Wellesley, April 8, 1921.

Cables to China, General, 1921. Sir J. Denison-Pender, Correspondence with Wellesley, Foreign Office, regarding Chinese Loans and Concessions, April 8, 1921. FO 262/1517.

PRO, Cables to China, Loans, 1925.

Cables to China, Loans. 1925. Japanese Embassy, Washington, Memorandum regarding Wireless Radio in China to Foreign Office, December 24, 1924. FO 262/1628.

PRO, Cables to China, Wellesley to Treasury, January 17, 1921.

Cables to China, General, 1921. Victor Wellesley, Foreign Office Correspondence with Treasury Office, January 17, 1921. PRO, FO 262/1517.

PRO, Cecil to Jones and Napier, December 8, 1916.

Reuters Ltd., Agreement with Foreign Office as to Services. Correspondence between Treasury Secretary (1917–22) Robert Cecil and Roderick Jones and Mark Napier outlining arrangements with Reuters, December 8, 1916. TS 27/90.

PRO, Clements Memorandum, August 16, 1920.

Memorandum from C. Carey Clements, Manager and Secretary Reuters: *Report on Supplementary Cable Service to Dominions and Colonies Prepared for the Foreign Office,* August 16, 1920. Treasury (1921), Communications: Telegraph: General: News Service to Overseas Dominions and Colonies, June 7, 1920, to November 2, 1923. T161/1.

PRO, Committee of Imperial Defense, 1920.

Committee of Imperial Defense, Imperial Communications Committee. Minutes of the Sixteenth Meeting held at the Colonial Office on Friday, 11th June, 1920. July 20, 1920. T162/135.

PRO, Dossier 1120 Telegraphs and Cables, 1922.

Dossier 1120 Telegraphs and Cables, Vol. 2. John Denison-Pender (n.d., probably January 12, 1922). Resolution of the Board of Directors of the Western Telegraph Company. FO 228/3416.

PRO, DUS Articles of Association, 1873.

Direct U.S. Cable Company Memorandum and Articles of Association. 1873. BT 31/1822/7058.

PRO, French Atlantic Cable Co., 1868.

Société du Câble Transatlantique Française, Ltd. Memorandum and Articles of Association. 1868. BT 31/1418/4105.

PRO, GPO, Purchase of Cable, 1920–23.

    GPO Purchase of Cable from DUS Cable Company and Temporary Lease, after Purchase, to the WU Company. June 16, 1920–November 23, 1923. T162/135.

PRO, Hamilton Memo, April 29, 1897.

    Correspondence from E. Hamilton, Treasury Office, to the Under Secretary of State, Colonial Office, April 29, 1897. CO 42/850.

PRO, Imperial Communications Committee, July 5, 1919.

    Imperial Communications Committee. International Conference to Consider International Aspects of Communications, July 5, 1919. T194/223.

PRO, International Conference Memo, May 27, 1921.

    Political Affairs, United States, 1921. General Post Office, Memorandum respecting the International Conference at Washington, May 27, 1921. FO 371/5631.

PRO, Lord Burnham Statement to the Newspaper Proprietors Association, December 1916.

    December 1916 draft statement to be made by Lord Burnham to the Newspaper Proprietors Association. TS 27/90.

PRO, Mercer Memo, April 29, 1897.

    W. Mercer, Memorandum on the Pacific Cable Report, etc., April 29, 1897. CO 42/850.

PRO, *Miscellaneous Correspondence between Treasury Office and Various HMG Officials*, 1870.

    Miscellaneous Correspondence between Treasury Office and Various HMG Officials as well as Copy Memorial to the Honorable the Senate and House of Representatives of the U.S., March–May 1870. T1/6978B.

PRO, South American Cable Company, 1891.

    South American Cable Company, Ltd. Memorandum and Articles of Association. 1891. BT 31/5102/34363.

PRO, South American Cables, Brazil Extract.

    South American Cables, Brazil Extract. File 65. Extract from annual Foreign Office report on Brazil. BT, post 33/1092A.

PRO, South American Cables, Brown to Under Secretary of State, September 15, 1921.

    South American Cables—General Papers. Correspondence from F. J. Brown, General Post Office, to Under Secretary of State, Foreign Office, September 15, 1921. BT, post 33/1092A, file 42.

PRO, South American Cables, Denison-Pender to Foreign Office, November 7, 1918.

    Correspondence from John Denison-Pender to Foreign Office, November 7, 1918. BT, post 33/1092A, file 29, file 3.

PRO, South American Cables, Kelly to Curzon, January 19, 1921.

    South American Cables—General Papers. Letter from D. Victor Kelly (HM Minister, Buenos Aires) to Earl Curzon, Foreign Office, January 19, 1921. BT, post 33/1092A, file 23.

PRO, South American Cables, Nye to Chief Censor, March 28, 1918.

    South American Cables—Personal correspondence from F. Nye, Captain, Royal Navy, to Chief Censor, March 28, 1918. BT, post 33/1091, file 3.

PRO, South American Cables, Paget to Curzon, May 8, 1920.

    South American Cables—General Papers. Letter from Ralph Paget to Lord Curzon from Embassy in Rio, May 8, 1920. BT, post 33/1092A, file 29.

PRO, South American Cables, Sperling to Foreign Office, July 18, 1921.

South American Cables—General Papers. Memo from R. Sperling, General Post Office to Foreign Office, July 18, 1921. BT, post 33/1092A, file 29.

PRO, South American Cables, Tower to Balfour, February 15, 1919.

Memorandum Regarding Argentina to Arthur Balfour from Reginald Tower, United Kingdom. Embassy, Buenos Aires, February 15, 1919. BT, post 33/1092A, file 29, file 4.

PRO, Telegraphs and Cables—China Related, Hurd, July 12, 1921.

Dossier 1120 Telegraphs and Cables, Vol. 2, China Related, April 1921 to July 1922. Hurd, question in House of Commons to Post Master General, July 12, 1921. FO 228/3416.

PRO, Telegraphs and Cables, Foreign Office Cypher to Balfour, December 22, 1921.

Dossier 1120 Telegraphs and Cables, Vol. 2, April 1921 to July 1922. Foreign Office Cypher telegraph to Balfour, Washington Delegation, December 22, 1921. FO 228/3416.

PRO, Treasury, *Memoranda of the Imperial Communications Committee re the Washington Conference*, 1920.

Treasury, *Memoranda of the Imperial Communications Committee regarding the Washington Conference on the disposal of German Submarine Cables*, 1920. T194/223.

PRO, *Treasury Minutes*, 1916.

"Purchase of the Undertakings of the Anglo-American Telegraph and the Direct United States Company by the Western Union Telegraph Company," *Treasury Minutes*, 1916. T 1/11980.

PRO, Treasury Secretary (1917–22), *Reuters Ltd., Agreement.*

Treasury Secretary (1917–22), *Reuters Ltd., Agreement with Foreign Office as to Services.* TS 27/90.

University College (London) Special Collections, Western Telegraph Company (WTC): Concessions, Correspondence, and Other Documents

WTC, Brazil, box 1.

July 12, 1885. Superintendent's Letters from Rio 1883–1893. Brazil files, Brazil, box 1.

WTC, Brazil, box 2.

General Memo, the Western WTC, Brazil, box 2. Letters from General Superintendent at Santos to E. Steer Hodson, Secretary to the Board, London, 1905–10. Brazil files, Brazil, box 6.

WTC, Brazil, box 6.

General Superintendent to Secretary, 1909–13. Brazil files, box 6.

United States Reports, Minutes of Evidence, Conferences, Correspondence, and Other Documents

U.S., *Cable-Landing Licenses.*

U.S. Congress. *Cable-Landing Licenses Hearings.* 66th Cong., 3rd sess., 1921.

U.S., Conference on International Communication, 1921.

United States, Conference on International Communication, *Report* (No. 387). House of Representatives, 66th Cong., 1st sess. 1921.

U.S., Department of Labor, *Investigation of Western Union*, 1909.
United States, Department of Labor, *Investigation of Western Union and Postal Telegraph Companies*. Washington, D.C: U.S. Government Printing Office, 1909.

U.S., *Digest of International Law*.
John Bassett Moore, *A Digest of International Law as Embodied in Diplomatic Discussions, Treaties and Other International Agreements, International Awards, the Decisions of Municipal Courts, and the Writings of Jurists, and Especially in Documents, Published and Unpublished, Issued by Presidents and Secretaries of State of the United States, the Opinions of the Attorneys-General, and the Decisions of Courts, Federal and State*. 8 vols.; Washington, D.C: U.S. Government Printing Office, 1906.

U.S., FTC, *Report on the Radio Industry*, 1923.
United States, Federal Trade Commission, *Report on the Radio Industry*, 67th Cong., 4th sess., December 1, 1923.

U.S., *Hearing on S. 6*, 1929.
United States, Senate, Committee on Interstate Commerce, *Hearings before the Committee on Interstate Commerce of the United States Senate, May 1929–February 1930 on S. 6, A Bill to Provide for the Regulation by Permanent Commission on Communications, of Transmission of Intelligence by Wire or Wireless* ("Couzens Bill").

U.S., H.R., *Pacific Cable*, 1900.
United States, House of Representatives, Committee on Interstate and Foreign Commerce. *Pacific Cable*. 1900. Library of Congress HE7731.

U.S., H.R., *Radio Stations*, July 1919.
United States, House of Representatives, *Radio Stations under Control of the Navy Department for Commercial Purposes*. 66th Cong., 1st sess., July 23, 1919, and July 24, 1919.

U.S., *Inter-American Committee on Electrical Communications*, 1924.
United States, *Inter-American Committee on Electrical Communications*. Mexico City, May 27 to July 22, 1924. Washington: Government Printing Office, 1927.

U.S., *International Communications*, 1923.
United States, Department of Commerce, *International Communications and the International Telegraph Convention*, 1923. Miscellaneous Series, no. 121, 8.

U.S., *International Communications*, 1945.
United States, Senate. *Hearings on Senate Resolution 187: A Resolution Directing a Study of International Communications by Wire and Radio*. 78th Cong., 1945.

U.S., *Memo*, 1920.
Memorandum Concerning Cable and Radio Telegraphic Communication with Mexico, Central and South America and the West Indies. Second Pan American Financial Conference, January 19–24, 1920. Washington, D.C.: Government Printing Office.

U.S., Message of President, 1887.
Message of President—landing of transatlantic telegraph companies, contracts of these with other cable and/or telegraph companies. Report of Secretary of State. March 3, 1887. U.S. Senate Executive Document 122 (49–2) 2449, 4–10.

U.S., *Navy in Peace*.
United States, Navy Department. *The United States Navy in Peace-Time*. Washington, D.C.: Government Printing Office, 1931.

U.S., Noyes, Associated Press.

F. Noyes, The Associated Press (Doc. 27), Presented to the United States Senate, 63rd Cong., 1st sess., Washington, May 14, 1913.

U.S., *Pacific Cable Bills*, 1902.

United States, House of Representatives, *Pacific Cable Bills*. 1902. Library of Congress HE7731.D250.231.

U.S., *Preliminary Report on Communication*, 1934.

United States, Congress. Committee on Interstate and Foreign Commerce, *Preliminary Report on Communication Companies*, Vol. 5. 72nd Cong., April 18, 1934.

U.S., *Regulation of International Services*, 1933.

United States, Senate, *Regulation of Rates for Messages in the International Service*. 73rd Cong., 2nd sess., November 15, 1933, doc. no. 127.

U.S., *Report*, 1923.

United States, House of Representatives, *Report on the Radio Industry in Response to House Resolution 548*. 67th Cong., 4th sess., 1923, published 1924.

U.S., *Report on Communication Companies*, 1935.

United States, Congress, Committee on Interstate and Foreign Commerce. *Report on Communication Companies*, Vol. 6, pt. 5. 72nd Cong., 1935.

U.S. Senate, *Navy Pacific Cable*.

United States, Senate, Committee of Naval Affairs, *Hearings on the Bill to Provide for the Construction, Maintenance and the Operation under the Management of the Navy Department of a Pacific Cable (Document 116)*. 1900. Mf 3851, file 2.

U.S. Senate, *Report on Magnetic Telegraph Companies*, 1858.

United States, Senate, Committee on the Judiciary, *Report on Magnetic Telegraph Companies*, June 9, 1858, 35th Cong., 1st sess., Rep. Com. no. 313.

SECONDARY SOURCES

Ackerman, C. *Report of the Graduate School of Journalism*. 2nd ed. New York: Columbia University Press, 1943.

Ahvenainen, J. *The European Cable Companies in South America*. Helsinki: Suomalaisen Tiedeakatemian, 2004.

——. *The Far Eastern Telegraphs*. Helsinki: Suomalaisen Tiedeakatemian, 1981.

——. *The History of the Caribbean Telegraphs before the First World War*. Helsinki: Suomalaisen Tiedeakatemian, 1996.

Aitken, H. *The Continuous Wave: Technology and American Radio, 1900–1932*. Princeton: Princeton University Press, 1985.

All-America Cables, Inc. *A Half Century of Cable Service to the Americas*. New York: All-America Cables, Inc., 1928.

Ambrosius, L. E. *Wilsonianism*. New York: Palgrave Macmillan, 2002.

Associated Press. *M. E. S. in Memoriam: A Tribute to the Life and Accomplishments of Melville E. Stone, General Manager of the Associated Press, 1893–1921*. New York: Associated Press, 1921.

Baark, E. *Lightning Wires: The Telegraph and China's Technological Modernization, 1860–1890*. Westport, Conn.: Greenwood Press, 1997.

Barty-King, H. *Girdle Round the Earth*. London: Heinemann, 1979.

Bektas, Y. "The Sultan's Messenger: Cultural Constructions of Ottoman Telegraphy, 1847–1880." *Technology and Culture* 41 (spring 2001): 669–96.

Berthold, V. *History of the Telephone and Telegraph in Argentina*. New York: AT&T, 1924.

——. *History of the Telephone and Telegraph in Brazil, 1851–1921*. New York: AT&T, 1924.

——. *History of the Telephone and Telegraph in Chile, 1851–1921*. New York: AT&T: 1924.

——. *History of the Telephone and Telegraph in Uruguay, 1866–1925*. New York: AT&T, 1925.

Bilby, K. *The General: David Sarnoff and the Rise of the Communications Industry*. New York: Harper and Row, 1986.

Blondheim, M. *News over the Wires*. London: Harvard University Press, 1994.

Boehmer. E. *Empire, the National and the Post Colonial, 1890–1920*. London: Oxford University Press, 2002.

Boemeke, M. F., ed. *The Treaty of Versailles: A Reassessment after 75 Years*. Cambridge: Cambridge University Press, 1998.

Boyce, R. "Canada and the Pacific Cable Controversy, 1923–28," *Journal of Imperial and Commonwealth History* 26 (January 1998): 72–92.

——. "Imperial Dreams and National Realities." *English Historical Review* 65 (2000): 460.

Boyd-Barrett, O., and T. Rantanen, eds. *The Globalization of News*. London: Sage, 1998.

Bright, C. "Cable versus Wireless Telegraphy." *The Nineteenth Century* 74 (June 1912).

——. *Imperial Telegraphic Communication*. London: P. S. King and Son, 1911.

——. *The Story of the Atlantic Cable*. New York: D. Appleton and Company, 1903.

——. "Telegraphs in War-Time." *The Nineteenth Century* 77 (1915).

——. *Telegraphy, Aeronautics and War*. London: Constable, 1918.

Britton, J. A., and J. Ahvenainen. "Showdown in South America." *Business History Review* 78 (spring 2004): 1–27.

Brown, F. J. *The Cable and Wireless Communications of the World*. London: Pitman and Sons, 1927.

Bullard, W. H. "The Naval Radio Service." Reprinted from *The Proceedings of the Institute of Radio Engineers* 3, no. 1 (March 1915).

Cameron, R. and V. I. Bovykin, eds. *International Banking, 1870–1914*. New York: Oxford University Press, 1991.

Carter, S., III *Cyrus Field: Man of Two Worlds*. New York: G. P. Putnam, 1968.

Cassidy, J. *dot.con*. New York: Harper Collins, 2002.

Central and South America Telegraph Company. *Via Galveston*. New York: Central and South America Telegraph Company, 1895.

Chandler, A., Jr. *Scale and Scope: The Dynamics of Industrial Capitalism*. Cambridge, Mass.: Harvard University Press, 1990.

Checkland, O. *Britain's Encounter with Meiji Japan, 1868–1912*. London: Macmillan, 1989.

Clark, K. *International Communications*. New York: Columbia University, 1931.

Coates, V., and B. Finn. *A Retrospective Technology Assessment: Submarine Telegraphy*. San Francisco: San Francisco Press, 1979.

Codding, A. and G. Rutkowski. *The International Telecommunications Union in a Changing World*. Dedham, Mass.: Artech House, 1982.

Coggeshall, I. S. *An Annotated History of Submarine Cables and Overseas Radiotelegraphs, 1851–1934.* 1934; reprint ed., 1992, edited and published by Donald de Cogan.

Cole, J. R. *Colonialism and Revolution in the Middle East.* Princeton: Princeton University Press, 1993.

Colley, L. *Captives: Britain, Empire and the World, 1600–1850.* New York: Pantheon, 2002.

Comor, E., ed. *The Global Political Economy of Communication.* New York: St. Martin's Press, 1994.

Conant, C. "The Economic Basis of Imperialism." *North American Review* 167 (1898): 326–46.

Cookson, G. "Ruinous Competition: The French Atlantic Telegraph of 1869." *Enterprises et Historie*, no. 23 (December 1999).

Cooper, K. *Barriers Down.* New York: Farrar and Rinehart, 1942.

Cooper, R. "The New Liberal Imperialism." *Guardian*, April 7, 2002.

Cox, R. "Global Perestroika." In *New World Order?* edited by R. Milliband and L. Panitch. New York: New Left Review, 1992.

——. *Production, Power and World Order.* New York: Columbia University Press, 1987.

Cowderoy, B. *Direct Telegraphic Communication with Europe and the East.* Melbourne: Stillwell and Knight, Printers, 1869.

Dayer, R. *Bankers and Diplomats in China, 1917–1925: The Anglo-American Relationship.* London: F. Cass, 1981.

Denny, L. *America Conquers Britain.* New York: Knopf, 1930.

Desbordes, R. "Western Empires and News Flows in the 19th Century: International News Agencies in South America, 1858–1918." Paper presented at the Conference of the International Association of Media and Communication Research, Porto Alegre, Brazil, July 24–30, 2004.

Dewey, J. "Imperialism Is Easy." *New Republic*, March 23, 1927.

Donald, R. *The Imperial Press Conference in Canada.* Toronto: Hodder and Stoughton, 1921.

Donaldson, F. *The Marconi Scandal.* London: Rupert Hart-Davis, 1982.

Douglas, S. *Inventing American Broadcasting, 1899–1922.* Baltimore: John Hopkins University Press, 1987.

Dunlap, O. *Marconi.* New York: Macmillan, 1941.

Duus, P., ed. *The Cambridge History of Japan*, Vol. 6. New York: Cambridge University Press, 1986.

"An Extraordinary Monopoly: How Newspaper Enterprise Is Throttled in New Zealand," *The Citizen*, January 15, 1909, 9–10.

Field, J. A. "American Imperialism: The Worst Chapter in Almost Any Book." *American Historical Review* 38, no. 3 (1978).

Feis, H. *Europe: The World's Banker, 1870–1914.* New Haven: Yale University Press, 1934.

Ferguson, N. *Empire.* New York: Basic Books, 2003.

Findley, C. V. *Bureaucratic Reform in the Ottoman Empire, 1789–1922.* Princeton: Princeton University Press, 1980.

Furth, C. "Intellectual Change." In *An Intellectual History of Modern China*, edited by L. O. Lee and M. Goldman. New York: Columbia University Press, 2002.

Gallagher, J. and R. Robinson. "The Imperialism of Free Trade." *Economic History Review* 6, no. 1 (1953): 1.

Giddings, F. H. *Democracy and Empire.* New York: Macmillan, 1900.

Gordon, A. *A Modern History of Japan.* New York: Oxford University Press, 2003.

Gramling, O. AP: *The Story of News.* New York: Farrar and Rinehart, 1940.

Griset, P. *Enterprise, Technologie et Souveraineté: Les Télécommunications TransAtlantiques de la France, XIXe–XXe Siècles.* Paris: Éditions Rive Droite, 1996.

——. "L'Etat et les Télécommunications Internationales au Debut de XXème Siècle en France." *Histoire, Économie et Société* 2, 2nd trimester (1987).

Hannah, L. "Visible and Invisible Hands in Great Britain." In *Managerial Hierarchies,* edited by A. Chandler. Cambridge, Mass.: Harvard University Press, 1980.

Harlow, A. F. *Old Wires and New Waves.* 1936; reprint. ed., New York: Arno Press, 1971.

Harvey, D. *The New Imperialism.* New York: Oxford University Press, 2003.

Headrick, D. *The Invisible Weapon.* New York: Oxford University Press, 1991.

Headrick, D., and P. Griset. "Submarine Telegraph Cables." *Business History Review* 75 (autumn 2001).

Heaton, J. Henniker. "The Cable Telegraph Systems of the World." *The Arena* 38, no. 214 (September 1907), 225–36.

——. "How to Smash the Cable Ring," *The Arena* 38, no. 212 (July 1907), 1–8.

——. "The Imperial Conference and Imperial Communications." *Nineteenth Century* 70, no. 414 (1911).

——. "Postal and Telegraphic Reforms." *Contemporary Review,* March 1891.

Held, D., A. McGrew, D. Goldblatt, and J. Perraton. *Global Transformations.* Stanford: Stanford University Press, 1999.

Hills, J. *The Struggle for Control of Global Communications.* Urbana: University of Illinois Press, 2002.

Hoare, J. "The Era of Unequal Treaties, 1858–1899." In *The History of Anglo-Japanese Relations, 1600–1930,* Vol. 1, edited by J. Nish and Y. Kibata. New York: St. Martin's, 2000.

Hobsbawm, E. *The Age of Empire.* London: Wiedenfeld and Nicolson, 1995.

——. *The Age of Extremes.* London: Penguin, 1994.

Hobson, J. A. *Imperialism,* 1904; 2nd rev. ed., London: George Allen and Unwin, 1954.

Hogan, M. *Informal Entente.* London: University of Missouri, 1977.

Hsiao, K. C. *A Modern China and a New World: K'ang Yu-wei, Reformer and Utopian, 1858–1927.* Seattle: University of Washington Press, 1975.

Ignatieff, M. *Empire Lite.* Toronto: Penguin Canada, 2003.

Iriye, A. "Japan's Policies towards the U.S." In *Japan's Foreign Policy 1869–1941,* edited by J. Morley. New York: Columbia University Press, 1974.

Johnson, J. *Pioneer Telegraphy in Chile, 1852–1876.* Stanford: Stanford University Press, 1948.

Johnson, R. *The Peace Progressives and American Foreign Relations.* Cambridge, Mass.: Harvard University Press, 1994.

Jones, G. *Multinationals and Global Capitalism.* Oxford: Oxford University Press, 2005.

Jones, R. *A Life at Reuters.* London: Hodder and Stoughton, 1951.

Kang Youwei. *Da Tong Shu* (The Great Cosmopolitan Community), 1935; new ed., Zhongzhou: Zhongzhou Ancient Books, 1998.

Kedourie, E. *Politics in the Middle East.* New York: Oxford University Press, 1992.

Kennedy, P. "Imperial Cable Communications and Strategy, 1870–1914." *English Historical Review,* October 1971, 728–53.

Keylor, W. R. "Versailles and International Diplomacy." In *The Treaty of Versailles: A Reassessment after 75 Years*, edited by M. F. Boemeke et al. Cambridge: Cambridge University Press, 1998.

Kieve, J. *The Electric Telegraph*. Newton Abbot, Devon: David and Charles, 1973.

Kindleberger, M. *Manias, Panics, and Crashes*. London: Macmillan, 1978.

Knock, T. J. "Wilsonian Concepts and International Realities at the End of the War." In *The Treaty of Versailles: A Reassessment after 75 Years*, edited by M. F. Boemeke et al. Cambridge: Cambridge University Press, 1998.

Koskenniemi, M. *The Gentle Civilizer of Nations*. Cambridge: Cambridge University Press, 2002.

Lange, O. *Partner og Rivaler: C. F. Tietgen, Eastern Extension and Store Nordiske, Expansion i Kina, 1880–86*. Copenhagen: Gyldendal, 1980.

Lee, L. O., and M. Goldman, eds. *An Intellectual History of Modern China*. New York: Columbia University Press, 2002.

Levin, N. G. *Woodrow Wilson and World Politics*. New York: Oxford University Press, 1968.

Lewis, P., and C. Pearlman. *Media and Power: From Marconi to Murdoch*. London: Camden, 1986.

Liang Q. C. *Yinbingshi Heji-Wenji* (Collected Writings from the Ice-Drinker's Studio), Vol. 6. 24 vols.; Shanghai: China Books, 1936.

Lippmann, W. *Liberty and the News*, 1920. Reprint ed., History of Ideas Series, New Brunswick, N.J.: Transaction Publishers, 1995.

Lippmann, W., and C. Merz. "A Test of the News." *New Republic (Supplement)*, August 4, 1920.

Livingston, K. T. *The Wired Nation Continent*. Melbourne: Oxford University Press, 1996.

Lukes, S. *Power: A Radical View*. London: Macmillan, 1974.

MacMillan, M. *Paris 1919*. New York: Random House, 2003.

Marconi, G. "The Progress of Wireless Telegraphy." *Transactions of the New York Electrical Society*, no. 15, 1912.

Marichal, C. *A Century of Debt Crisis in Latin America*. Princeton: Princeton University Press, 1989.

Mattelart, A. *Mapping World Communication: War, Progress, Culture*. Trans. by Susan Emanuel and James A. Cohen. Minneapolis: University of Minnesota, 1994.

Munro, D. *Intervention and Dollar Diplomacy in the Caribbean*. Princeton: Princeton University Press, 1964.

Ninkovich, F. *Modernity and Power*. Chicago: University of Chicago Press, 1994.

——. *The United States and Imperialism*. London: Blackwell, 2002.

Ollivier, M. *The Colonial and Imperial Conferences from 1887 to 1937*, Vol. 1. Ottawa: E. Cloutier, Queen's Printer, 1954.

O'Rourke, K., and J. Williamson. *Globalization and History*. Cambridge: MIT Press, 1999.

Oslin, G. P. *The Story of Telecommunications*. Macon, Ga.: Mercer University Press, 1992.

Pacific Cable Board, *The All Red Route Via Pacific*. London: Pacific Cable Board, 1924.

——. *Via Pacific: Some Notes on the Pacific Cable*. London: Pacific Cable Board, 1911.

Pacific Cable Company of New York. *Pacific Cable*. New York: Pacific Cable Company of New York, 1900.

Parsons, F. *The Telegraph Monopoly*. Philadelphia: C. F. Taylor, 1899.

Peterson, L. "The Moguls Are the Medium." *Foreign Policy*, November 2004.

Pike, R. "National Interests and Imperial Yearnings: Empire Communications and Canada's Role in Establishing the Imperial Penny Post." *Journal of Imperial and Commonwealth History* 26, no. 1 (1998).

Pletcher, D. M. *The Diplomacy of Trade and Investment: American Economic Expansion in the Hemisphere, 1865–1900*. Columbia: University of Missouri Press, 1998.

Polanyi, K. *The Great Transformation*. New York: Beacon Press, 1957.

Putnis, P. "Reuters, Competition and the Australian Press in the Late-19th Century." Paper presented at the Conference of the International Association of Media and Communication Research, Porto Alegre, Brazil, July 24–30, 2004.

——. "The Integration of Reuters into the Australian Media System in the Late 19th and Early 20th Centuries." Paper presented to the International Association of Media and Communication Researchers, Glasgow, July 1998.

Rantanen, T. "The Struggle for Control of Domestic News Markets." In *The Globalization of News*, edited by O. Boyd-Barrett and T. Rantanen. London: Sage, 1998.

Read, D. *Reuters: The Power of News*. New York: Oxford University Press, 1992.

Reinsch, P. *Intellectual and Political Currents in the Far East*, 1911. Reprint ed., Freeport, N.Y.: Books for Libraries, 1971.

Richeson, D. *Sandford Fleming and the Establishment of a Pacific Cable*. Ph.D. diss., University of Alberta, 1972.

Rippy, F. "Notes on Early Telephone Companies of Latin America." *Hispanic American Historical Review* 26, (1946): 116–18.

Rodgers, D. T. *Atlantic Crossings*. Cambridge, Mass.: Harvard University Press, 1998.

Rogers, W. "Tinted and Tainted News." *Saturday Evening Post*, July 21, 1917.

Rosenberg, E. *Financial Missionaries to the World*. Cambridge, Mass.: Harvard University Press, 1999.

Sarnoff, D. *Looking Ahead: The Papers of David Sarnoff*. New York: McGraw-Hill, Ryerson, 1968.

Scrymser, J. A. *Personal Reminiscences of James A. Scrymser in Times of Peace and War*. Easton, Pa.: Eschenback Printing Company, 1915.

Sharp, E. W. "International News Communications." *University of Missouri Bulletin* 28, no. 3 (1927).

Shreiner, G. *Cable and Wireless and the Role in the Foreign Relations of the United States*. Boston: Stratford Company, 1924.

Sklar, M. *The Corporate Restructuring of American Capitalism, 1890–1916*. Cambridge: Cambridge University Press, 1988.

Sobel, R. *I.T.T.: The Management of Opportunity*. New York: Times Books, 1982.

——. *RCA*. New York: Stein and Day, 1986.

Standage, T. *The Victorian Internet*. New York: Walker Publishing, 1998.

Stewart, W. "Notes on an Early Attempt to Establish Cable Communication Between North and South America." *Hispanic American Historical Review* 26, no. 1 (1946): 118–24.

Streeter, T. *Selling the Air*. Chicago: University of Chicago Press, 1996.

Tang, X. *Global Space and the Nationalist Discourse of Modernity*. Stanford: Stanford University Press, 1996.

Tang, Z. C. *Shanghai Shi*. Shanghai: Shanghai People's Press, 1989.

Thompson, R. L. *Wiring a Continent*. Princeton: Princeton University, 1947.

Thussu, D. *International Communication: Continuity and Change*. New York: Oxford University, 2000.

Triana, S. P. *The Pan-American Financial Conference of 1915*. London: Heinemann, 1915.

Tribolet, L. *The International Aspects of Electrical Communication in the Pacific*. Baltimore: Johns Hopkins University Press, 1929.

Tulchill, J. *The Aftermath of War*. New York: New York University Press, 1971.

Twain, M. [Samuel Clemens]. *Following the Equator*. Hartford: American Publishing Company, 1897.

Vittinghoff, N. "Unity vs. Uniformity: Liang Qichao and the Invention of a 'New Journalism' for China." *Late Imperial China* 23, no. 1 (2002).

Wagner, R. G. "The Role of the Foreign Community in the Chinese Public Sphere." *China Quarterly* 142 (1995): 432.

Wardl, M. W. *Joint Purse*. 1897. London: British Telecom Archives. Post 83/56.

——. *Telegraphic Communication with India*. 1897. London: British Telecom Archives, Post 83/56.

Widenor, W. C. "The Construction of the American Interpretation." In *The Treaty of Versailles: A Reassessment after 75 Years*, edited by M. F. Boemeke et al. Cambridge: Cambridge University Press, 1998.

Williams, R. *The Sociology of Culture*. Chicago: University of Chicago Press, 1995.

Wilson, G. G. *Submarine Telegraphic Cables in Their International Relations*. Lecture delivered at the Naval War College, Library of Congress, August 1901, 3–5.

Winseck, D. *Reconvergence*. Cresskill, N.J.: Hampton, 1998.

Winston, B. *A History of Communication Technology*. Thousand Oaks, Calif.: Sage, 1998.

Yang, D. Q. "Submarine Cables and the Emerging Japanese Empire, 1884–1915." Paper presented at the Academy of International Business Annual Meeting in Charleston, S.C., 2000. Unpaginated.

Index

Page numbers in italics refer to illustrations.

DWAYNE WINSECK is an associate professor in the School of Journalism and Communication at Carleton University. He is the author of *Reconvergence: A Political Economy of Telecommunications in Canada* (1998) and coeditor (with Mashoed Bailie) of *Democratizing Communication?: Comparative Perspectives on Information and Power* (1997), and (with Annabelle Sreberny, Jim McKenna, and Oliver Boyd-Barrett) of *Media in Global Context* (1997).

ROBERT PIKE is professor emeritus of the Department of Sociology at Queen's University in Kingston. He is the author of many articles on the social and political history of postal services, the telephone, and broadcasting.

Library of Congress Cataloging-in-Publication Data
Winseck, Dwayne Roy, 1964–
Communication and empire : media, markets, and globalization, 1860–1930 /
Dwayne R. Winseck and Robert M. Pike.
    p. cm. — (American encounters/global interactions)
Includes bibliographical references and index.
ISBN 978-0-8223-3912-0 (cloth : alk. paper)
ISBN 978-0-8223-3928-1 (pbk. : alk. paper)
1. Telegraph, Wireless—History—19th century. 2. Telegraph, Wireless—History—20th century. 3. Telecommunication systems—Technological innovations—History—19th century. 4. Telecommunication systems—Technological innovations—History—20th century. I. Pike, Robert M., 1937– II. Title.
TK5711.W56 2007
384.109′034—dc22
2006036830